Large-Scale Peculiar Motions

Matter in Motion

Other Related Titles from World Scientific

The WSPC Handbook of Astronomical Instrumentation
(In 5 Volumes)
Volume 1: Radio Astronomical Instrumentation
Volume 2: UV, Optical & IR Instrumentation: Part 1
Volume 3: UV, Optical & IR Instrumentation: Part 2
Volume 4: X-Ray Astronomical Instrumentation
Volume 5: Gamma-Ray and Multimessenger Astronomical Instrumentation
Editor-in-chief: David N Burrows
edited by Alex Wolszczan and Anna M Moore
ISBN: 978-981-4644-31-0 (Set)
ISBN: 978-981-4644-35-8 (Vol. 1)
ISBN: 978-981-4644-36-5 (Vol. 2)
ISBN: 978-981-4644-37-2 (Vol. 3)
ISBN: 978-981-4644-38-9 (Vol. 4)
ISBN: 978-981-4644-39-6 (Vol. 5)

Black Holes, Cosmology and Extra Dimensions
Second Edition
by Kirill A Bronnikov and Sergey G Rubin
ISBN: 978-981-123-344-9

A Cabinet of Curiosities: The Myth, Magic and Measure of Meteorites
by Martin Beech
ISBN: 978-981-122-491-1

Large-Scale Peculiar Motions

Matter in Motion

Gary A Wegner

Emeritus Professor of Physics & Astronomy
Dartmouth College

World Scientific

NEW JERSEY • LONDON • SINGAPORE • BEIJING • SHANGHAI • HONG KONG • TAIPEI • CHENNAI • TOKYO

Published by

World Scientific Publishing Co. Pte. Ltd.
5 Toh Tuck Link, Singapore 596224
USA office: 27 Warren Street, Suite 401-402, Hackensack, NJ 07601
UK office: 57 Shelton Street, Covent Garden, London WC2H 9HE

Library of Congress Cataloging-in-Publication Data
Names: Wegner, Gary A., author.
Title: Large-scale peculiar motions : matter in motion / Gary A. Wegner, Emeritus Professor of Physics & Astronomy, Dartmouth College.
Description: Hackensack, NJ : World Scientific Publishing Co., [2022] | Includes bibliographical references and index.
Identifiers: LCCN 2021033765 | ISBN 9789811211805 (hardcover)
Subjects: LCSH: Galaxies--Motion in line of sight. | Large scale structure (Astronomy) | Gravitation.
Classification: LCC QB857.5.M67 W44 2022 | DDC 523.1/12--dc23
LC record available at https://lccn.loc.gov/2021033765

British Library Cataloguing-in-Publication Data
A catalogue record for this book is available from the British Library.

For any available supplementary material, please visit
https://www.worldscientific.com/worldscibooks/10.1142/11594#t=suppl

Desk Editor: Ng Kah Fee

Typeset by Stallion Press
Email: enquiries@stallionpress.com

To Cindy (Cynthia Kay), who wouldn't let me quit, and to my border collie, Bugsy, my constant Muse.

Acknowledgements

The material is based on lectures I gave at Dartmouth College in the Department of Physics & Astronomy 1982–2016, while the Margaret Anne and Edward Leede '49 professor. I owe a debt of gratitude to my many generous colleagues wherever they are and to all those connected with the MDM (formerly Michigan-Dartmouth-MIT) Observatory. I wish to thank my friend and colleague, Professor Roberto Onofrio, for reading an early version of the manuscript and suggesting many valuable comments and corrections. Dr. Richard Watkins sent correspondence and Dr. Michael Hudson provided valuable information. It is a pleasure to thank the authors of papers and publishers who corresponded and generously gave me permission to reproduce their diagrams in this book and to others who made their work publicly available.

Cover design: Kristen Wegner (adapting Fig. 14.6 from Zaroubi *et al.*, 2001).

Foreword

This book is for the scientific reader interested in how peculiar motions of the galaxies fit in with modern extragalactic astrophysics and cosmology. It brings together details of the subject that are not currently in one place.

All bodies in the universe move, from the tiniest subatomic particles to the largest clusters of galaxies and the expansion of the universe itself. It is a fundamental fact and from studying matter in motion, science has unravelled, what we call the Laws of Nature. Even before the times of Galileo and Newton, astronomical investigations, growing in scale initially from planetary motions to binary star orbits to motions in the Milky Way led to galactic dynamics, and discovery of the universe's expansion. Along with the host of atomic and particle physics experiments this has progressively produced greater generalizations and knowledge extending the limits of understanding of the physical world. Less discussed or publicized are motions on the largest scales of the universe, the extragalactic peculiar motions.[1]

Here the development and present state of the subject are covered. This includes details of measurements and their interpretation. Necessary corrections to peculiar velocity data are presented. The techniques are important and are common to other related measurements. I have endeavoured to piece together details of peculiar velocity measurements to show the

[1]As used here peculiar motions refer to intrinsic velocities. A galaxy's 'peculiar velocity', $V_{\rm pec}$, is obtained from its redshift, cz along the line of sight as a deviation from the smooth expansion of the universe, H_0D, where D is its distance and H_0 is the Hubble constant. Neglecting the observer's velocity and other corrections, this redshift is $cz = H_0D - V_{\rm pec}$ giving $V_{\rm pec} = H_0D - cz$, as explained in greater detail further on.

complex chain of observations, corrections and calibrations needed to obtain answers. These are common to interrelated fields and it reveals some of the problems and complexity encountered in general extragalactic endeavours. This illustrates the complexity and difficulties in obtaining accurate and reliable answers from many types of astronomical observations.

The numbers of researchers, groups, and publications giving results about peculiar motions have multiplied. To compile a complete list of all the work that is appearing is beyond the scope of this book, but it might provide an outline and source for the subjects included. For this reason, the reader is urged to consult the bibliographies of the referenced papers for further details of ongoing projects as they develop.

In the late 20th century, the discovery of the peculiar motions of galaxies was an early result of modern cosmology. It grew from one of the important developments in astronomy, the improvement of accurate redshift independent distance indicators. Some of these are described here. Other important discoveries associated with the cosmic microwave background (CMB) and large-scale structure subsequently added overwhelming impact and forever changed all of astronomy in a few years. In the 1980s, peculiar motions helped solve questions about the mean matter density of the universe. Large redshift surveys showed the void and filament features of the large-scale structure. Studies of the peculiar motions found flows of galaxies and attractors in the large-scale structure providing a comparison with motion derived from the cosmic microwave dipole and other methods. Peculiar motions provide one facet of current astrophysics and cosmology. The whole field has expanded with the measurements of the cosmic microwave background, increasingly large redshift surveys from new optical telescopes, space satellites that detect radiation across the electromagnetic spectrum, and massive radio telescope arrays in size up to the earth's diameter. The list of developments grows and to these are added the startling recent discoveries in gravitational wave and neutrino astronomy and ever larger and elaborate numerical simulations which are taking us into previously unexplored physical conditions.

The description of peculiar motions employs several intertwined disciplines of physics and astronomy. It is apparent that several individual volumes on each subject would be necessary for detailed explanations. The literature on all fields expands exponentially, almost weekly, and it seems impossible to include and discuss all of the excellent books and papers that have been published in recent years. The NASA Astrophysics Data System (ADS) is a powerful tool to get a grip on this information (e.g., M. J. Kurtz *et al.*, 2000) and can be used to find additional papers from references to these papers.

I have assumed that the reader has a knowledge of astronomy. There are now so many excellent books on astrophysics and cosmology, at every level that it is difficult to choose but the following are some that might be helpful for background and develop the topic in more detail:

Dressler A. (1994), *Voyage to the Great Attractor*. New York: Knopf.
Harrison E. R. (2005), *Cosmology*, 2nd ed., Cambridge: Cambridge University Press.
Weinberg S. (2008), *Cosmology*, Oxford: Oxford University Press.
Peebles P. J. E. (1993), *Principles of Physical Cosmology*, Princeton: Princeton University Press.
Narlikar J. V. (1993), *Introduction to Cosmology*, 2nd ed., Cambridge: Cambridge University Press.
Bradt H. (2007), *Astronomy Methods*, Cambridge: Cambridge University Press.
Ryden B. (2014), *Introduction to Cosmology* 2nd ed., Cambridge: Cambridge University Press.
Schneider P. (2015), *Extragalactic Astronomy and Cosmology*, 2nd ed., Berlin: Springer-Verlag.

The advances on solar and stellar astrophysics are not covered. They are equally exciting and strongly impact extragalactic astronomy and cosmology. The spectacular discovery of thousands of extra-solar system planets and the ability to probe their physical characteristics has provided insight into planetary formation. It has also opened avenues into exobiology with the possibility of discovering extraterrestrial life and even civilizations. This will certainly have untold significance in the future as a great deal remains in formative stages. Were signals or some hitherto unrecognized sign from an intelligent artificial source thousands of light years distant received, it would be reminiscent of the early archaeologists endeavouring to translate inscriptions from anonymous authors millennia in the past. This discussion, however, is clearly beyond the scope of this book.

This book's outline roughly follows a timeline. Chapters 1 and 2 briefly sketch a bit of history to remind one of the efforts of many earlier astronomers working without modern instruments who laid the foundations of today's astronomy. They achieved this through persistent efforts and labour that has brought us to the present time.

In Chapters 3 and 4, the pervasive role of cosmological simulations in interpreting all aspects of astronomy and cosmology and the details of galaxy evolution through modelling and observations are mentioned. Much of the current theoretical interpretation assumes the ΛCDM model as a framework. Understanding how objects employed to determine distance scale, peculiar motions, the Hubble constant, and other parameters change with look-back time and distance is of great importance for obtaining precise answers.

In Chapter 5, the development of observations and analysis resulting in the mapping and quantifying of the large-scale structure is considered.

Gravity drives motions and dominates on many scales in the universe. Chapter 6 discusses some of the questions about its nature and the searches for dark matter. Even though current research is very intensely pursued, the answer remains a mystery. Although General Relativity is a powerful theory, dark matter and energy, quantum gravity, and large-scale considerations suggest that modifications are needed. It has not been established what they might be.

The next five chapters discuss measurement of quantities needed for determining peculiar motions: Chapters 7 and 8 explain determination of the optical brightnesses of sources and corrections to the measurements needed to estimate distances. Chapter 9 presents the bane of astronomy, biases, and some additional corrections needed to different types of data. Chapters 10 and 11 cover distance indicators. Although many are available, those, including scaling relations, employed for peculiar motion investigations are detailed.

Chapters 12–14 include peculiar motions in the nearby universe, starting from the Milky Way, the Local Group, and Virgocentric infall to greater distances including the Great Attractor and dipole motions as established to approximately the year 2000. Here we momentarily leave peculiar motions to describe additional important cosmological developments in the 21st century. These are new discoveries that have made fundamental changes and enhanced our understanding.

Chapters 15 and 16 briefly outline the cosmic microwave background and the baryonic oscillations. Chapter 17 describes gravitational lensing and its applications. Chapter 18 gives an account of galaxy clusters and the motions of interacting double galaxy clusters, which can provide information on dark matter and gravity. Chapter 19 touches on gravitational radiation and some of the associated phenomena related to black holes and neutron stars as well as nucleosynthesis.

Returning to peculiar motions in the 21st century, Chapter 20 is an account of new developments and the changes in peculiar motion studies from the emergence of new generations of projects; resulting from these new discoveries and combining the information with peculiar motions and dipoles. Observational databases are becoming larger, more detailed, and sophisticated.

Chapter 21 concludes about the consistency of the current picture of cosmology. These are based on many varied pieces of the puzzle. The agreement and discrepancies of the ΛCDM model are discussed as well as some basic philosophical questions relating to cosmology.

Research activity on peculiar motions continues to expand in parallel with these other discoveries. It is becoming increasingly recognized as important to cosmology and provides an additional method to gain information on large-scale structure. If some important results have been left out, it is due to the numbers of papers, rather than personal prejudice. One must keep in mind that science primarily progresses in small increments from the accumulation of hard efforts by many dedicated workers. This book relies on some of the work that I have been fortunate to be involved in and the fine people I met along the way, so it is also a personal perspective.

Walpole and Vinalhaven, Maine

Contents

List of Figures

List of Tables

Chapter 1

Historical Summary

The fundamental question to be addressed in this book is where are we, and where are we heading? Since ancient times, people have asked this and given answers, often based on their religious beliefs. In modern times, this is a question for the science of cosmology. Cosmology has been discussed throughout the ages and in many instances some aspects of today's possible theories were anticipated. If we really knew the true nature of the universe, it would obviously have deep philosophical and religious consequences. Although many old timers in the past century referred to cosmology, which then was based on few concrete facts, as 'religion,' since then it has become a true science. Emerging in the 20th century are developments that were unknown in ancient times that have placed this on a quantitative basis. These include quantum theory, special and general relativity, and the expansion of the universe. In the past few decades of the 21st century, there are great discoveries and strides of progress thanks to the development of powerful new instruments permitting detailed investigations of the cosmic microwave background (CMB), the distant and earlier universe at high redshift, galaxy evolution and modelling and gravity waves. The more specialized study of the intrinsic or the 'peculiar' motions of galaxies, the subject of this book, is one of the windows into the large-scale structure of the universe and requires an understanding of the nature and processes of galaxy formation, structure, and evolution.

1.1. Ancient Astronomy

Ancient people did not contemplate cosmology as we do, but one may trace in their thoughts notions related to our modern ideas. Many considered the earth to be fixed in space at the centre of their world, but widespread belief in magic and superstition existed. The ancient Egyptians considered that the world formed from a 'primeval abyss,' from a kind of water unlike any in the sea. Often there was a lack of words to express such notions. Nevertheless, as quaint as they may seem, many of their ideas can be argued to contain a grain of truth.

Speculations about the deeper nature of the universe have existed since ancient times. Some ancient Greek and Indian philosophers adhered to notions of atomism. The atomists Leucippus, Democritus, and Epicurus (4th and 5th centuries BC) held that the universe consists of the void containing an infinity of eternal individual particles they called the atoms. These have a variety of sizes and shapes and comprise all matter. They fill the endless space of the universe and due to their collisions over gigantic intervals of time, they form larger bodies, for example our earth, but infinite other worlds are possible. These bodies endlessly dissolve and reform other worlds. In Indian philosophy, time consists of unending cycles in which, just one day of Brahma exceeds four billion years, roughly the age of the earth. Biblical cosmology implies the different standpoint that the universe was created and has an ending and the answers to these questions are still being debated. The ancient philosophers lacked the scientific method and had no instruments for making detailed observations and experiments and although some might identify their ideas with ours about fundamental particles and forces, they did not possess the concepts of quantum mechanics and relativity which became the basis of scientific thought in the 20th century.

One might wonder that in ancient times someone would question whether all worlds move through space, but to many ancient and medieval philosophers, the earth was fixed and unmovable. Aristotle (384–322 BC) and others argued for a geocentric universe, mostly based on incorrect concepts of mechanics in basic physics. However, some astronomical arguments, correct in principle, reached incorrect conclusions. They noted the lack of observable annual and secular parallaxes of the stars. To detect this requires telescopic observations. Secular parallax is produced if the earth moves; the apparent directions of the nearer stars will shift more than those farther away. This and parallax cannot be observed without a telescope. The largest known stellar trigonometric parallax is about 0.77 arcsec whereas the naked eye only resolves about one arcmin or 60 arcsec and this observation was only achieved in the 19th century. Aristotle knew of the Egyptian star maps that were made nearly a thousand years earlier and used the apparent absence of secular changes in the constellations of the stars as evidence that the earth stands still.

Limited by naked eye observations and an incorrect view of physics, some of the ancient Greek philosophers tried to answer questions based on astronomical observations. Aristarchus of Samos (ca. 270 BC) and others deduced the relative sizes and distances of the earth, moon, and sun, albeit inaccurately. Aristrchus found the distance to the sun (or one astronomical unit) to be 400 earth radii ($R_\oplus$) away. Eratosthenes (ca. 230 BC) measured the circumference of the earth accurate to within about 5% of the modern value. Posidonus (ca. 90 BC) had independently determined a similar value for the earth's circumference and estimated the sun's distance to be nearly 10 thousand $R_\oplus$. The accepted value is about 23 thousand $R_\oplus$. Hipparchus (fl. 146–127 BC) made accurate astronomical observations of planetary motions and among his many contributions are the epicycles to explain their seemingly irregular motions and a star catalogue. The geocentric theory of Ptolemy (ca. 120 AD) was based on these ideas. Although it failed to explain the reason for the physical cause, the motions were constrained to move on mechanical tracks as in a clockwork using circular deferents and epicycles to explain the apparent motions of the sun, moon, and known planets around the stationary earth, and predominated well into the 16th century.

1.2. The Emergence of Modern Ideas

Detailed calculations of the positions of the planets then known, or ephemerides were carried out in the Middle Ages based on the Ptolemaic theory. One of the best known, the Alfonsine tables, was computed for Alfonso X of Castile for years following 1252 by a team of over 50 mathematicians. Although earlier skepticism of this work had sometimes been expressed, by the 15th century inadequacies began to surface. The Ptolemaic system is complicated and Ptolemy required 80 circles; further accuracy could be obtained by adding additional epicycles, but in 1543 Copernicus published his book *De revolutionibus orbium coelestium* in which he demonstrated that with a heliocentric system, placing the sun at the centre only required 34 circles and was adequate for the same accuracy. This produced a major revolution in astronomy and gave rise to the 'Copernican Principle,' that we occupy no special position in the universe and is a fundamental pillar of modern cosmology and many other sciences.

Copernicus's theory was not universally accepted and it did not reveal the true elliptical shapes of the planetary orbits. One distinguished astronomer who disbelieved the heliocentric theory was the colourful

Tycho Brahe (1546–1601) who had developed his own 'Tychonic theory'. Brahe had constructed the first modern astronomical research institute, Uranaborg, based on the Danish Baltic island of Hven where he strove to construct instrumentation and to make the most precise pre-telescopic positional measurements (Christianson, 2003). Among his many achievements, Brahe was interested by Mars and he secured nearly 20 years of observations of the planet. Losing favour with the new Danish king Christian IV in 1600, he obtained the post of imperial astronomer of the Holy Roman emperor Rudolph II in Prague near the end of his life. Brahe's sudden death the following year led to his assistant, Johannes Kepler (1571–1630) analyzing Brahe's Mars data, from which he empirically derived the three laws of planetary motion, the first two in 1609 (*Astronomia nova*) and the third in 1619 (*Harmonices Mundi*).

In this era, Galileo Galilei (1564–1642) accomplished several discoveries in physics and astronomy that changed science. His first telescopic observations of, e.g., the sequence of the phases of Venus and the four largest (Galilean) satellites of Jupiter, helped verify the heliocentric theory. Galileo's studies of mechanics were also substantially valuable and were incorporated by Newton (1643–1727) in his three laws of motion. Using these three laws, Newton demonstrated that Kepler's laws and the moon's motion result from an inverse square (r^{-2}) gravity force, the 'universal law of gravitation' or today Newtonian gravitation, in which the attracting force between two particles of masses m_1 and m_2, separated by a vector distance $\mathbf{r_{12}}$ is

$$\mathbf{F_{12}} = -G\frac{m_1 m_2}{|\mathbf{r_{12}}|^3}\mathbf{r_{12}} \tag{1.1}$$

(Newton, 1687). Innumerable discussions of this formula are given in elementary textbooks on astronomy and physics and Maxwell (1877) gives an excellent account. This remarkably simple result grows vastly more complicated when enormous numbers of particles, such as stars in galaxies are considered. As seen later, Newtonian physics and gravity are remarkably accurate for weak gravity forces and velocities slow compared to the speed of light. Until the latter 19th century this was generally considered capable of explaining all known observations.

The development of spectroscopy in the 19th century permitted chemical analysis of the sun and stars and the measurement of radial velocities. It also contributed to the development of quantum theory. When this was applied to the galaxies, this led to the discovery of the expansion of the universe, which relates the recession expressed as a redshift[1] velocity, $V = cz$, to the object's distance by $V = H_0 D$ where H_0 is the Hubble constant. This provides a means for mapping the galaxies. Measuring V yields the distance along the line of sight, $D = V/H_0$. In modern times this has become increasingly technical as will be described in following chapters.

1.3. Early Developments in Space Motions

In the 18th century, it was considered that Newtonian gravity would cause the stars to move resulting from their mutual attraction. In a way, thinking the same as Aristotle, Tobias Mayer (1760) showed how the sun's motion could be revealed by determining the solar apex using proper motions (Hoskin, 1980). In Fig. 1.1, W. Herschel (1783) used proper motion measurements by N. Maskelyne (1782–1811) and J. Lelande (1732–1807) to determine that the solar apex lies near λHerculis, not far from the modern value towards the bright star Vega (αLyr) with a velocity $\sim$20 km s^{-1}. Detailed explanations of the stellar case for determining the solar apex from proper motions and radial velocities and the results for different classes of stars can be found in many places including Smart (1962) and Binny & Merrifield (1998).

1.4. Peculiar Motions of Galaxies

It was only in the last half of the previous 20th century that the analogue of the solar motion for galaxies was tackled using peculiar motions. The implementation of this required not only the technological developments of instrumentation, sophistication of cosmological theories, and cataloguing of galaxies, but also the development of accurate astronomical distance indicators and statistical methods. Almost all but the nearest galaxies can only have the radial component of motion measured from the redshift at the present time. The transverse velocity perpendicular to the line of sight is usually well below the resolution of any existing telescopes. The redshift, $V = cz$, divides into two components: (1) the cosmological expansion of the universe, which at low redshifts is $V_{\text{cosmic}} = H_0 D$ where H_0 and D are the Hubble constant and distance, and (2) the component of the independent

[1] Redshift is defined as $z = (\lambda - \lambda_o)/\lambda_o$ where λ is the observed wavelength of a spectral line and λ_o is its terrestrial laboratory wavelength. Also called the Doppler shift, in most extragalactic sources $\lambda > \lambda_o$ so lines have longer wavelength or are shifted redward.

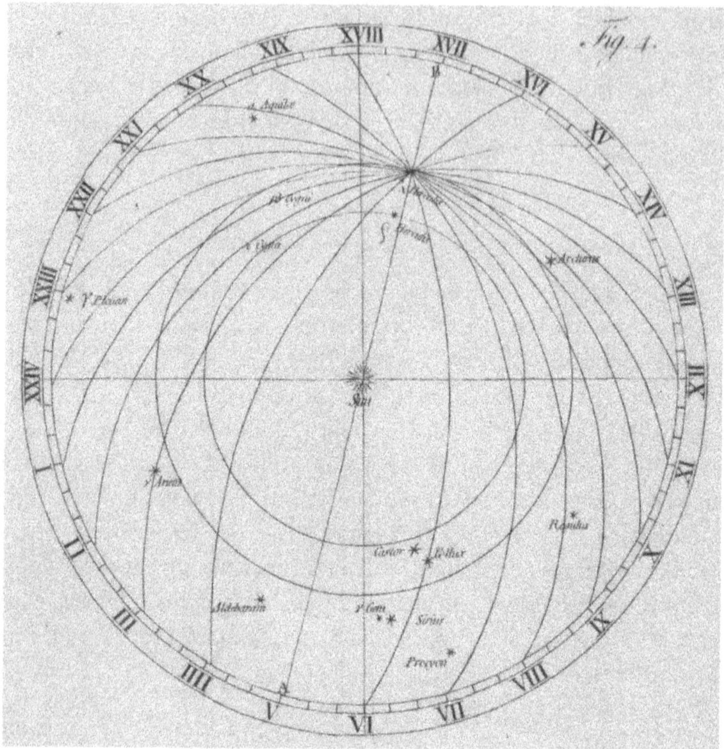

Fig. 1.1. Herschel's map of the portion of the celestial sphere showing the apex of the solar motion near λ Her using about 15 stars that had proper motions determined at the time. (*Source*: Herschel W. (1783), *Phil Trans*, lxxiii, 247, "On the proper motion of the sun and solar system..." Fig. 4, open access.)

or peculiar velocity of the galaxy along the line of sight $V_{\rm pec}$

$$cz = H_0 D + V_{\rm pec}(D) - V_{\rm pec}(0), \tag{1.2}$$

and $V_{\rm pec}(0)$ is the observer's peculiar velocity.

Rubin *et al.* (1973; 1976a,b) attempted the pioneering investigation of the motion of the Milky Way, known as the velocity anisotropy using the Sc I spiral galaxies and determined the dipole of their motions. These observations appear primitive in light of modern techniques, but this investigation encountered the complications that occur in subsequent peculiar motion work that will be described in this book.

At about the same time a powerful new observation became available, the CMB. The measurement of the dipole anisotropy, produced by the

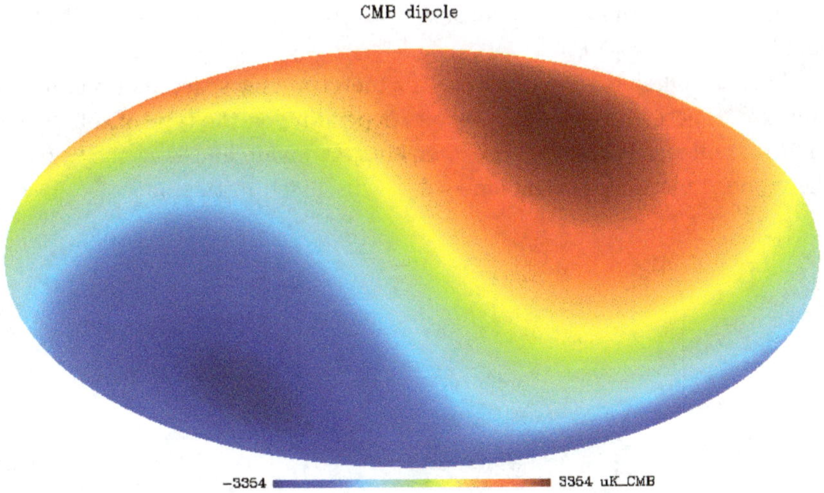

Fig. 1.2. The CMB dipole as it would appear from the PLANCK satellite after removal of all local galactic structures. Colours represent red- and blueshifts corresponding to the velocity and directions given in the text. This is from a simulation. (*Credit*: Delabrouille J. *et al.* (2013), *A&A*, 553, A96, "A pre-launch planck sky model," Fig. 1.1, reproduced by permission ©ESO.)

Doppler shift due to the sun's motion through the CMB was first attempted by Penzias & Wilson (1967) and Partridge & Wilkinson (1967) who set upper limits on the peculiar motion of the Milky Way of order $\sim$300 km s^{-1}. Rubin *et al.* found a dipole velocity of about 454 km s^{-1} relative to the sun toward galactic coordinates[2] ($l = 165°, b = +32°$). The CMB dipole is now well determined (e.g., Planck Collaboration, 2014): $V_{\rm CMB} = 384$ km s^{-1} ($l = 264°$, $b = +48°$), directed near the constellation Crater. Figure 1.2 shows a sky map from data by the PLANCK satellite and illustrates the CMB dipole after all galactic sources of emission by dust and gas have been removed. It can be seen to be smooth with the red and blue sides indicating recession and approach due to our motion through the CMB. Thus, the Rubin *et al.* result from the relatively nearby Sc I galaxies compared to the CMB arising from a far distant region was surprising and alternative explanations were put forth. Either the galactic extinction law

[2]As defined by the IAU (Blaauw *et al.*, 1960), the sun is at the centre of the coordinate system. The equator is defined by the plane of our Milky Way galaxy. Galactic longitude, l is measured in degrees clockwise along the galactic equator with $l = 0°$ pointing to the galactic center. Galactic latitude, b, is measured N or S from the galactic equator.

was incorrect or the Hubble constant was anisotropic (Lynden-Bell *et al.*, 1988).

There are now several methods for determining space motions in the universe. These give different answers depending on the type of source and their distances. How these measurements are made and how they fit into the modern picture of cosmology will be discussed in this book.

Chapter 2

The Beginnings of Modern Astrophysics

Starting about two centuries past, observational astronomy began the discovery of additional components of the large-scale structure of the universe in addition to the stars that had been known since antiquity. The first component is the nebulae or the galaxies as they are now more commonly called. In the 20th century, this was increased by the discovery of the phenomena now called 'dark matter' and followed by 'dark energy' at the turn of the 21st century that further dominate the picture of the structure of the universe. Although its importance was not fully realized until the 20th century, the nearest large galaxy in the constellation Andromeda (M 31, NGC 224) is our neighbour. Already noted by the astronomer al-Sufi in the 10th century this is a spiral galaxy, about 2.5 million light years distant is roughly as large as our Milky Way, but of the millions of known galaxies, it is the only galaxy easily visible to the naked eye as a large fuzzy region. With the development of ever-larger telescopes, revealing fainter and fainter sources, observers became aware of many fixed faint fuzzy sources or 'nebulae'. The comet hunter Messier (1771–1781) compiled a catalogue of 102 pesky stationary nebulae, to avoid confusion with comets. Many of these are bright nearby galaxies, but some are planetary nebulae and unresolved star clusters in our Milky Way, such as M13, which is a globular cluster.

In the 19th century, increasing numbers of nebulae were discovered. Wm. Herschel (1738–1822) in his sweeps of the sky in the northern hemisphere with telescopes up to 40 inches (1 m) in diameter, aided by his sister Caroline (1750–1848), published a catalogue of 2500 objects in 1802. His son John Herschel (1792–1871), completed work in the southern hemisphere compiling a catalogue of 5079 objects in 1864. The large light grasp of Lord Rosse's 6 foot (1.8 m) 'Leviathan' telescope revealed additional nebulae as well as discovered the spiral structure of some, e.g., the Whirlpool galaxy (M51A). Towards the end of the 19th century 7840 objects were compiled by Dreyer (1888) in the New General Catalogue (NGC) and in Dreyer (1895; 1908) 5386 additional objects were added in the Index Catalogue (IC).

This work was done by visual observers, painstakingly recording these observations by hand, and it formed the basis for extragalactic astronomy in the 20th century. Nevertheless, at the time these catalogues were completed, the nature, distributions, and motions of the nebulae were not generally understood. Kant (1755), Herschel, and others had speculated that the nebulae might be distant vast star systems like the Milky Way and Proctor (1897) had called attention to their projection on the sky in what is today recognized as the supergalactic plane, but the nature of the nebulae was debated.

Some experts in the 19th century considered the spiral nebulae to be huge vortices inside the Milky Way condensing into star systems. Lockyer (1890) thought the nebulae to be systems of meteors. Huggins & Miller (1864) had noted that the spectra of some nebulae showed bright emission lines of hydrogen and a mysterious element nebulium, now known to be from doubly ionized oxygen (O III). However, other nebulae showed continuous spectra like those of stars indicative of their being distant unresolved star systems. Although suggestive, these observations are near the limit of what can be achieved using visual techniques and the small telescopes then in existence, but at this time astronomy received a new impetus starting with the development of photography to be followed by the many more kinds of detectors now in use.

2.1. The Development of Photography

Photography developed from the early 19th century adapting the camera obscura for manually recording images to the daguerreotype, tintype, and later the wet plate. John Herschel, already mentioned, was one of the pioneers of photography. All of these early processes were too cumbersome and insensitive to be useful in astronomy for studying faint sources. A typical daylight portrait required an exposure of typically several minutes.

However, this changed with the invention of the dry plate in 1871 by R. L. Maddox. The increased light sensitivity permitted long integrations. The dry photographic plate consists of a glass plate base coated by a gelatin film containing tiny grains (typically 15–30 μ[1]) of light sensitive silver halide. For many popular applications, a cellulose or plastic backing was employed rather than glass. Exposure to light followed by developing and fixing produces a negative picture of the exposed image on the plate, which is a permanent record that can be examined and measured when convenient using microscopes, densitometers, etc. Some technical details on the properties of photographic plates are in Selwyn (1950). Unlike the human eye, which tires after a few minutes from staring into the dark, the plates can be exposed for long periods of time, hours, even up to weeks, integrating and detecting light from faint sources well below what the eye can see. Photographic plates were manufactured throughout the 20th century, notably by the Eastman Kodak, Ilford, and Agfa companies with increasing improvements in size, sensitivity, and wavelength coverage, but by the end of the century they had been replaced by Charge Coupled Devices (CCDs).

In addition to their myriad other scientific employments outside of astronomy and in solar, stellar, and planetary physics, photographic plates allowed astronomers to conduct surveys of nebulae and to obtain spectra. In the case of stars, the spectra of over 225 thousand stars were obtained and compiled in the Henry Draper (HD) Catalogue between 1918 and 1924 (Cannon & Pickering, 1918–1924).

Several large-scale sky surveys were undertaken using photographic plates. Among the earliest was the international plan of 206 selected areas under the leadership of J. C. Kapteyn started in 1906. Most of these early surveys employed relatively small telescopes and were used to study the statistics of stars. Nevertheless, they remain a valuable resource for today's investigations. Shapley (1957), from his photographic surveys, considered that the galaxies belong to a large structure that he called the 'metagalaxy'.

Examples of surveys relevant to extragalactic astronomy that were completed in the last half of the 1900s include the Lick Survey and the Palomar Sky Surveys. The Carnegie Double Astrograph at the Lick Observatory had dual 20-inch (0.5 m) objective lenses and employed 17 × 17-inch photographic plates covering a 6 × 6-degree field on the sky, one objective lens optimized for blue light and the other in the yellow (Shane, 1947). From 1947 to 1957, C. D. Shane and C. A. Wirtanen covered most of the northern sky with the blue sensitive plates and scanned them visually, cataloguing nearly a million galaxies. Papers published in 1967 revealed that galaxies

[1]The non-standard abbreviation μ will be often used to denote the micron or 10^{-6} meter.

belong to clusters and were arranged in filaments separated by voids. Peebles (1971) and Peebles & Hauser (1974) used these data for quantitative determination of the distribution function of the galaxies (see below).

The National Geographic Society — Palomar Sky Survey, known as the POSS-1 was conducted from 1949 to 1958 using the 48-inch (1.2 m) Schmidt telescope on Mount Palomar in the northern hemisphere using 14×14 inch plates and covering 6×6 degrees. Each field was photographed in the red and the blue. The POSS-1 was published as a series of prints or glass copies (935 in each colour) and for decades, studies of galaxies were based on them. Some of the important early studies using it include the Abell (1958) and Zwicky *et al.* (1961–1968) catalogues for galaxy clusters, Vorontsov-Vellyaminov, Krasnogorskaya, & Arkhipova (1962–1974), Nilson (1973) for detailed galaxy types.

To cover both hemispheres of the sky, the POSS-2 and UK Schmidt (UKS), and European Southern Observatory (ESO) surveys were undertaken utilizing the development of better photographic plates. The POSS-2, a second sky survey was started using the Palomar Schmidt, renamed the Oschin Schmidt with improved optics. To cover the southern sky, the UK Schmidt telescopes at Siding Spring Observatory in Australia and the European Southern Observatory (ESO) in Chile started collaborative observations in 1973 (West, 1974). These telescopes are close in design to the Palomar Schmidt. Observations were completed with improved plates. The images on these plates have been digitized and are available online (e.g., The ESO Online Digitized Sky Survey, The STScI Digitized Sky Survey, and several other sites). These are most likely the last photographic sky surveys as the higher quantum efficiency of mosaics of CCDs has surpassed photographic plates. Some observatories maintain collections of their older data, which continue to contain useful information. Notable is the Harvard College Observatory collection, which has over 500 thousand glass photographic plates from the years 1885 to 1993, which are in the process of digitization.

The large areas covered by the Schmidt surveys were utilized for galaxy and cosmology studies around the turn of this century. These involved digitization of the photographic plates and development of methods for automatically extracting and classifying all the star and galaxy images, which are still useful for modern CCD images. Examples of two large surveys in the southern sky using these techniques are the SERC Automatic Plate Measuring (APM) Survey and the Münster survey. The APM survey is described in an extensive series of papers (Maddox *et al.*, 1990) published between 1990 and 1997, measured 185 UKSTU blue sensitive Eastman Kodak IIIa-J plates utilizing the SERC Automatic Plate Measuring facility in Cambridge, covered 4300 square degrees, and extracted about 20 million images. The Münster survey (Ungruhe *et al.*, 2003) used 271 ESO Sky Atlas R plates and covered 5000 square degrees.

2.2. Development of Astronomical Spectroscopy

The development of spectroscopy since the 19th century permits chemical analysis of the sun, stars, nebulae, and extragalactic objects and gives information on their physical conditions. Equally important, their radial velocities and redshifts can be measured. As well, laboratory measurements of atomic spectra contributed to the development of quantum mechanics. In the 20th and 21st centuries, the discovery of Hubble's law from redshift measurements unlocks the basis for mapping the galaxies as will be discussed further.

The nature of light was unknown in ancient times. Since the 17th century, the speed of light and other properties have continued to be measured with increasing precision. The realization that light has a fixed finite speed is a fundamental discovery of physics. Newton (1730) in his book *Opticks* described experiments breaking sunlight into the rainbow colours employing glass prisms. He used circular entrance apertures and missed the discovery of spectral lines, for this requires a narrow entrance slit. Wollaston (1802) had noted a few dark lines in the spectrum of sunlight, but Fraunhofer (1814–1815) discovered and mapped over 500 dark lines in the visible light region. Kirchoff and Bunsen (Kirchoff, 1861) found that bright emission lines in the laboratory have the same wavelengths as dark absorption lines which enabled identification of elements in astronomy. Young (1873) observed the reversal of the solar lines at the limb of the sun during eclipse. Mitchell (1932) gives an excellent account of early solar spectroscopy.

In the 19th century, numerous observers conducted visual observations of the spectra of bright stars. Secchi (1866) classified stars by their colour and observed lines. Measurements of radial velocities using the Doppler effect (e.g., Huggins, 1868) were attempted, but accurate results required photographing the spectra pioneered by Vogel (1889) and Keeler (1890).

Early spectroscopy utilized glass and liquid prisms, but most modern spectrographs employ diffraction gratings. Although gratings were known for a long time, they became widely used at the end of the 1800s when their construction by ruling the required grooves was perfected, notably by Rowland (1882) who catalogued the solar spectrum 1895–1897 (Moore, Minnaert, & Houtgest, 1966; Pierce & Breckenridge, 1973). Extensive laboratory measurements are also highly important (e.g. Meggers, Corliss, & Scribner, 1975). Sawyer (1948) describes some earlier spectrographs and reduction techniques as does Hearnshaw (1996). Descriptions of modern astronomical instruments can be found detailed in the web pages of the major observatories.

2.3. Development of Electronic Detectors

With the discovery of the photoelectric effect, new devices for detecting photons became available and had profound influence on the development of quantum mechanics and many other fields utilizing photons. The photoelectric effect arises when light induces an electrical influence in different substances of matter. Although some photoelectric effects were known earlier in the 19th century, experiments by H. Hallwachs in 1887 reveiled that a zinc plate illuminated by ultraviolet light becames positively charged. This discovery was rapidly developed and the first photoelectric cells were built by J. Elster and H. Geitel in 1890. By 1920 photoelectric photometry was regularly conducted on the brightest stars.

The earliest electronic detectors were one dimensional and employed phototubes. For many years, accurate photometric or photoelectric measurements were obtained using photomultiplier tubes. Like photographic plates, these have now been mostly superseded by CCDs, but a large amount of astronomical photometry is based on the foundation provided by these devices. Although phototubes have no positional information (i.e. one pixel), they are otherwise to be considered almost ideal detectors. They have high quantum efficiency, up to QE $\approx 45\%$, depending on the composition of the cathode, and linear response over a large range of intensity. The original photon hitting the photocathode ejects an electron, which is focused at high energy onto an electrode called a dynode. This produces a number of electrons that are accelerated by the electrical field onto another dynode, which now emits a number of electrons for each of the incoming electrons. These electrons are accelerated by the electrical field onto another dynode. This process is repeated typically about ten times, multiplying the number of electrons by several thousand. The charge from this cascade of electrons is collected on an anode and stored by charging a capacitor used for measurement. The number of photons recorded in an interval of time is a measurement of the intensity of the incident light.

From these processes, television systems were developed. For application in astronomy, a number of two dimensional detectors known as image tubes were developed. In general, such image intensifiers are amplifying devices and need to be combined with an information storage device. An early example was developed by A. Lallemand (Lallemand, Duchesne, & Walker, 1960), and contained a photocathode on which the light image was focused. The resulting photoelectrons were focused using electronic optics onto a special high contrast photographic plate. The Spectracon (McGee & McMullan, 1969) used a mica window to transmit the photoelectrons onto a nuclear emulsion. Another widely used device was the Carnegie DTM image tube (Ford, 1968), which used normal photographic plates for recording. Although these devices had higher quantum efficiencies (QEs) and delivered increased sensitivity, they still suffered the disadvantage of photographic plates as they required the additional step of digitization.

Subsequently, a number of detectors were developed that replaced photographic plates with a television camera and digital storage. Examples of some of these are the Image-dissector-scanner (IDS) (Robinson & Wampler, 1972), the Image Photon Counting System (IPCS) (Boxenberg & Coleman, 1979), and the Faint Object Camera (FOC) for the Hubble Space Telescope (Machetto, 1980). These devices delivered exceptional performances, but they were complicated to build, use, and maintain.

This situation changed with the development of solid-state imagers, constructed of MOS (Metal–Oxide–Semiconductors) and MIS (Metal–Insulator–Semiconductor), which are small photosensitive storage devices and FET (field effect transistors) that serve as switches. These are combined into arrays to obtain spatial coverage.

The Reticon was an early application of this technology. This is a linear array of pixels and extensively used for spectroscopy. One of these devices used in astronomy attached to two three-stage image intensifiers was the photon-counting Reticon TM detector at the MDM Observatory on Kitt Peak (Schechtman & Hiltner, 1976). The Reticon has now been replaced by the two-dimensional CCD.

2.4. Development of the CCD

The development of the CCD created a revolution in all branches of optical astronomy. The marked advance over photographic plates was due to the higher quantum efficiency (QE), where QE = 100% indicates that each incident photon is detected. Operating in their optimum wavelength ranges, QE $\approx$ 5% for a photographic plate, whilst QE $\approx$ 90% for a CCD. This means that a telescope with a 48-inch (1.2 m) aperture with a CCD will be as powerful as the 200-inch (5 m) telescope, one of the largest instruments in the mid-1900s, with photographic plates. Compared to photographic plates, which are nonlinear in their response to photons, CCDs are linear and have a larger dynamic range, which also implies that higher photometric accuracy can be obtained with CCDs, although the plates were excellent for positional measurements. More importantly however, CCDs have digital output and for similar data, the plates have to be scanned. The earliest CCDs had small physical sizes and for many applications such as the POSS surveys, the large sizes of the photographic plates made them preferable. However, with the development of large format CCDs and placing them in arrays, CCDs are now used entirely in optical astronomy.

The first CCDs were invented by W. Boyle and G. E. Smith, at AT&T Bell Labs in 1969 (Boyle & Smith, 1970) for which they were awarded the Nobel Prize. Development of CCDs by several companies was rapid (Lesser 2015) and as well as in astronomy have sweeping applications for imaging. At present, the largest CCDs have dimensions of about 150 mm with near 10^8 pixels or 100 Megapixels. Mosaics of CCDs increase the sizes of detectors to cover the focal planes of the instruments to use them (e.g., Tonry *et al.*, 2005; Zacharias *et al.*, 2007; Jorden *et al.*, 2014). Technical details on CCDs can be found many places (e.g., Bradt, 2007; Cooper *et al.*, 2004).

A CCD is an array of rectangular pixels (typically $15 \times 15\mu^2$). These sit on a lower layer of p-type silicon with a middle layer of n-type silicon with a top layer of silicon dioxide. Each pixel is a polysilicon gate electrode sitting above these. There are two varieties of CCD: top illuminated or 'thick chips' and bottom illuminated or 'thin chips'. The thin chips are the more sensitive and thus more desirable, but more difficult to manufacture and thus expensive. Along with the development of the detectors were accompanying advances in computing.

The reduction of an astronomical CCD image requires the following steps after exposure to light and readout: bias reduction, flat fielding, and removal of cosmic ray (actually muon) hits. The latter can be alleviated by taking three or more tandem exposures. Some publicly available suites of programs are available for imaging and spectroscopic data reductions Starting with early development of the Starlink project in the UK, these currently include IRAF (Tody, 1986) and ESO-MIDAS (Banse, Grosbøl, & Baade, 1991) and STSDAS developed for reducing HST data.

Additional types of detectors are being developed such as the Superconducting camera (SCAM) by the European Space Agency (ESA). This device is a superconducting tunnel junction, runs at 0.3 K and will produce an image plus generating spectroscopic information for it.

2.5. Construction of Multi-object Spectrographs

The invention of large format CCDs was accompanied by developments in optics and diffraction grating dispersers, which made new innovations in instrumentation available. In addition to imaging, this has had a huge impact on spectroscopy. Although low resolution objective prism spectra were secured photographically by mounting a thin prism in front of the telescope in the early 1900s (e.g., for the HD catalogue) or using CCDs (Salzer *et al.*, 2000) to simultaneously observe multiple objects, spectra of stars and galaxies were generally obtained one at a time by placing the image of a star or galaxy on the slit of a spectrograph. In his description of the expansion, Hubble (1936) mentions that redshifts of 46 galaxies were available in 1929. In the classical paper of Humason, Mayall, & Sandage (1956), about 800 redshifts of galaxies are reported and this number continued to mount to about 57,536 redshifts by 1995 in the ZCAT catalogue of the Cfa (Huchra *et al.*, 1992).

Obtaining redshifts continues to be a prime activity in astronomy, but in the 1980s their measurement was accelerated by the development of multi-object spectrographs (MOS) in which many entry slits to the spectrograph, connected by optical fibers, could be moved to simultaneously observe numbers of targets. An early instrument used on the 2.4 m telescope at the MDM Observatory in Arizona was the 'decaspec' (Fabricant & Herz, 1990), which was used to obtain 1019 spectra of galaxies for the Century Survey (Wegner *et al.*, 2001). Much more powerful instruments rapidly became available. The still operating Sloan Digital Sky Survey begun in 2000 (Newman *et al.*, 2004) was capable of observing up to 640 objects simultaneously using a plugboard system could observe up to 5760 objects per night and has provided over 3 million galaxy spectra. The two degree field (2dF) Galaxy Redshift Survey (Colless *et al.*, 2001b), which relied on the APM Survey for imaging, collected data from 1998 to 2003 on the 4 m Anglo-Australian telescope (AAT) used 400 movable fibers and obtained about 232 thousand galaxy redshifts. The hectospec (Fabricant *et al.*, 2005) on the 6.5 m Multiple Mirror Telescope (MMT) has 300 movable fibres. The SAURON integral-field spectrograph (Peletier *et al.*, 2001) on the 4.2 m William Herschel Telescope (WHT) has an array of many fixed slitlets for detailed spectroscopy of single objects. Even larger MOS instruments are being brought into the field and larger and deeper surveys are underway such as the Dark Energy Spectroscopic Instrument (DESI), which will be capable of observing 5000 spectra simultaneously (Aghamousa *et al.*, 2016a,b). Several instruments of this type are now available around the world.

2.6. Growth of the Telescopes

The development of telescopes parallels learning about the nature of light, advances in physics giving the materials to construct them, and their detectors throughout the electromagnetic spectrum, as well as neutrinos, and gravity waves. Until the 19th century, the eye was the only detector. Mirrors, glass, and transparent crystals, such as quartz, were known in ancient times and several early authors mention optical devices. Examples of lenses and mirrors have been found by archaeologists, although their purposes are unknown. One well-known example is the Layard/Nimrud lens (750–710 BC) in the British Museum, which is a ground and polished plano-convex lens of crystal. The Greeks seem to have used 'burning glasses' for starting fires. The burning of the Roman fleet at the siege of Syracuse using a giant mirror at the hands of Archimedes in 212 BC is a famous story. According to Seneca the Younger (1st century AD), magnifying lenses can be constructed from water filled glass globes and Pliny the Elder states in his *Natural History* that Nero used a lens of smaragd (possibly emerald) for viewing. Geometric optics, crucial for telescope design, can be traced to Ptolemy's book *Optics* (1st century AD) that discussed reflection, refraction, and colour. This was used by Ibn Sahk (940–1000) who discussed what is now known as Snell's Law of refraction plus curved lenses and mirrors. This was followed by the influential *Book of Optics* by Alhazen (ca. 965–1040).

Besides burning glasses, 'reading stones' were used in Europe in the 11th to 13th century. A reading stone is roughly half of a sphere that can be set on a book page to help discern the writing. The 11th or 12th century Visby lenses, found on Gotland, Sweden, are sophisticated aspheres and made of quartz crystal. They may have been employed as reading stones, but some are of sufficient quality to have been used in a telescope. Eye glasses evolved from reading stones about 1250 and by the 16th century manufacturers were grinding and polishing lenses in Holland and Germany. Given the history of these artefacts, it is possible to speculate that in earlier times, someone devised a telescope. Da Vinci (1998) described a project to study the moon's surface with a magnifying optical device, but the details are lost.

Hans Lippershey, the Dutch spectacle manufacturer, is often credited with inventing the refracting telescope in 1608, although it is certain that other contemporaries knew about these instruments. It is well known that Galileo learned of these devices and designed and fashioned his own telescopes for astronomical observations in 1609, starting with an objective lens of 1.5 cm aperture and certainly inferior to a good pair of binoculars today. At nearly the same time, other astronomers including Simon Marius and Thomas Harriot also began making astronomical observations, even preceding Galileo, with telescopes. Since that time, there has been the race to construct the largest and best telescopes and competition to make the first observational discovery.

The two fundamental types of telescopes appeared in the 17th century: the refractors and the reflectors (although in the 20th century a third lens plus mirror catadioptric system, such as the Schmidt telescope, was added). Following Lippershey for the first type, Newton invented the second, the reflecting telescope in 1668 with a 3.3 cm (1.3 inch) parabolic mirror. Over the years, a large number of telescopes of different designs have been built, but there are some basic developments that have contributed to telescopes in the 20th and 21st centuries.

Early refractors utilizing singlet objective lenses suffer from chromatic aberration, whereby not all wavelengths of light reach the same focus. This can be improved by constructing extremely long aerial telescopes, e.g., Hevelius's telescope built around 1641 was 46 m in length. These instruments were notoriously unwieldy and difficult to use. Refractors were improved by the development of the achromatic lens by Hall in 1733 by the construction of a doublet lens, one element composed of crown glass and the other of flint glass.

Reflecting telescopes are free from chromatic aberration, but early mirrors were made of speculum metal, an alloy, known since ancient times, of about two-third copper and one-third tin with possibly small amounts of arsenic and other metals, depending on the builder's formula. Speculum metal is hard, white, and brittle and can be highly polished. However, for large telescope mirrors it is difficult to use; it is hard to cast, tarnishes fast requiring frequent re-polishing, and at best only reflects about 66% of the incident light. Although some large speculum reflectors were constructed, notably W. Herschel's 48-inch (1.2 m) 'forty foot' telescope in use 1787 to 1815 and Lord Rosse's 72-inch (1.8 m) 'Leviathan' in use 1845 to 1908, and the 48-inch (1.2 m) Great Melbourne Telescope erected in 1868, refractors were easier to utilize and maintain and preferred by many investigators throughout the 19th century, which saw the construction of several great refractors, the largest being the Yerkes 40-inch (1.0 m) refractor still in use. Clarke (1908) gives an excellent account of 19th century telescopes.

For the reflecting telescope to gain dominance in the 20th century, in addition to improvements in glass technology, two developments enabled the construction of efficient glass mirrors. The first was the invention of the silvering process, independently by K. A. von Steinheil and L. Foucault about 1856–1857 for telescope mirrors. Glass has great advantages over speculum in that it is much cheaper and easier to cast and work. The second advance was the invention of aluminizing by thermal vacuum evaporization by J. D. Strong in 1932. Reflecting telescopes of increasing size have continued to be constructed. A. A. Common constructed a 36-inch (0.9 m) telescope in 1879, which he sold to E. Crossley in 1895 and was rebuilt at Lick Observatory and used until 2010. Reflecting telescopes of 1.5 m aperture soon appeared on the scene and these were followed after 1900 by several larger instruments, such as the famous 100 inch (2.5 m) on Mount Wilson, and the 200 inch (5.0 m) on Mount Palomar, which contributed to the opening up of extragalactic astronomy and cosmology. By the close of the 20th century, there were nearly three dozen optical telescopes with apertures between 10.0 and 3.0 m in use at world observatories with dozens only somewhat smaller.

In the 21st century, the building of ever larger optical telescopes continues. A major factor is the impracticality of casting glass mirrors with diameters above about 10 m. At present, the largest single mirrors are 8.4 m in diameter and an impressive number of new telescopes up to this size have come into use. To construct larger mirrors has led to the development of multiple and segmented mirror telescopes, although liquid mercury and gallium mirrors have also been seriously considered. The multiple mirror concept has circular mirrors placed in proximity and their light brought to a common focus. The Multiple Mirror Telescope (MMT) in Arizona using six 1.8 m mirrors employed this design in 1979. Segmented mirrors are comprised of hexagonal components placed nearly in contact to form the mirror surface. The construction of these mirrors required overcoming several challenges, which include cutting and figuring the hexagonal segments, mounting and keeping them aligned, and construction of the associated mechanical equipment. The 10 m Keck telescope on Hawaii, completed in 1993 employed the first large segmented mirror. These technical complications have been largely overcome and telescopes of this type give excellent

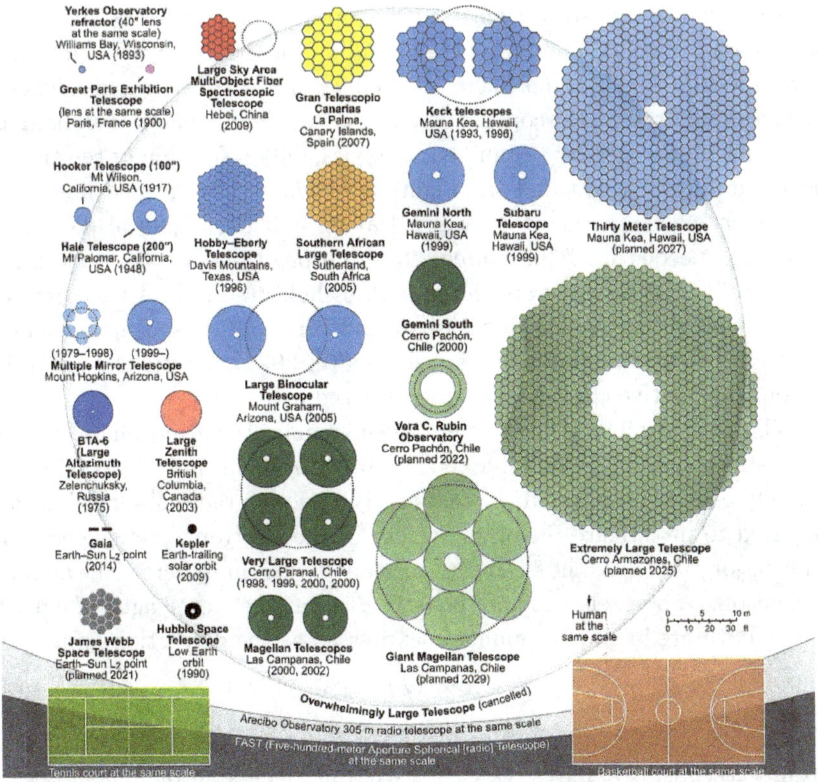

Fig. 2.1. Comparison of some existing and proposed primary mirrors for optical telescopes along with the curves at the bottom showing outlines of the Arecibo and FAST radio dishes. Tennis and basketball courts are also shown. (*Source*: Cmglee, "Data on holes in mirrors provided by an anonymous user from IP 71.41.210.146," https://commons.wikimedia.org/wiki/File:Comparison_optical_telescope_primary_mirrors.svg, Wikipedia, Creative Commons Attribution-ShareAlike 3.0 (CC BY-SA).)

performance. Several segmented mirrors of the 10 m class are now in service or under construction and the sizes of some different telescope apertures are compared in Fig. 2.1.

Larger telescopes under construction include the Extremely Large Telescope (ELT), 39.3 m, to be completed in Chile in 2024, the Thirty Meter Telescope (TMT), 30 m, which might be finished in Hawaii by 2027, and the Giant Magellan Telescope (GMT), 24.5 m, expected by 2025 in Chile. Doubtless even larger telescopes will eventually be built, e.g., the Colossus Telescope (74 m) proposed by the Planets Foundation and immense hypertelescopes either in the earth or in space (Xie *et al.*, 2020).

Although ground-based optical telescopes seem only to be limited in size by their available budgets (\$1245 million for the ELT), they are hampered by the earth's atmosphere or 'seeing', and with the rapid advances in space and radio telescopes, their eclipse as the primary astronomical instruments has already begun. For galaxies, only a fraction of the time is usable due to daylight and the nights are divided into bright time when the moon makes the sky too bright, leaving half the nights in dark time which are needed for faint sources. In addition, there are the vagaries of the weather. In recent years the growth of lighting in cities have created the problem of light pollution for observing faint sources and the launching of thousands of communication satellites after 2020 constitutes a potential serious destructive challenge to ground-based astronomy.

The optical window due to the transmission of the atmosphere extends from approximately 3000 Å due to absorption by ozone in the ultraviolet to roughly 5 μ in the infrared. Consequently, space or radio observations are required to investigate the other regions of the electromagnetic spectrum. The major problem is the 'seeing' or motion of the atmosphere. The angular resolution, α, of a telescope of diameter d cm at the wavelength of light λ, where both are in cm, is commonly expressed as the diffraction limit,

$$\alpha = 1.22\lambda/d \text{ radians.} \tag{2.1}$$

For visible light ($\lambda \approx 5500\,\text{Å}$ or $5.5 \times 10^{-5}\,\text{cm}$), $\alpha \approx 13.8/d$ arcsec. For a 10 m telescope, $\alpha \approx 0.01$ arcsec, but without some kind of compensation, atmospheric motion usually limits the resolution to about 0.5 arcsec. In space, there is no atmosphere, eliminating the seeing, which is a major factor in the superiority of space observations, e.g., the HST. On the ground, sometimes resolution approaching the diffraction limit has been achieved using adaptive optics techniques that are now being widely used, including speckle interferometry (e.g., Ghez, 2004) and laser guide stars (e.g., Corsini *et al.*, 2017).

2.7. Space Telescopes

Starting with the launch of the first artificial satellite, Sputnik 1, in 1957 by the USSR, it is an understatement to mention the impact that subsequent growth of satellite technology has had on society and it is impossible to give a detailed account of these developments here. By the year 2019, many nations and some private organizations had launched satellites and nearly 5000 of these devices are orbiting the earth. The launching of tens of thousands of communications satellites is now being facilitated by private firms and can create severe hindrance for some optical and radio observations. Although not remotely approaching this, a substantial number of satellites and space probes intended for astronomical observations have been constructed making use of the advances in satellite technology. The development, construction, and organization of the large teams of collaborators, the tracking, and data handling of these experiments are as remarkable as the observations they have produced. There are two obvious advantages for these devices: First, to explore the electromagnetic spectrum outside the optical and radio windows transmitted by our atmosphere, the infrared, ultraviolet, X-ray, and γ-ray regions, and second to avoid the blurring effects from peering through the moving sea of air at the earth's surface, or seeing.

A few of the many recent satellites that have produced spectacular data and helped revolutionize different fields of astronomy and their years of launch include: Hipparcos (1989), the Hubble Space Telescope (HST) (1990), the Chandra X-ray Observatory (1999), XMM-Newton (1999), GALEX (2003), Spitzer Space Telescope (2003), Fermi (2004), Herschel Space Observatory (2009), Kepler (2009), and Gaia (2013). These and other instruments are well documented on the internet as are the results of their observations, which are generally publicly available. Some of these instruments are generally not ideal for studies of large-scale structures and peculiar motions because they have small fields of view or do not cover the entire celestial sphere for galaxies.

Some satellites cover the celestial sphere and have produced data used for large-scale structure, bulk motions, and the cosmic microwave background (CMB): IRAS (1983), COBE (1989), WMAP (2001), AKARI (2006), Planck (2009), and Wise (2009) and the data from these will be discussed in more detail.

Even more powerful and ambitious space telescopes are under construction, including the James Webb Space telescope, JWST (2021), Euclid (2022), and the Wide Field Infrared Space Telescope, WFIRST (mid-2020s). Detailed descriptions of all of the projects mentioned in this chapter can be found in their respective webpages.

2.8. Development of Radio Astronomy

The development of radio telescopes has reached a stage where they are equally important as optical telescopes and surpass them with higher resolutions. Radio telescopes have contributed to a large degree in large-scale structure and peculiar motion studies, notably by the discovery of the

Tully–Fisher relation and galaxy structure observations. Radio telescopes are also applied to a diverse range of applications from planet and star formation, black holes, to searches for extraterrestrial signals.

After the mid-19th century J. C. Maxwell (1865) predicted the existence of electromagnetic waves and they were first detected by H. R. Hertz by 1888. With the development of the wireless in the early 20th century there was rapid and extensive development of radio technology. Several investigators, e.g., O. Lodge in 1897–1900 and N. Tesla attempted to detect radio waves from the sun and other celestial sources. Since the technology and knowledge of the pertinent physics had not been developed, this proved unsuccessful until the 1930s when the first radio telescopes were constructed. K. G. Jansky (1933) detected the galactic centre and G. Reber conducted the first survey for astronomical sources in 1937 (Tyson, 2003). Ground-based radio observations utilize the transmission of the earth's atmosphere in the 'radio window' that extends in approximately in wavelength $\lambda = 1\,\text{cm}$ ($3 \times 10^{10}\,\text{Hz}$) to 11 m ($3 \times 10^{7}\,\text{Hz}$) and since that time numerous increasingly extensive surveys at different frequencies have been conducted. Of the earlier surveys, the Cambridge 2C survey that lists 1936 sources and the 3C survey, now extending to the 10 C survey (Shakeshaft *et al.*, 1955; Edge *et al.*, 1959), greatly expanded knowledge of known radio sources. The observations were made with the one-mile interferometer telescope in Cambridge and M. Ryle and A. Hewish who headed up the project were awarded the Nobel Prize in 1974. Additional early work was done by J. Bolton in Australia, the UK and US (Kellermann, 1996). The Ohio State Survey (Kraus, 1977) found about 19,000 radio sources.

The large expansion in radio astronomy has followed these efforts in the latter half of the 20th century in several countries and research projects have expanded from the cataloguing of radio sources. Many increasingly large radio telescopes have been constructed. Earlier designs often utilized a fully steerable single parabolic dish. The Jodrell Bank 250 ft (76.2 m) completed in 1957 was the first large one and still is in use. Due to flexure of building materials as well as prohibitive cost, the largest of this class of instrument is the Robert C. Byrd Green Bank telescope (GBT) with a diameter of 330 ft (100 m) and earlier plans to construct a 600 ft (183 m) diameter steerable instrument were abandoned. The Xingjiang Qitai 110 m radio telescope (QTT) scheduled to be completed in 2023 would become the largest steerable single dish instrument.

Because of these problems, larger fixed parabolic reflectors have been used, notably the 1000 ft (305 m) Arecibo radio telescope, which ended in a fatal collapse in 2020 December, and the Five-hundred-meter Aperture Spherical-radio Telescope (FAST) completed in 2016 in Guizhou Province, China, which is 1640 ft (500 m) in diameter. Other designs are utilized, such as the decommissioned Ohio State Big Ear, a flat design equivalent to a 53 m diameter mirror and the RATAN-600 radio telescope made up of rectangular radio reflectors set in a 576 m diameter circle.

Radio observations are capable of producing much higher resolutions than currently possible with optical telescopes. From Eq. (2.1), since λ is typically cm in size, much larger effective diameters, d, are necessary. Besides operating at the shortest observable microwave wavelengths transmitted by the atmosphere, the sizes of radio telescopes have been increased by arrays of single dishes and the resulting images synthesized using interferometry. Since the entire aperture is not filled, they are less sensitive than the large full dish instruments, but they achieve significantly higher resolutions. It is possible that this mode of observation could be applied to future optical telescopes. Many instruments of this type have been constructed; here we mention just a few. The Very Large Array (VLA) in New Mexico, USA was begun in 1975 and upgraded in 2011. It is an interferometer and is comprised by 28 moveable 25 m diameter antennae. It operates at wavelengths 0.6–410 cm and is capable of resolutions of 0.02 arcsec, compared to the HST's of 0.05 arcsec. Several newer radio telescope arrays have been built or are under construction.

The Atacama Large Millimeter Array (ALMA) was fully operational in 2013. Sited on Chile's Atacama Desert by an international partnership, ALMA consists of 66 antennae that can be deployed in an array 150 m to 16 km across and operates at wavelengths 9.6–0.3 mm. The resolution that can be achieved is about 0.01 arcsec, about five times better than HST and velocity accuracies are better then $\pm 50\,\text{km s}^{-1}$.

In order to achieve higher resolution, a number of radio telescope arrays have been placed at different points on the earth's surface using the technique of Very Long Baseline Interferometry (VLBI). Crucial to this method has been the development of atomic clocks with which the data from each distant telescope component can be precisely synchronized as they cannot be simultaneously connected due to their large distances. The largest VLBI in regular use is the Very Long Baseline Array (VLBA) completed in 1993, it is made of 10 dedicated radio telescopes reaching 8612 km across the USA. Operating at a wavelength of 0.7 cm, a resolution of 170 μarcsec can be achieved.

The Event Horizon Telescope (EHT) uses VLBI in a global network of ten radio telescopes covering both the northern and southern hemispheres. It was designed by the Event Horizon Telescope Consortium to study the black holes in SgrA* and M87 (Akiyama *et al.*, 2019). A resolution of about 15μarcsec has been achieved. Imaging of the shadow of the black hole in M87 concluded that it contains a $(6.5 \pm 0.7) \times 10^9$ $M_\odot$ spinning Kerr black hole.

The next step towards increasing the baseline of VLBI telescopes would be to develop arrays of radio telescopes in space. This concept has already been tested with limited instruments. Recent satellites carrying radio telescopes include the Japanese HALCA and the Russian Spektr-R satellites.

The Square Kilometre Array (SKA) is a joint project between South Africa and Australia and many consortium members. It would combine both size and collecting area. Construction is planned to commence in 2020 with completion after 2027. Thousands of small antennae will be spread from New Zealand and Australia to South Africa and it will have a total collecting area of about 1 km^2. This ambitious project will need to overcome several technical problems including data handling and international construction. Several precursor experiments related to the development of the SKA already exist.

As is the case in optical astronomy, there has been substantial development of publicly available software for the reduction of radio telescope data, such as the Astronomical Image Processing System (AIPS), which has evolved towards the Common Astronomy Software Application (CASA) described by e.g., Jaeger (2008).

Chapter 3

Cosmological N-body Simulations

Paralleling observational techniques for gathering data, the development of cosmological N-body simulations has provided widely used methods in astrophysics and cosmology for understanding these data. Numerical simulations are widely employed throughout science and engineering, and particularly in other branches of physics such as plasma physics, as well as stellar atmosphere and evolution codes, which will not be discussed here. The gravitational force is generally considered to dominate on large scales and even for the seemingly simple Newtonian inverse square law, simulating and understanding the dynamics, formation, and time evolution of galaxies and larger scale systems using large numbers, N, of components is beyond the ability of analytic techniques (see, e.g., Binney & Tremaine (2008) for discussions of some of the known analytic solutions) and manual calculations with mechanical calculators only available a few generations ago. Even for the Newtonian case, complex nonlinear dynamics and processes are encountered and this becomes increasingly difficult when including General Relativity (GR). In all of astronomy, it is impossible to experiment on the behavior of extragalactic systems, stars, and black holes, but in a sense this can be be done with the increasing complexity of numerical calculations as the capabilities and speeds of computers have grown.

Holmberg (1941) conducted a pioneering analogue N-body simulation of two tidally interacting galaxies (Wegner, 2007a). Since light, like gravity, obeys an inverse square law, Holmberg represented two flattened 'nebulae', each by 37 light bulbs with luminosities proportional to mass, which he measured with a photocell. Current computer simulations use billions of particles and include the expansion, gas dynamics, and stellar evolution effects.

The invention of the digital computer in the 1940s and the discovery of the transistor have led to the many manifest technological developments in all areas of science and society. N-body simulations have benefited from the increase in computing power, following Moore's law (Moore, 1965). Early N-body simulations utilized a few hundred mass points (e.g., Toomre & Toomre, 1972; Wright, 1972), while some current simulations exceed 8×10^9 particles.

In these simulations, or numerical experiments, the constituents, including forms of gravity, cosmology, matter properties, and initial conditions are included and varied, and the results compared with Nature. As current interpretations seem to indicate that about 95% of the mass of the universe is dark matter (DM), unknown, and collisionless, and normal or baryonic matter comprising the visible stars, planets, galaxies, and gas is the rest. These DM particles appear 'cold' in the sense that their velocities are non-relativistic hence denoted 'cold dark matter' (CDM) opposed to relativistic hot dark matter (HDM) particles.

Each type of DM has a different power spectrum, as illustrated in Fig. 3.1, but it became apparent that those of the CDM type agree better with data. The model power spectra of density fluctuations have the form

$$P(k) = Ak^n T^2(k/\Gamma), \tag{3.1}$$

where $n \approx 1$ is the exponent in the primordial Harrison–Zeldovich power spectrum. There is no analytic function and it is fit numerically

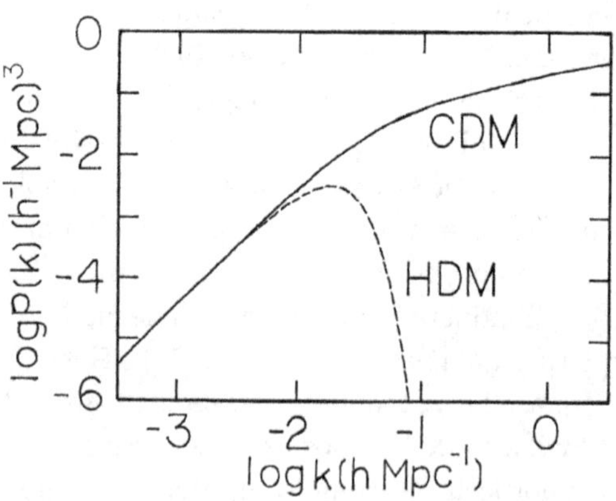

Fig. 3.1. Sketch of power spectra for adiabatic models of CDM and HDM particles, where $k \propto 1/\ell$. Although they are similar for large scales, HDM does not produce power and thus structure at the smaller scales.

to simulations (e.g., Sugiyama, 1995),

$$T(q) = \ln \frac{1.234q}{2.34q} \left[1 + 3.89q + (16.1q)^2 + (5.46q)^3 + (6.71q)^4\right]^{-1/4}, \quad (3.2)$$

is the transfer function for CDM (Bardeen *et al.*, 1986) with $q = k/\Gamma h$ and $\Gamma = \Omega_0 h \exp[-\Omega_b - (2h)^{\frac{1}{2}}\Omega_b/\Omega_0]$ is the 'shape parameter'. Other parametrizations of $P(k)$ have been employed, e.g., the 'Γ model',

$$T(k) = (1 + [ak/\Gamma + (bk\Gamma)^{3/2} + (ck/\Gamma)^2]^{\nu})^{-1/\nu} \quad (3.3)$$

(e.g., Efstathiou, Bond, & White, 1992).

Figure 3.2 compares matter distributions generated for CDM and HDM. In this case the difference is easily seen by eye.

In most current analyses matter contributes to about 30% of the expansion dynamics and 70% is governed by 'dark energy' or Λ. Consequently, these constituents have been included in many N-body simulations and are ΛCDM models. Comparison between observation and theory have arrived at a more precise description of the constituents and eliminated several cosmological theories including the steady state and HDM models.

The N-body simulations seek to describe the dynamics of galaxies in large volumes of sizes typically ~ 10–10^3 Mpc (where $1\,\text{Mpc} = 3.08 \times 10^{24}$ cm or 3.26×10^6 light years), including the global cosmological expansion given by the scale factor $S(t)$. If there are no other outside forces, two particles a distance $\mathbf{r}_1$ apart at time t_1 will be apart $\mathbf{r}_2 = \mathbf{r}_1 S(t_2)/S(t_1)$ at time t_2. From Hubble's law, $V = Hr$ where $H(t) = \dot{S}/S$ is the Hubble parameter. Using 'comoving' coordinates, $\mathbf{x} = \mathbf{r}/H(t)$ and Newton's second law for a particle is

$$\frac{\mathrm{d}^2\mathbf{x}}{\mathrm{d}t^2} + 2H\frac{\mathrm{d}\mathbf{x}}{\mathrm{d}t} = -\nabla\phi, \quad (3.4)$$

where ϕ is the gravitational potential of the perturbation. When the motions are non-relativistic and gravity is weak, this Newtonian equation is correct, even for general relativity. Using the spatially varying part of the mass density field, $\delta\rho/\rho = [\rho(x,t) - \bar{\rho}]/\bar{\rho}(t)$, where $\bar{\rho}(t)$ is the mean density, this can be used to write Poisson's equation

$$\nabla^2\phi = 4\pi G\,[\rho(x,t) - \bar{\rho}(x,t)] = \frac{3}{2}\Omega H^2 \frac{\delta\rho}{\rho}, \quad (3.5)$$

and $\Omega(t) = 8\pi G\bar{\rho}/3H^2$ is the cosmological density parameter. If $\Omega < 1$ the universe is open and expands; if $\Omega > 1$ it is a closed recollapsing universe, and $\Omega \equiv 1$ for a flat Euclidian universe. Current measurements employing

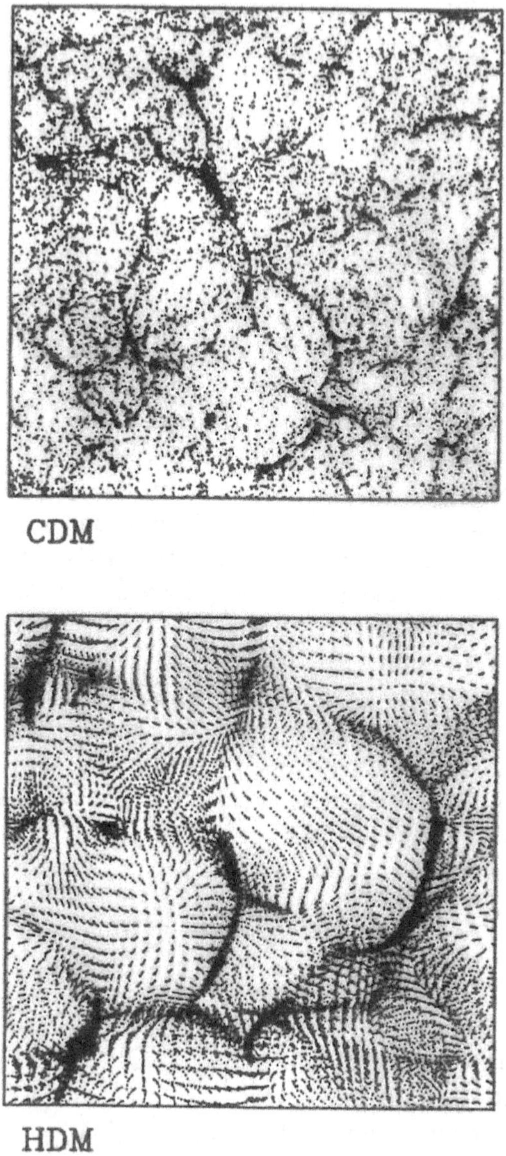

Fig. 3.2. Comparison at $z = 0$ of large-scale structure resulting from cold dark matter (CDM) (upper panel) and hot dark matter (HDM) simulations (lower panel). (*Source*: Bromley B. C. (1994), The large-scale structure of the universe: wavelet analysis of the distribution of galaxies. PhD thesis, Dartmouth College, with permission.)

a number of different methods to be described including the CMB, redshift surveys, and peculiar motions give $\Omega \approx 1$, but its value is undecided for very early or old ages when Λ would play an important role.

For initial conditions, the positions and sizes of the particles at the starting need to be given. The assumption of the cosmological principle and the observed CMB indicate that matter was initially distributed extremely smoothly and during the inflation era (first 10^{-33}–10^{-32} sec), quantum zero-point fluctuations or some other early mechanism generated tiny density fluctuations described by a power spectrum, $P(k)$, where k is the wave number. Often this is taken as the Harrison–Zel'dovich form $P(k) \sim k^n$ with n close to one. These fluctuations are expected to have generated a Gaussian random field. That is, at a given time the variance of $\delta\rho/\rho$ from plane waves in a volume element d^3k is Gaussian. In some theories involving cosmic strings or textures, these assumptions are modified, but $P(k)$ and the Gaussian random field completely describe the initial conditions of the density.

The initial positions and velocities of the particles are obtained from this using a linear approximation assuming $|\delta\rho/\rho \ll 1|$ in Lagrangian coordinates (Zel'dovich, 1970). Small displacements $\mathbf{q}$ from a grid of comoving coordinates take the form

$$\mathbf{x}(\mathbf{q}, t) = \mathbf{q} + \psi(\mathbf{q}, t) \approx \mathbf{q} + D(t)\psi_0(\mathbf{q}), \tag{3.6}$$

$$\frac{\mathrm{d}\mathbf{x}}{\mathrm{d}t} = \frac{\partial \psi(\mathbf{q}, t)}{\partial t}, \tag{3.7}$$

where ψ is the displacement field and $D(t)$ is the growth factor. For a given cosmology these are known, e.g., for a flat universe, $D(t) \sim S(t) \sim t^{2/3}$.

In general, N-body simulations can be solved by integrating a system of ordinary differential equations of the form

$$m_i \frac{\mathrm{d}^2\mathbf{q}_i}{\mathrm{d}t^2} = \sum_{j \neq i}^{N} G m_i m_j \frac{\mathbf{r}_i - \mathbf{r}_j}{|\mathbf{r}_i - \mathbf{r}_j|^3}. \tag{3.8}$$

Supercomputers are required for large N. Figure 3.3 shows the evolution of structure in the Millenium ΛCDM simulation (Springel *et al.*, 2005), where smoother structure at earlier time grows into larger density contrasts at the later time. The Millenium XXL simulation required the equivalent of 300 years of CPU time using 1200 cores on the JUROPA supercomputer. Difficulties arise when N becomes large. For direct integration, the order of magnitude of the computational effort increases as N^2. For this reason,

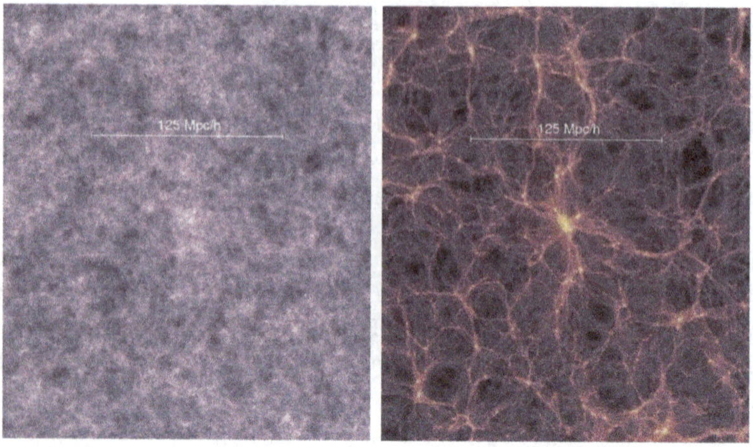

Fig. 3.3. Portions of structure computed by the Millenium ΛCDM simulation in a 15 Mpc thick layer. This shows the growth of structure from an early time, $t = 0.21$ Gyr (redshift $z = 18$) to the present ($z = 0$). (*Credit*: Springel V. *et al.* (2005), *Nature*, 435, 629, "Simulations of the formation, evolution and clustering of galaxies and quasars," and the Millennium Simulation Project, https://wwwmpa.mpa-garching.mpg.de/galform/virgo/millennium/.)

several methods have been devised to reduce the required computation times. Some of the basic methods include:

- Direct integration. This is the slowest method and is used for smaller simulations (Aarseth, 2003).
- Hierarchical tree codes. Distant particles are clumped to form separate massive particles. This generally reduces computation time to order $N \log N$.
- Fast multipole methods.
- Particle mesh methods.

Several international groups have generated and continue to improve cosmological simulations focused on ΛCDM. Many continue to be active and continually update and make results freely available. The Virgo Consortium computed the Millenium simulations. Millenium-I (Springel *et al.*, 2005) included 2160^3 $10^9 M_\odot$ particles corresponding to 20 million galaxies and a $(2 \times 10^9$ LY$)^3$ volume cube using the GADGET program, which is a hierarchical tree algorithm that uses smoothed particle hydrodynamics to include gas (Springel, 2005). The Millenium-II simulation used the same number of particles, but with $6.9 \times 10^6 M_\odot$ particles and a $(400 \times 10^6$ LY$)^3$

cube. The Millenium XXL run used an enlarged volume $(13 \times 10^9\ \text{LY})^3$ (Angulo *et al.*, 2012).

The Bolshoi cosmological simulation (Klypin, Trujillo-Gomez, & Primack, 2011; Klypin *et al.*, 2016) used adaptive mesh refinement (AMR). AMR uses an adaptive mesh refinement tree (AMT), uses dynamic gridding, adapt accuracy according to density higher or lower where needed. This uses a $(10^9\ \text{LY})^3$ cube and up to about 6×10^{10} particles.

The Illustris project (Vogelsberger *et al.*, 2014) is based on the AREPO code that uses a Voronoi tessellation moving mesh, which has advantages over others for including gas dynamics. About 2×10^{10} particles are used in a $(106.5\ \text{Mpc})^3$ cube.

The Eris simulation (Guedes *et al.*, 2011) was intended to simulate the physics of the Milky Way and other spiral galaxies and uses a smaller distance scale.

Hundreds of papers have been written using the results of these calculations. These simulations cannot be expected to give a detailed fit to the actual galaxy or the universe, but statistical methods for treating the calculations and observational results are employed. Also peculiar velocities will be generated and discussed in later chapters.

Chapter 4

Galaxy Evolution and Semi-analytic Models

In the determination of the peculiar velocities of the galaxies, as well as other aspects of astrophysics, it is important to study how the measurements of distance indicators are affected by environment, age, metal content, and other factors governing the properties of galaxies at their observed redshift. Since most of the distance estimators are not based on first principles, their observation and calibration need examination. Important examples are the scaling relations (e.g., TF and FP) for large samples of spirals and ellipticals, relied upon as distance indicators for peculiar motions, but this applies to the other distance methods as well. If one wishes to determine distances good to a few percent or better, these considerations become important.

Some questions can be answered by models of the galaxies and their constituents including the stars, gas, dark matter, etc. Although considerable advances are being pursued using detailed hydrodynamical models combined with N-body simulations to directly calculate the evolution of galaxies employing basic physics, this modelling is labourious and has lacked small-scale spatial resolutions. The semi-analytic models have been developed to attack these problems based on merging histories combined with empirical laws and adjustable parameters to model the different physical processes.

4.1. Galaxy Evolution Models

Simple models of galaxy evolution start assuming a burst of star formation in which the distribution of masses is described by an initial mass function

(IMF), ξ, using,

$$dN = N_0 \xi \left(\frac{M}{M_\odot} \right) \frac{dM}{M_\odot}, \tag{4.1}$$

where dN is the number of stars in the mass interval $(M + dM)$, and N_0 is a normalization factor. Subsequent evolution of the different masses with time is calculated by evolutionary tracks from stellar evolution codes and stellar atmosphere models to give the spectroscopic and photometric properties. Among early models of galaxy evolution are those by Tinsley (1972) and Larson & Tinsley (1978) for the evolution of the broadband UBV filters. using the Salpeter (1955) initial mass function (IMF),

$$\xi(M/M_\odot) = (M/M_\odot)^{-2.35}. \tag{4.2}$$

Figure 4.1 shows the colour changes with age in these models in the $(U - B), (B - V)$ two-colour diagram compared with some observed elliptical galaxies. Even as crude as these early models now seem, this shows a reasonable fit.

The IMF is most important to all aspects of galaxy evolution. It forms the basis of understanding stellar populations. The single power law in equation (4.2) predicts increasing large numbers of low mass stars, while modern data show that their numbers decrease for masses below about a solar mass. For a more detailed fit to known data, several versions of the IMF have been put forward (e.g., Kroupa, 2002). One of these, often used, is Kroupa's (2001) IMF, which is a broken power law of the form

$$\begin{aligned} \xi(M/M_\odot) &= (M/M_\odot)^{-0.3}; (M/M_\odot \leq 0.08) \\ &= (M/M_\odot)^{-1.3}; (M/M_\odot = 0.08 - 0.50) \\ &= (M/M_\odot)^{-2.3}; (M/M_\odot \geq 0.50). \end{aligned} \tag{4.3}$$

For a long time one important question has been whether the IMF and the DM fraction in galaxy centres is universal or is set by a galaxy's environment or formation history. Increasingly, detailed semi-analytical models of galaxy evolution have been constructed utilizing factors that include the IMF, the density structure of the galaxy, stellar evolution models, dynamics of the stars and DM, starbursts, accretion and recycling of the gas component.

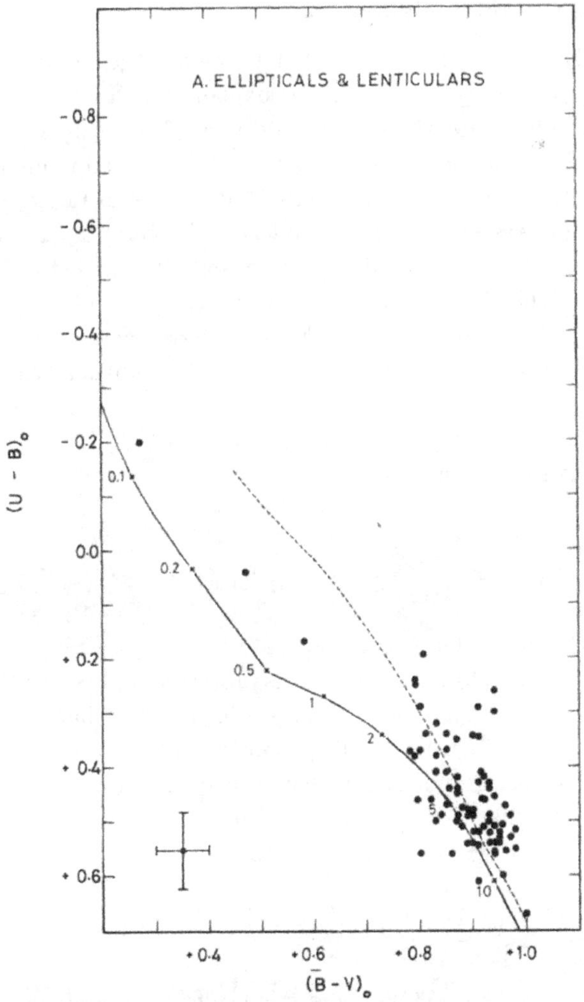

Fig. 4.1. Comparison between observed reddening corrected broadband *UBV* photometry of the inner regions of early-type galaxies and some of the first simple galactic evolution models of Larson & Tinsley (1978). Solid curves are loci of galaxies with a 10^7 year starburst followed by indicated ages in Gyr. The dashed curves show 10^{10} year old systems with decreasing rates of star formation. It is interesting to note that very red ($(B - V)_o > 1.0$) ellipticals do not appear much beyond the limit of ∼10 Gyr, which would give a lower limit on the age of their formation and the universe. (*Source*: Wegner G. (1979), "UBV photometry of 262 southern galaxies," Fig. 4, *Ap&SS*, 60, 15.) [Reprinted by permission from Springer Nature].

4.2. More Complex Galaxy Models

Several early investigators attempted to model the heavy element production in galaxies (Pagel, 2001) to explain the limited data available at the time. Of course these efforts occurred before dark matter was a major consideration and are now superseded, but remain illustrative of the complexity involved. Lynden-Bell (1975) describes examples of the early models and derives some general models with accretion and recycling to explain the observed metals distribution and their structures in forming galaxies. The interplay of several factors governs this and includes the mass loss of the evolving stars and accretion of intergalactic gas. Assume that a galaxy has total mass $M(t)$ at a time t. The total masses in stars and interstellar gas are $s(t)$ and $g(t)$. These depend on the rates at which mass forms into stars, $F(t)$, and the stellar mass lost to gas, $L(t)$, as stars evolve, so the conservation of mass gives:

$$\dot{s} = F - L, \tag{4.4}$$

$$\dot{s} + \dot{g} = \dot{M} = A, \tag{4.5}$$

where A is the intergalactic accretion rate.

Some of the mass formed into stars is processed and returns to the interstellar medium via supernovae, planetary nebulae, stellar winds, etc. The mass loss from these effects will be proportional to the star formation rate at an earlier 'cooking' time Δ for each process, $F(t-\Delta)$. For nuclear processes, assuming that the mass loss after formation decreases exponentially with an e-folding time τ, L^* the mass lost will be

$$L^* = \beta \int_0^t \exp(-\Delta/\tau) F(t-\Delta) d\Delta/\tau, \tag{4.6}$$

where as $\tau \to 0$, $L^* \to \beta F(t)$, where β is the fraction without nuclear processing. If the abundance of a metal Z_i were returned to the interstellar medium unchanged, then its amount would be $Z_i L^*$, but since nuclear reactions have gone on in the stars for the cooking time, each element's abundance will be enhanced and an amount of the element y_i will be returned. Consequently the mass lost is the sum of the enriched material by nuclear reactions and the material returned unchanged:

$$L_i^* = y_i \int_0^t \exp(-\Delta/\tau_i) F(t-\Delta) d\Delta/\tau_i + \beta \int_0^t \exp(-\Delta/\tau) Z_i(t-\Delta) d\Delta/\tau. \tag{4.7}$$

Also there are contributions of slow processes that require times comparable to the galaxy's age and are approximately constant so that in total

$$L = L^* + \omega s, \tag{4.8}$$

$$L_i = L_i^* = \omega z_i s, \tag{4.9}$$

where z_i is the abundance of the slowly returned element and ωs the steady mass-loss rate of stellar material

Equations (4.4) and (4.5) along with

$$\frac{d}{dt}(gZ_i) = L_i - Z_i F, \tag{4.10}$$

relate the evolution of s, g, and Z_i. The total mass of element i will be $Z_i = g_i/g$ once these quantities are defined. The system of equations (4.4) to (4.10) can be applied to rings or shells in the galaxies weighting them by the masses of these regions. Secondary and tertiary elements produced from the production of the first generation of elements, e.g., by the r and s processes, would be included setting their yield proportional to the abundance of the parent, Z_i, by $y_{II}^* = Y_{II}^* Z_i$.

This system of equations becomes increasingly complicated depending on the choices of processes and their mathematical description. The simplest model due to Schmidt (1963) assumes no accretion ($A = 0$) and instantaneous cooking time ($\Delta = 0$), and disagrees with observed metal abundances in G stars as it produces too many low metal stars. Lynden-Bell (1975) describes two models that are soluble, where the details are set out and will not be elaborated here. The first is the 'best accretion model'. The gas accretion rises to a maximum and then ends; this produces a fast initial enrichment and few stars with low metals and most stars being in a narrow range of metal abundances. The second is the 'complicated linear model', which assumes a linear relationship between star formation rate and gas mass and considers how lags in star formation affect element production.

Considerably more observational data are now available and many refinements and complexity have been introduced to fit the detailed observations making analytic methods difficult and these models are handled using numerical methods. Cole *et al.* (2000), Almeida *et al.* (2014), Lagos *et al.* (2018) have described some of these.

4.3. Semi-analytic Models

Semi-analytic models have been developed with increasing refinements with the purpose of predicting the evolution of galaxy properties in

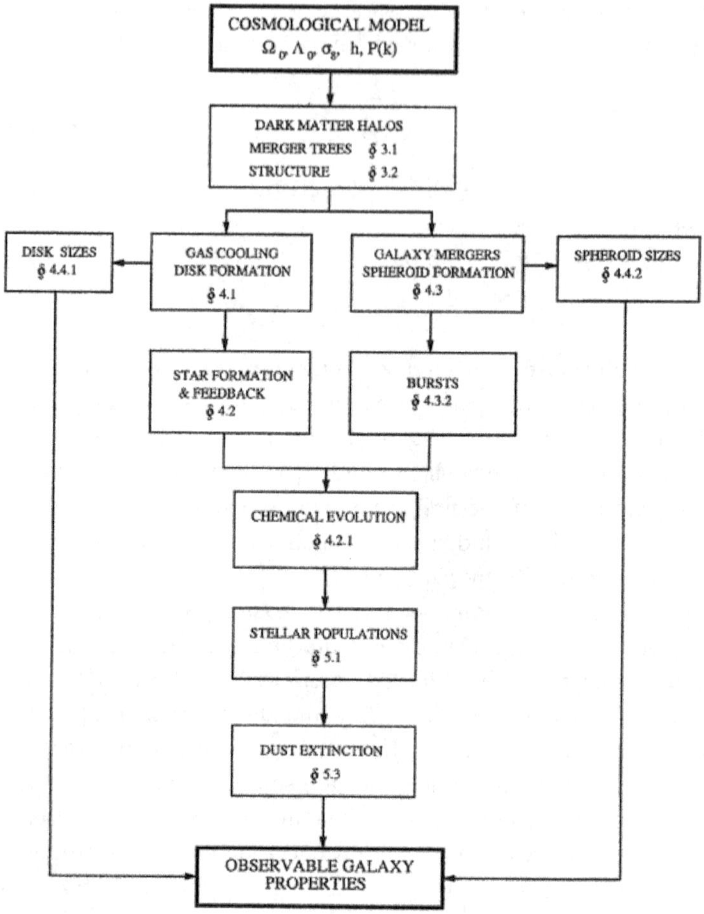

Fig. 4.2. A flowchart of how different physical processes, starting with an assumed cosmological model, combine to model observable parameters of galaxies. Numbers in the boxes refer to the original paper. (*Source*: Cole S., Lacey C. G., Baugh C. M., Frenk C. S. (2000), *MNRAS*, 319, 168, "Hierarchical galaxy formation," Fig. 1, Reproduced by permission of Oxford University Press on behalf of the Royal Astronomical Society.)

hierarchical clustering (e.g., Lacey *et al.*, 2001; Somerville *et al.*, 2008; Hopkins *et al.*, 2008), but other cosmologies, self-interacting dark matter, etc., can be considered. Figure 4.2 shows a flow chart of different steps involved in the semi-analytic models of Cole *et al.* (2000). Formulae are adopted to describe each process as shown in the diagram. Most current semi-analytic models assume standard ΛCDM cosmological parameters

($\Omega_\Lambda \approx 0.7, \Omega_m \approx 0.3, \Omega_b \approx 0.05, H_0 \approx 70\text{km s}^{-1}\,\text{Mpc}^{-1}$, etc.). The density fluctuations start from a power law of the $P(k) \sim k^n$ Harrison–Zel'dovich form and a Gaussian random field as outlined in Chapter 3. The dark matter (DM), usually considered to be collisionless, dominates so the subsequent histories of the DM density fluctuations are driven by DM, which grow into dark matter halos. As described in Chapter 18, the evolution can be understood based on a spherical collapse model where regions that are dense enough,

$$\rho/\rho_0 > 1.69/(1+z), \tag{4.11}$$

decouple from the Hubble expansion and become gravitationally bound to form DM halos. As different mass halos merge (e.g., large + small, small + small masses, etc.), the distribution of masses, $N(M)$, can be found from the Press–Schechter formula (Press & Schechter, 1974; Longair, 1998; and Chapter 18) or Monte-Carlo methods and halo merger trees are built up, which is the distribution of halo masses with redshift as the smaller halos merge and form increasingly large halos with time.

Numerous examples of a merger tree can be found in the literature (e.g., Lacey & Cole, 1993). The density profiles of the DM halos are taken to be representative of DM simulations, e.g., the NFW profile (Navarro, Frenk, & White, 1996). More recently, detailed merger trees are produced using N-body simulations.

Although the details involving collisionless DM are well studied, the situation increases in complexity for the baryonic matter component, initially in the form of a diffuse gas, which ultimately is observed as galaxies. Several complicated processes involving the baryons are not as well understood. They are not calculated from first principles and are used in the models as free parameters or follow empirical relations. Here some of the major processes are described.

When the DM halo turns around from the expansion and begins to contract, the diffuse gas falls towards its centre and the gas forms a rotating disk. The gas is shock heated to the halo virial temperature $kT = \mu m_H V_c^2/2$ where V_c is the disk's rotation rate. The disk cools by radiative processes that include photoionization and radiative cooling via atomic lines. The cooled gas can form stars at a rate $\dot{M}^\star = M_{\text{cold}}/t^\star$ where M_{cold} is the cold gas mass in the halo and $t^\star$ is a characteristic time, both being model parameters. The mass distribution of the stars is set by the IMF.

Subsequently, supernova feedback is an important process. The massive stars formed evolve quickly and explode as supernovae. The radiation and the expanding shells' kinetic energy reheats some of the gas and pushes some of it into the halo so that $\dot{M}_{\rm eject} \propto \dot{M}^{\star}$ reducing the available gas and regulates the star formation so that a galaxy is unable to form all the gas into stars. Recycling of the gas also occurs and results in chemical evolution of the gas, such as metals enrichment already mentioned. AGN accretion and feedback are also included.

The chemical enrichment produces dust, which is subsequently included in the models. Stars and gas in different parts of the formed galaxies evolve differently and bulge + disk models are constructed. Mergers and starbursts are included. If two halos merge, the result depends on their mass ratio. When the masses are nearly equal, these are major mergers, and the disks will be destroyed with gas consumed in a starburst, ultimately resulting in elliptical galaxies. If the mass ratio of the two galaxies is largely unequal, the smaller halo begins to orbit its larger companion and dynamical friction causes it to spiral into the companion with time. Tidal stripping will occur and the smaller system may be tidally disrupted.

The normalization of the free parameters in the semi-analytic models is facilitated by comparing the predictions of the models with a number of observed galaxy relations. These include, e.g., properties of the Milky Way, the Tully–Fisher relation, the star formation rate, and the stellar mass density vs. redshift (Madau, Pozzetti, & Dickenson, 1998; Madau & Dickenson, 2014), galaxy morphologies at low ($z < 1$) and high ($z = 2$–3) redshifts, etc. Figure 4.3 illustrates how some of these observations are utilized to constrain the parameters in models constructed by Cole *et al.* (2000). GALACTICUS (Benson *et al.*, 2012) is a publicly available semi-analytic code that is in the Astrophysics Source Code Library (ASCL.net).

The semi-analytic models can explain several properties of galaxies. They predict evolution of the slope and zero point of the Faber–Jackson relation (Gallazzi *et al.*, 2006). Covington *et al.* (2011) find that dissipation during mergers accounts for reduced scatter of the size-mass relation observed in disk galaxies and the tilt of the Fundamental Plane (the relationship for elliptical galaxies between the effective radii, velocity dispersions, and surface brightness described in Chapter 11). The semi-analytic models predict the evolution of the red and blue sequences in the colour-magnitude diagrams of galaxy clusters described in Chapter 18. Galaxies in the blue cloud are generally less massive and have more gas. They form from 'wet mergers' and as their gas is depleted by star formation and mergers, they move into the more massive red cloud which is formed by 'dry mergers'. The star formation history and stellar mass buildup can be fit (Madau *et al.*, 1998; Madau & Dickenson, 2014) as well as the extragalactic

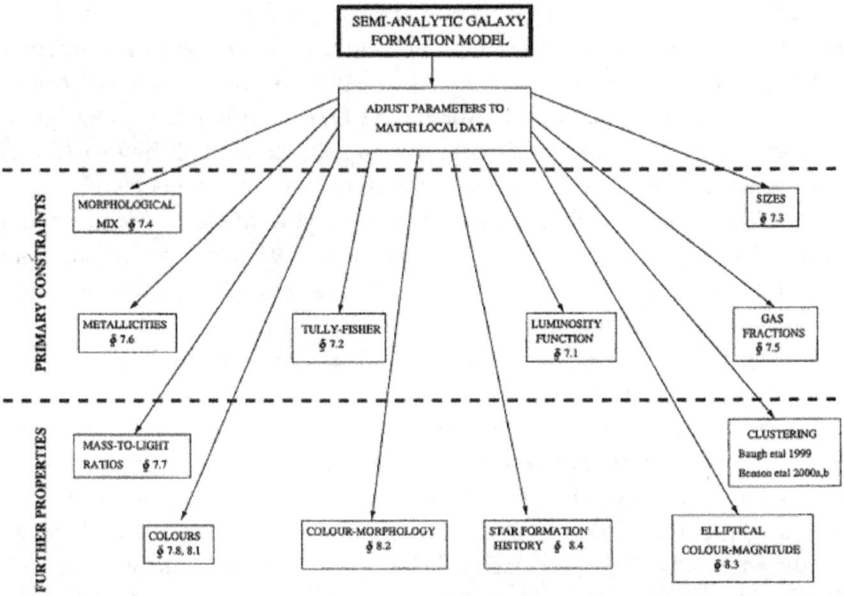

Fig. 4.3. Illustration of how the galaxy properties predicted by the model in Fig. 4.2 are compared with observations and used to adjust the model. (*Source*: Cole S., Lacey C. G., Baugh C. M., Frenk C. S. (2000), *MNRAS*, 319, 168, "Hierarchical galaxy formation," Fig. 2, Reproduced by permission of Oxford University Press on behalf of the Royal Astronomical Society.)

background light and galaxies at higher redshifts ($z = 2$–3). Although the semi-analytic models continually improve with time, they are based on empirical relations with many free parameters and they are not fully predictive, so many details of galaxy evolution remain to be explained.

4.4. Colour and Age Models

Crucial to the investigation of galaxy ages and chemical composition used in all the models are the comparison of the observed spectra with stellar population models. Evolving from the simple Tinsley models that compare broad band photometric indices, these have developed in detail and several groups have produced libraries of these models utilizing compilations of stellar spectra. Starting either from computed stellar atmosphere models, (e.g., Kurucz, 1992) or empirical spectra such as from the Multi-Unit Spectroscopic Explorer (MUSE) library (Ivanov *et al.*, 2014) or the MILES stellar library (Falcón-Barroso *et al.*, 2011; Röck *et al.*, 2016), choices of the IMF and reddening law are made. Bruzual & Charlot (2003) produced software that has been extensively applied that can be used with different IMFs, reddening, and aging functions to calculate the spectra and colour indices of stellar populations. Worthey (1994) describes how these models can be constructed and calibrated. Thomas, Marston, & Bender (2003) calculated Lick indices (Chapter 7) for stellar evolution models based on theoretical stellar population models with varying metal abundances and ages 1–15 Gyr. With the advances in instrumentation, the entire observed spectrum can be compared with the theory. Ge *et al.* (2019) have compared the results of five different libraries.

Using these models to interpret the observations of galaxies, correlations between age, metals, and dynamics can be found. Nelan *et al.* (2005) measured Lick indices and velocity dispersions, σ, for 4097 early-types in 93 galaxy clusters and used the Thomas *et al.* (2003) population models to determine ages and element abundances, with the results that age, $\tau \propto \sigma^{0.59}$, metals, $Z/H \propto \sigma^{0.53}$, and alpha elements (i.e. lighter elements produced by the α-process fusing helium into heavier elements, e.g., C, O, Ne, Mg, Si, S) relative to iron, $\alpha/Fe \propto \sigma^{0.31}$, which agreed with older studies generally using smaller samples referenced in that paper. The most significant result was the measurement of the age–mass relation for red sequence cluster galaxies. Since σ correlates with mass, the most massive are older and have more metals. This gives weight to the 'downsizing' picture that the cluster galaxies in the red sequence grow with time starting with the more massive galaxies forming at ~13 Gyr and the lower mass galaxies emerging later at ~4 Gyr. McDermid *et al.* (2015) find similar results using the ATLAS^{3D} data. Galaxies in different density environments were compared. In higher density regions, such as the environs of the Virgo cluster, galaxies are older, more alpha enhanced, and have shorter star formation durations than galaxies in lower density regions.

4.5. Dynamical Analyses of Elliptical Galaxies

Early-type galaxies (morphological types E and S0) span a wide mass range, from the most massive to dwarfs. In contrast to the spirals, their stars are old, they contain only small amounts of gas, and represent some of the most ancient galaxies. In addition to their application as a distance indicator in Chapter 11, they have been the subject of many extensive observational investigations to study the dynamics of their stars, ages, and metal contents. In this active field, many theoretical studies have been described (e.g., Osipkov, 1979; Merritt, 1985; Corollo, de Zeeuw, & van der Marel, 1995; Binney & Tremaine, 2008) and by results of N-body simulations such as described in Chapter 3.

Observations of velocity dispersions, rotation, and metal line strengths at points across the projected surfaces of the early-types as well as their light profiles provide information that can be employed to determine the dynamical structures of these objects. Measurements of σ, the Gauss–Hermite coefficients, H_3, H_4, plus the brightness profile at points on the galaxy can be used to determine the luminous and DM content.

Flattened axisymmetric early-type galaxies were studied in the Coma cluster and the less dense cluster Abell 262, comparable to the Virgo cluster, for nearly a decade (Mehlert *et al.*, 1998, 2000, 2003; Wegner *et al.*, 2002; Corsini *et al.*, 2008). Spectroscopy was secured by placing a long slit along the major and minor axes of such a galaxy, as shown in Fig. 4.4, as well as

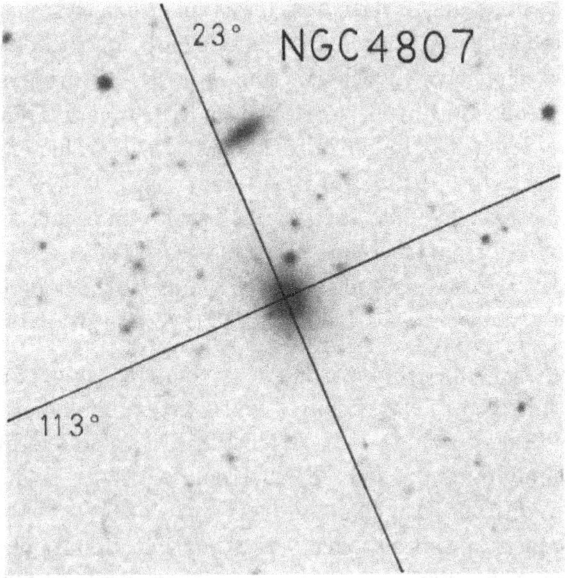

Fig. 4.4. Observed longslit positions on NGC 4807 as used in Thomas *et al.* (2005) from the POSS-2 *R*. (The Digitized Sky Surveys were produced at the Space Telescope Science Institute under U.S. Government grant NAG W-2166. The images of these surveys are based on photographic data obtained using the Oschin Schmidt Telescope on Palomar Mountain and the UK Schmidt Telescope. The plates were processed into the present compressed digital form with the permission of these institutions.) The image is approximately 5 arcmin on edge with N on the top and W to the right. The position angle of the major axis was taken to be 23° and the minor axis 113°.

other positions so that σ, H_3, H_4, and metal line strengths could be obtained at points between the centre and extending outwards to typically 1–3 R_e. The line strength index measurements were used for determining ages and metal abundances (Mehlert *et al.*, 2003).

Of the many spectroscopic campaigns of galaxies, the larger ATLAS^{3D} project observed 260 early-type galaxies, although not specifically focused on flattened systems. This used the SAURON integral-field spectrograph on the William Herschel Telescope (Cappellari *et al.*, 2011 and references therein) with a 1431 grid of 0.94 arcsec pixels in a $41 \times 33\,\mathrm{arcsec}^2$ field. As shown in Fig. 4.4 for the galaxy NGC 4807, the long slit placed at different position angles covers axially symmetric targets and provides a longer linear region of ± 30 arcsec with a scale of 0.5–0.6 arcsec along its length. Ma *et al.* (2014) and Greene *et al.* (2015) have conducted a similar investigation of the 100 most massive early-type galaxies within a distance, $D < 108$ Mpc using integral field spectroscopy.

Dynamical modelling of the ellipticals as spheroidal systems uses the two dimensional long slit and photometric data is based on the 'orbit superposition technique' pioneered by Schwarzschild (1979) who originally applied it to triaxial systems and further developed by Richstone & Tremaine (1988), Thomas *et al.*(2004, 2005b, 2007a,b, 2009a,b, 2011). This employs Jeans' theorem (Jeans, 1915, 1919; Binney & Tremaine, 2008), wherein the stationary distribution functions are described by the integrals of motion. For an axisymmetric system, there are three integrals, the energy E, the angular momentum along the symmetry axis, L_z, plus I_3, the non-classical 'third integral'. These major steps are included in the application of the method:

- Assume a density distribution. For a system containing DM, $\rho = \rho_* + \rho_{\rm halo}$, where ρ_* represents the stars and matter that follow their profile and $\rho_{\rm halo}$ is taken to be spherical. Typically, a cored profile given by the logarithmic potential (Binney & Tremaine, 2008) or a NFW profile (Navarro *et al.*, 1996) might be used.
- Use these density profiles to calculate the gravitational potential.
- Calculate an orbit library in the potential, typically 13,000 to 19,000 orbits including both direct and retrograde orbits.
- Calculate the integrals of motion.
- Check to recover the assumed density distribution.

A maximum entropy technique was employed for the fitting, where the quantity $\hat{S}$ related to the entropy, S,

$$\hat{S} = S - \alpha\chi^2_{\rm LOSVD}, \tag{4.12}$$

is maximized (Richstone & Tremaine, 1988), where α denotes a constant that quantifies the relative importance of maximizing the entropy and minimizing χ^2_{LOSVD}, determined from the observations along the line of sight. Empirically $\alpha = 0.02$.

Fits to the observed spectroscopic kinematical data, σ, V, H_3, and H_4 along with the optical light profiles of a galaxy can then be used to fix its light profile, assuming a LF, which gives mass-to-light ratio. Then the density is the sum of the luminous and dark components,

$$\rho = \Upsilon_* \nu + \rho_{DM}. \tag{4.13}$$

Υ_* denotes the stellar mass-to-light ratio and ν is the stellar number density so $\Upsilon_* \nu$ describes the mass in stars. Generally, the kinematics provide a larger mass than just the stars and the difference is made up from a DM component.

A series of calculations using the orbit superposition technique was carried out using data for the Coma and A262 ellipticals. Thomas *et al.* (2007) constructed dynamical modelling of 17 Coma early-types. Thomas *et al.* (2009a,b) studied the flattening and orbital structure of these galaxies and their DM scaling relations. Thomas *et al.* (2011) compared the derived dynamical masses of 16 Coma galaxies with masses using gravitational lensing, finding that the masses from the two methods agree within about 20–30%, although for the low σ galaxies if mass follows light the masses are higher than the lensing values but for the high σ objects, the two component models give masses in agreement with the lensing. Several galaxies in the Abell 262 cluster display Balmer H I lines, that provide independent rotational velocities and provide a consistency check with the photometric and kinematical data.

New and improved versions of the Schwarzschild orbit superposition technique as well as more extensive and better observations continue to appear such as the SMART program (Neureiter *et al.*, 2021). Here a few typical investigations will be described.

Halo masses correlate neither with luminosity nor population age, α-element abundances and metallicity as determined from line indices as explained in Chapter 7. The majority of the early-types surveyed halos 2–10 times denser than spirals in the work of Persic *et al.* (1996) as derived from the kinematical fits. Using the simple spherical collapse model, the formation redshift $(1 + z_{\mathrm{DM}})^3 \sim \rho_{\mathrm{DM}}$, spirals form at $z_{\mathrm{DM}} \sim 1$ and early-types earlier at $z_{\mathrm{DM}} \sim 2$–3. For the majority, $z_{\mathrm{DM}} \approx z$, but cases appear where the stars are older than the halo and some younger.

Figure 4.5 illustrates the analysis of the $V = 12.6$ mag flattened early-type galaxy NGC 4807 in the Coma cluster in Fig. 4.4 (Thomas *et al.*, 2005). This is typical of the features of other galaxies studied. Here one sees dark and luminous matter profiles where the DM dominates in the outer regions. Panel (a) shows the distribution function using $\alpha = 0.02$, (b) compares the measured major and minor axis rotational velocity, the velocity dispersion, H_3 and H_4, and the errors with the model fits. Section (c) shows the mass reconstruction from the stellar and DM contributions and in (d) the major axis stellar mass density is above that of DM until $r = r_{\rm eff}$ where they cross. For most objects, redshift $z_* \approx z_{\rm DM}$, while galaxies with $z_* > z_{\rm DM}$ may be from a gas-poor binary merger, which would move them down in the plot and galaxies with $z_* < z_{\rm DM}$ would have had an episode of star formation after the formation of the halo.

Some of the results found for the flattened early-type galaxies (Fig. 4.6) are summarized in Wegner *et al.* (2012) and Corsini *et al.* (2013, 2017). The majority, but not all, of objects exhibit significant evidence for DM halos. In the Coma cluster 8/17 and in the Abell 242 cluster, 4/8 the significance is greater than 95%. On average, the DM fraction of halo mass to total mass inside R_e is $\langle f_{\rm halo} \rangle = 0.2$, a result similar to earlier studies of spherical systems (e.g., Gerhard *et al.*, 2000). It should be noted that when dark and luminous matter closely follow similar radial profiles, it is difficult to separate the two.

The systematic trend found between the early-type galaxies with mass might result from a non-universal IMF or dynamical evolution as well as a combination. The processes of baryonic contraction and violent relaxation have been suggested to produce similar DM and light profiles (see Chapter 18). However, questions remain about the operation of these processes in the galaxies.

The ATLAS$^{\rm 3D}$ project obtained observational data for a 260-volume limited sample of early-type galaxies with the integral-field instrument already mentioned, and is described in a series of papers described in Cappellari *et al.* (2011). In the series of papers, Young *et al.* (2011) quantified the molecular gas content of the galaxies and Davis *et al.* (2011) studied the TF relation using the molecular lines. Bois *et al.* (2011) compared fast and slow rotators using numerical simulations of binary mergers and Khochfar *et al.* (2011) studied the formation of the formation of early-type galaxies using semi-analytic modelling. Detailed axisymmetric dynamical models were applied to get M/L, DM content, and the classical

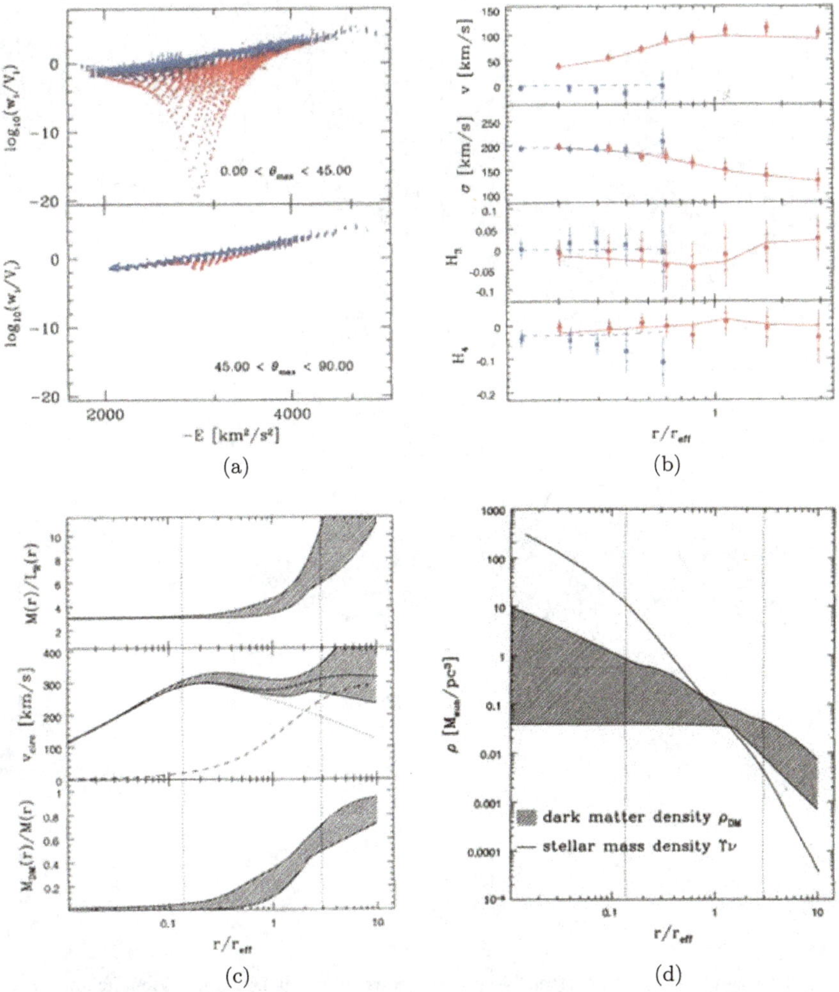

Fig. 4.5. Some results of the Schwarzschild orbit superposition modelling analysis of NGC 4807. (a) The distribution function for the best fitting orbit model with $\alpha = 0.02$ for orbits in 45° cones, top around the equator and lower remaining orbits. Red and blue dots are the phase-space density along separate orbits. (b) Fits to the observed kinematical data. Red dots: major axis, blue squares: minor axis. (c) Mass reconstruction aspects for the $\alpha = 0.02$ model. The thick areas represent 68% confidence levels. In the central panel, the contributions to the rotation curve from stellar matter and dark matter are shown as the dotted and dashed curves. (d) The mass densities for dark and stellar matter along the major axis. (*Source*: Thomas J., Saglia R. P., Bender R., Thomas D., Gebhardt K., Magorrian J., Corsini E. M., Wegner G. (2005), *MNRAS*, 360, 1355, "Regularized orbit models unveiling the stellar structure and dark matter halo of the Coma elliptical NGC 4807," Figs. 13, 15, 16, 20, Reproduced by permission of Oxford University Press on behalf of the Royal Astronomical Society.)

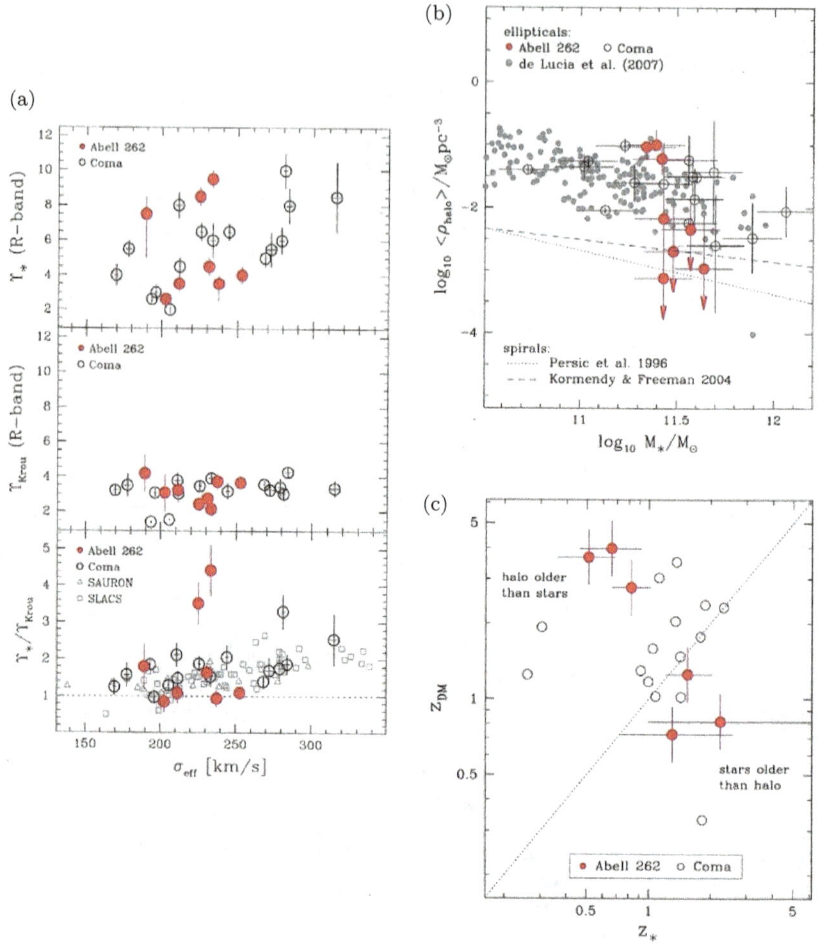

Fig. 4.6. Some properties of early-type galaxies in Abell 262 and Coma from fitting kinematics and photometry. (a) Mass-to-light ratios, Υ, in the $R_{\rm KC}$ band vs. velocity dispersion ($z_{\rm eff}$). Top: dynamical estimates, centre: Kroupa IMF, bottom: their ratio adding SLACS (Treu *et al.*, 2010; Cappellari *et al.*, 2006) and showing the trend with $\sigma_{\rm eff}$. (b) The average dark halo density within $2R_e$ from dynamical models vs. mass, M_* that follows light for Coma (open circles from Thomas *et al.*, 2009b) and Abell 262 (red circles), and DeLucia & Blaizot (2007) semi-analytic models (grey circles). Blue lines are from spiral galaxy scaling relations (Persic *et al.*, 1996; Kormendy & Freeman, 2004) and show the trend of higher densities in early-type galaxies. (c) Comparison of assembly redshifts for stellar populations (z_*) using a Kroupa IMF and dark matter halos (z_{DM}) in Abell 262 (red circles) and Coma (open circles from Thomas *et al.*, 2011). The dotted diagonal is for $z_* = z_{\rm DM}$. (*Source*: Wegner G. A., Corsini E. M., Thomas J., Saglia R. P., Bender R., Pu S. B. (2012), *AJ*, 144, 78, "Further evidence for large central mass-to-light ratios in early-type galaxies: the case of ellipticals and lenticulars in the A262 cluster," Figs. 11–13, ©AAS reproduced with permission.)

fundamental plane (Capellari *et al.*, 2013a,b). This study also finds that the IMF is not universal with a transition from the Kroupa relation to the Salpeter function in the velocity dispersion range $\sigma_{\rm eff} \approx 90$–$290\,{\rm km\ s^{-1}}$, which can be compared to Conroy & van Dokkum (2012). Measurements of the fundamental plane are discussed in Chapter 11.

Chapter 5

Early Developments on Large-Scale Structure

5.1. The Correlation Function

Discussions about the large-scale distribution of the galaxies have existed over two centuries and can be found e.g., in the works of Wright (1750), Kant (1755), and Lambert (1761), long before the many modern investigators have grappled with this problem. Charlier (1922a,b), Hubble (1934) and others had tried to map the galaxy distribution on the sky. Some early investigations presented qualitative indications that the universe is inhomogeneous on large scales, e.g., Hoffmeister (1962) described a region unusually low in galaxies in the southern sky. From counts of galaxies, Shapley (1957) used the term 'metagalaxy', and de Vaucouleurs (1958) described the local supercluster.

However, the analysis of the Lick galaxy counts by Neyman & Scott (1959) investigating the clustering of the galaxies pioneered the development of modern quantitative methods for measuring the properties of the observed galaxy distributions. Different investigators considered whether or not there are limitations to the size of the clustering, e.g., Zwicky (1958) held that structures' sizes do not extend beyond the clusters or if it is random or hierarchical.

At first sight, the distribution of the galaxies may appear random, but for analysis and comparisons with numerical simulations, statistical tests are needed. The n-point correlation function in two, three, or more dimensions is one of the earliest and a standard method that is extensively employed as an important tool in cosmology. In Fig. 5.1(a), if the density of objects is n, then the probability, δP, that an object falls in a small

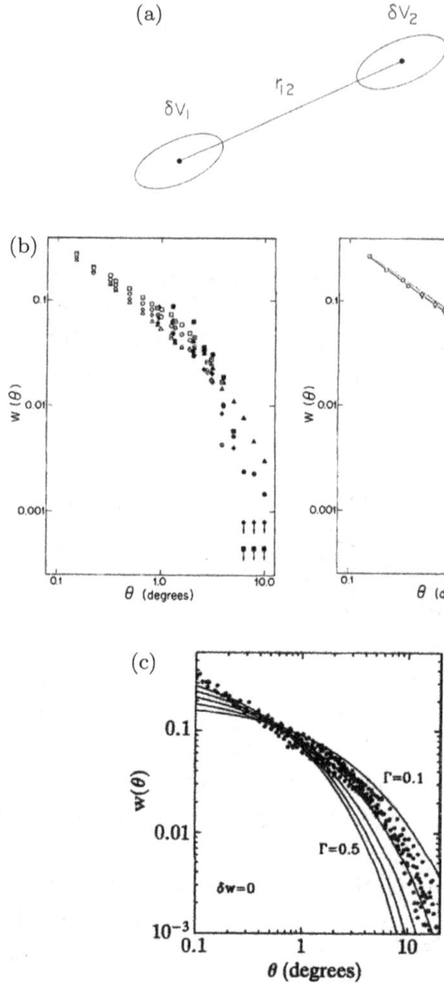

Fig. 5.1. Some 2-point angular correlation functions from galaxy counts. (a) Illustrating the definition of the 2-point correlation function. (b) Early 2-point angular correlation functions, $w(\theta)$ (Groth & Peebles, 1977). Left: Results derived from different surveys. Right: Averages for the estimates. (c) 2-Point angular correlation function from the APM Survey (Efstathiou *et al.*, 1990; Efstathiou, 1993). Curves are for ΛCDM simulations using Γs in Eq. (3.1). (*Source*: (a) Author, (b) Groth E. J., Peebles P. J. E. (1977), "Statistical analysis of catalogues of extragalactic objects. VII. Two- and three-point correlation functions for the high-resolution Shane-Wirtanen catalogue of galaxies," *ApJ*, 217, 385, Figs. 4 and 5, ©AAS reproduced with permission, (c) *Source*: Efstathiou G. (1993), Galaxy clustering on large scales, *PNAS*, 90, 4859, reproduced by permission ©1993, National Academy of Sciences, U.S.A.)

volume, δV, is

$$\delta P = n\delta V. \tag{5.1}$$

For objects at two different locations separated by a distance r_{12} each being in the small volumes, $\delta V_1, \delta V_2$, is

$$\delta P = n^2 \delta V_1 \delta V_2 \left[1 + \xi(r_{12})\right], \tag{5.2}$$

where ξ is the correlation function.

For measured positions of galaxies on the sky, the two-point angular correlation, $w(\theta)$, is

$$\delta P = n^2 \delta\Omega_1 \delta\Omega_2 \left[1 + w(\theta)\right], \tag{5.3}$$

where $\delta\Omega_1$ and $\delta\Omega_2$ are small solid angles on the sky. The two-point function, $w(\theta)$, is the simplest and can be found from imaging data.

The three-dimensional correlation function, $\xi(r)$, needs distance information, typically as given by redshifts, and requires a more extensive observational effort, but w and ξ can be related by Limber's equation (Limber, 1954; Rubin, 1954) if the galaxies' luminosity function is known,

$$w(x) = \frac{\int_0^\infty y^4 \phi(y)^2 \mathrm{d}y \int_{-\infty}^{\infty} \mathrm{d}u \xi(\sqrt{(u^2 + (xy)^2)})}{\left[\int_0^\infty y^2 \mathrm{d}y \phi(y)\right]^2}, \tag{5.4}$$

where $\phi(y)$ is the luminosity function. An early approximation of ϕ consisting of two broken lines is due to Abell (1962), but this is better fit by Schecter's (1976) function

$$\phi(y) = \phi_* y^\alpha e^{-y}, \tag{5.5}$$

where $y = L/L_*$, ϕ_* normalizes the number density, $\alpha \approx -1.07$, and the luminosity L_* is where $\phi(y)$ stops decreasing exponentially.

Numerical methods for calculating the correlation functions can be found in great detail elsewhere (Peebles, 1980; Efstathiou *et al.*, 1990; Dalton *et al.*, 1992). Under the simplest conditions, the correlation functions are power laws:

$$w(\theta) = A\theta^{1-\gamma}, \tag{5.6}$$

$$\xi(r) = \left(\frac{r}{r_0}\right)^{-\gamma}. \tag{5.7}$$

The value of $\gamma \approx 1.8$ is approximately valid over a wide range of objects.

The angular correlation function, $w(\theta)$, was found before extensive redshift surveys were available and the spatial function, $\xi(r)$, was estimated from Limber's (1954) equation. The higher order correlation functions are discussed, e.g., by Peebles (1980) and Dalton *et al.* (1992). This can be related to the power spectrum of the structure. Allen *et al.* (2005) is one example of the determination of the angular correlation function $w(\theta)$ from CCD observations.

It is well known from other fields, e.g., signal processing theory, quantum mechanics, etc., that the power spectrum and the correlation spectrum are related by a Fourier transform pair. In three dimensions, such a pair has the form

$$\begin{aligned} \varphi(\mathbf{k}) &= \int d^3x f(\mathbf{x}) e^{-i\mathbf{k}\cdot\mathbf{x}}, \\ f(\mathbf{x}) &= \int \frac{d^3k}{(2\pi)^3} \varphi(\mathbf{k}) e^{i\mathbf{k}\cdot\mathbf{x}}, \end{aligned} \tag{5.8}$$

where $\mathbf{k}$ and $\mathbf{x}$ denote frequency and spatial vectors and $k \sim 2\pi/L$.

For many astronomical applications, the power spectrum, $P(k)$ and $\xi(r)$ are related by the pair

$$\begin{aligned} P(k) &= 2\pi \int_0^\infty \mathrm{d}r r^2 \frac{\sin(kr)}{kr} \xi(r), \\ \xi(r) &= \frac{1}{2\pi^2} \int_0^\infty dk k^2 \frac{\sin(kr)}{kr} P(k). \end{aligned} \tag{5.9}$$

This enables determination of the power spectrum of different objects and the comparison with model power spectra, e.g., hot dark matter (HDM), cold dark matter (CDM) and the many others, e.g., hot, cold, fuzzy, tilted, etc. Bond *et al.* (1987) considered the correlation function in terms of CDM. Figure 5.1 has examples of early determinations of $w(\theta)$. Efstathiou *et al.* (1990) used galaxy counts from the APM survey (Maddox *et al.*, 1990) and showed that cold dark matter simulations can fit the observational data, which led to further development of ΛCDM. In Chapter 16, the application of newer CMB data and large redshift surveys for the determination of $\xi(r)$ and the newer power spectrum leading to Baryonic Acoustic Oscillations (BAO) is discussed.

5.2. Redshift Surveys

With the availability of sufficiently large numbers of redshifts, maps of the three-dimensional galaxy distributions can be constructed. These are not strictly true spatial maps, but actually maps of the galaxies in redshift space. The data can be displayed by constructing plots in relatively thin strips across regions of the sky. Galaxies are placed using a distance given by Hubble's law, $D = \frac{H_0}{cz}$ along the vector corresponding to their direction from earth. Galaxy peculiar motions will produce redshift space distortions (RSD), but at large redshift these are observed to be of smaller magnitude. These plots called 'cone' or 'pie' diagrams take the fan shape with the earth at the narrow vertex. An early picture of the galaxy distributions was by Lapparant & Geller (1985) based on the ZCAT data and centred on the Coma cluster. Such plots were followed by others, e.g., Fairall (1985) using other redshift surveys and collected from the literature.

Figures 5.2 and 5.3 are cone diagrams from the Century Survey (Wegner *et al.*, 2002). Figure 5.4 shows galaxies in the direction of the Perseus–Pisces region (Giovanelli & Haynes, 1985; Wegner *et al.*, 1993). The first view shows the positions of the galaxies as they appear on the sky. The following two plots give the same objects in redshift space corresponding to the sky distribution and the cone diagram. The structures approximate the spatial distribution of the galaxies, but as these are redshift measurements any space motions of the galaxies produce redshift distortions. The most notable feature of this effect is the elongated feature corresponding to the

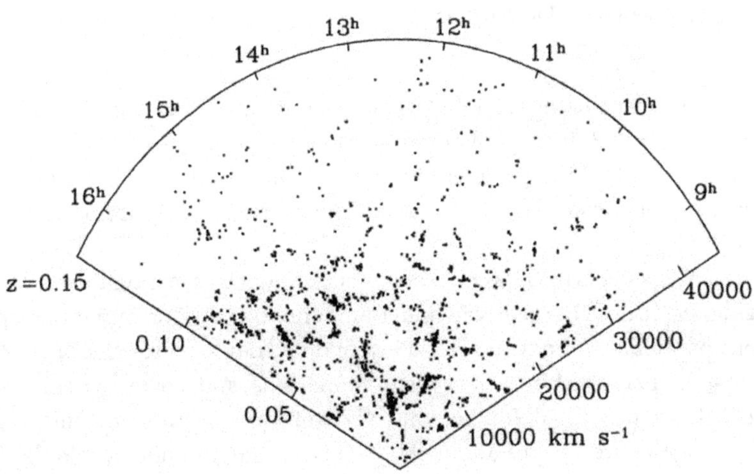

Fig. 5.2. Example of a wide cone diagram of galaxy distributions in redshift compiled from the Century Survey. (*Source*: Wegner G., Thorstensen J. R., Kurtz M. J., Brown W. R., Fabricant D. G., Geller M. J., Huchra J. P., Markzke R. O., Sakai S. (2001), *AJ*, 122, 2893, "Redshifts for 2410 galaxies in the century survey region," Fig. 7, ©AAS Reproduced with permission.)

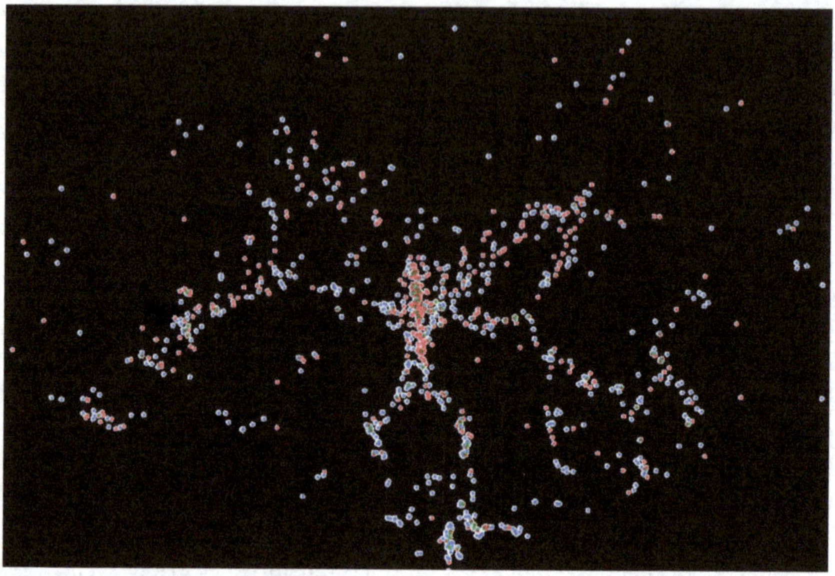

Fig. 5.3. Enlargement of the inner part of the region in Fig. 5.2 from the first slice of the CFA cone diagram extending to redshift $z \sim 0.1$ showing the distributions of the different morphological types measurements in Wegner *et al.* (2002). Red dots denote early type galaxies and blue dots the spirals. (*Credit*: courtesy Kurtz M. J., ©SAO.)

massive Perseus cluster (Abell 426) at redshift $cz = 5366$ km s^{-1}near right ascension, $\alpha = 3.2$ hours, colloquially known as a 'finger of God,' which is produced by the orbital motions of the cluster members. Accurate measures of these distortions contain additional information which are addressed in Chapter 13.

In the 21st century, MOS instruments and large arrays of CCDs for imaging are producing overwhelming amounts of data for numerous applications of stellar and extragalactic astronomy that are growing in size and reaching higher redshifts. For peculiar motions and cosmological results these will be described further on. The surveys are well documented in their respective web pages and most of the data are publicly available. The 2dF and related surveys in the southern hemisphere, measured over 400,000 redshifts. The Sloan Digital Sky Survey (SDSS) covers about 35% of the sky and obtained both photometry and about 4 million spectra in four data releases 2000–2020. The related Baryon Oscillation Spectroscopic Survey (BOSS) has about 800 thousand redshifts. The Dark Energy Survey (DES) currently getting underway in the southern hemisphere will vastly enlarge

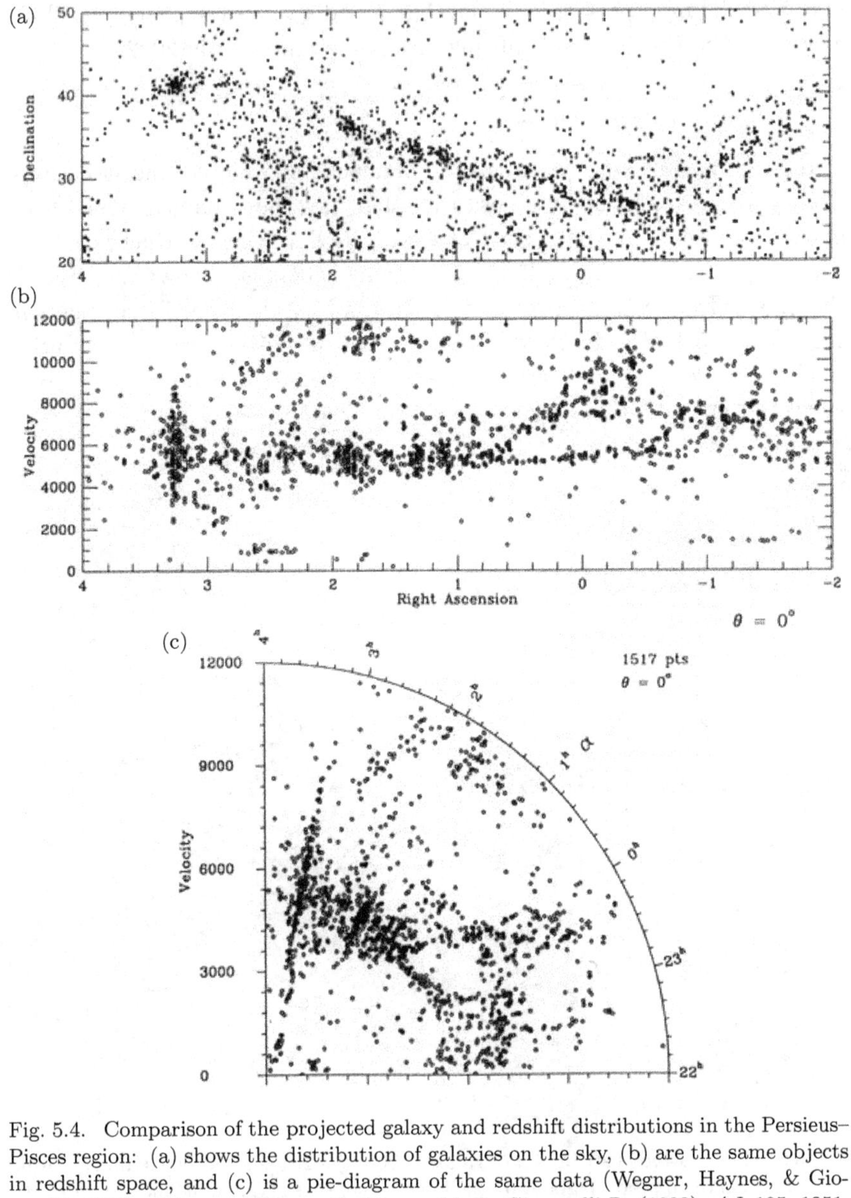

Fig. 5.4. Comparison of the projected galaxy and redshift distributions in the Persieus–Pisces region: (a) shows the distribution of galaxies on the sky, (b) are the same objects in redshift space, and (c) is a pie-diagram of the same data (Wegner, Haynes, & Giovanelli, 1993). (*Source*: Wegner G., Haynes M. P., Giovanelli R. (1993), *AJ*, 105, 1251, "A survey of the Pisces–Perseus. V: The declination strip +33.5 degrees to +39.5 degrees and the main supercluster ridge," Fig. 7, ©AAS Reproduced with permission.)

photometric and spectroscopic databases and reach even higher redshifts as well as the Euclid spacecraft planned to be launched in 2022.

5.3. Filament Structures

The large redshift surveys reveal the filament and void cosmic structures. Among the filament-like regions, a number have been named. One of the first to be recognized is the CfA2 Great Wall (Geller & Huchra, 1989). This can be seen in Figs. 5.2 and 5.3. Many others are known and include the Sloan Great Wall (Gott *et al.*, 2005) and the Perseus–Pisces filament seen in Fig. 5.4. The increased measurement of redshifts continue to bring the large-scale structures into sharper focus. Figure 5.5 shows the galaxies measured in the 2dF survey (Colless *et al.*, 2001b).

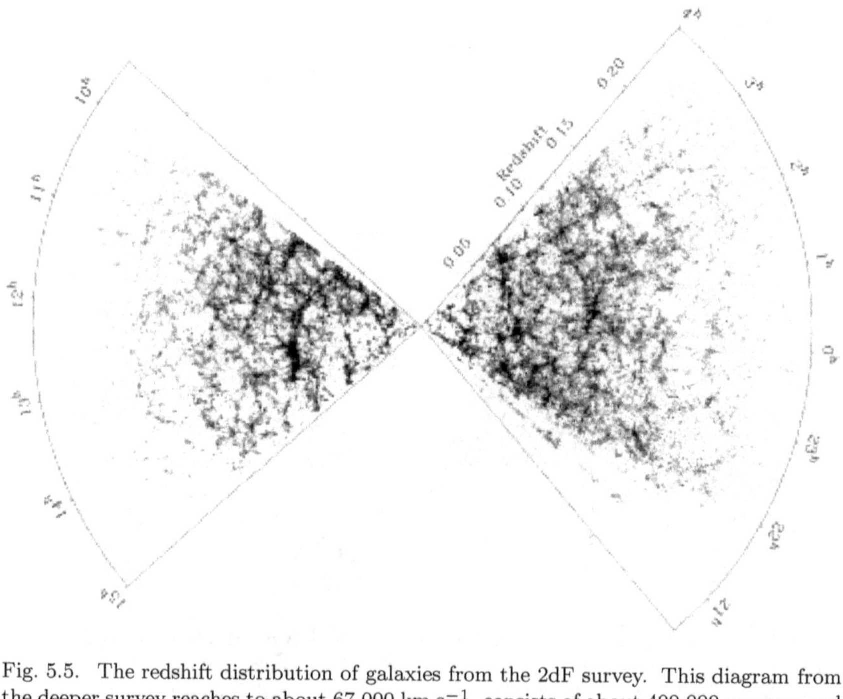

Fig. 5.5. The redshift distribution of galaxies from the 2dF survey. This diagram from the deeper survey reaches to about 67,000 km s^{-1}, consists of about 400,000 sources, and shows details of the filament and void structures. (*Source*: The 2dF Galaxy Redshift Survey team and Colless M. *et al.* (2001), *MNRAS*, 328, 1039, "The 2dF galaxy redshift survey: spectra and redshifts," Fig. 18, Reproduced by permission of Oxford University Press on behalf of the Royal Astronomical Society, www.2dfgrs.net.)

5.4. Void Structures

Voids are the other important feature in the large-scale structure. A typical large void has a diameter roughly 20–50 Mpc and a density of order 0.1 the mean cosmic density. Void properties are being intensely studied and with the ongoing redshift surveys, substantial numbers of voids are being located (e.g., Patiri *et al.*, 2006; Platen *et al.*, 2007; Pustilnik & Tepliakova, 2011; Pan *et al.*, 2012; Varela *et al.*, 2012; Pustilnik *et al.*, 2013; Mao *et al.*, 2017; Liebeskind *et al.*, 2018). Because of its lower density, a void expands faster than the Hubble flow and cosmological models (e.g., van de Weygaert & Platen, 2011; Einasto *et al.*, 2011; Cautun *et al.*, 2014) show that matter flows from the centre of the void and forms higher density walls leaving a complex internal structure of weak filaments with void galaxies (Aragon-Calvo & Szalay, 2013; Apaslan *et al.*, 2014). The voids should provide useful information on dark energy and non-standard gravity (e.g., Liebscher *et al.*, 1992; Moss *et al.*, 2011; Li *et al.*, 2012; Clampitt *et al.*, 2013; Lam *et al.*, 2015) and a test of the ΛCDM model (Peebles, 2001; Peebles & Nusser, 2010). The lower densities of the voids also provide different environments to study galaxy evolution, where simulations predict that galaxies formed there will have younger ages (e.g., DeLucia *et al.*, 2006; Krekel *et al.*, 2012).

A number of studies of the properties of void galaxies in the optical and with H I radio surveys in low-density regions seem to confirm the younger ages (e.g., Popescu *et al.*, 1997, 1999; Grogin & Geller, 2000; Wegner & Grogin, 2008; Hoyle *et al.*, 2012; Kreckel *et al.*, 2011, 2012, 2015; Ricciardelli *et al.*, 2014; Moorman *et al.*, 2014, 2015, 2016; Douglas *et al.*, 2018). Different investigators sought differences between void and field galaxies and although some evidence for lower masses and some metal-poor objects have been found, the general conclusion is that the lower densities for the voids have a rather small effect on their properties.

This is indicated by the metal abundances and star-formation rates (SFR) of emission-line galaxies in the Boötes Void. The Boötes Void is probably the largest nearby void at a redshift roughly $z = 0.04$ to 0.06 and ($\alpha = 206°$ to $240°$ and $\delta = +32°$ to $+57°$) on the sky (Fig. 5.6). Discovered by Kirshner *et al.* (1981, 1983, 1987) its contents seem to typify void objects. Studies of star-forming galaxies in the Boötes Void based on the Hα-selected KPNO International Spectroscopic Survey (KISS) of Salzer *et al.* (2000) by Wegner *et al.* (2019) found that the metallicity and SFR are nearly identical to comparison galaxies in density regions more than 16 times higher than in the void.

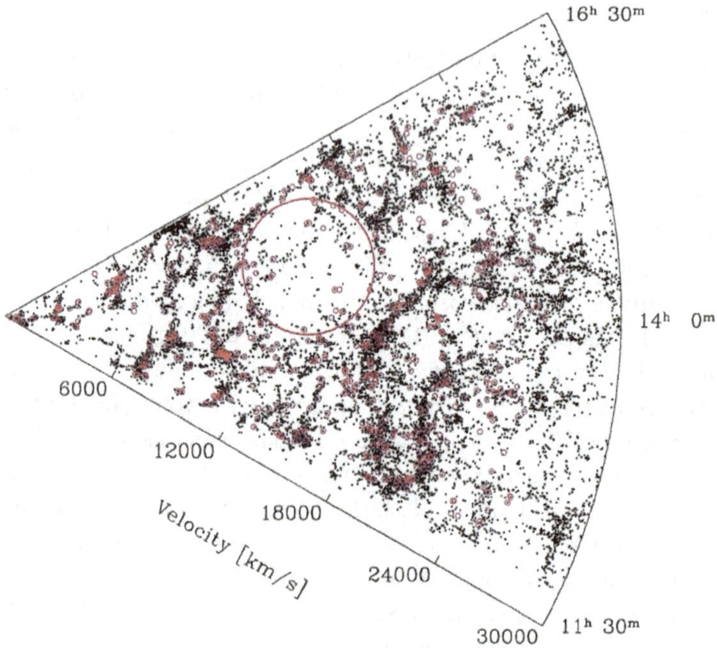

Fig. 5.6. Pie diagram in the region of Boötes Void. Red circles are KISS galaxies (Salzer *et al.*, 2000) and black dots are SDSS galaxies. The red circle outlines approximately the position of the Boötes Void (*Source*: Wegner G. A., Salzer J. J., Taylor J. M., Hirschauer S. A. (2019), *ApJ*, 883, 29, "Metal abundances and star formation rates of emission-line galaxies in and around the Botes void," Fig. 4, ©AAS Reproduced with permission.)

Future plans to study voids along with many references are given in the first year summary of the Dark Energy Survey (DES) by Fang *et al.* (2019).

5.5. The Lyman α Forest

The pattern of void and filament structures, e.g., shown in Fig. 5.7 can be detected by the 'Lyman α forest', produced by absorbing gas clouds in them. A bright quasar observed at large redshift is observed along a line of sight and shines through the intervening clouds. The cumulative effect superimposes patterns of spectral lines from these clouds at different redshifts. At redshifts above about ~ 6, the Gunn–Peterson effect is produced as so much neutral hydrogen accumulates that the many lines merge and the quasar's light is cut off and gives information on reionization (Fan, Carilli, & Keating, 2006). This provides a probe to studying the intergalactic medium

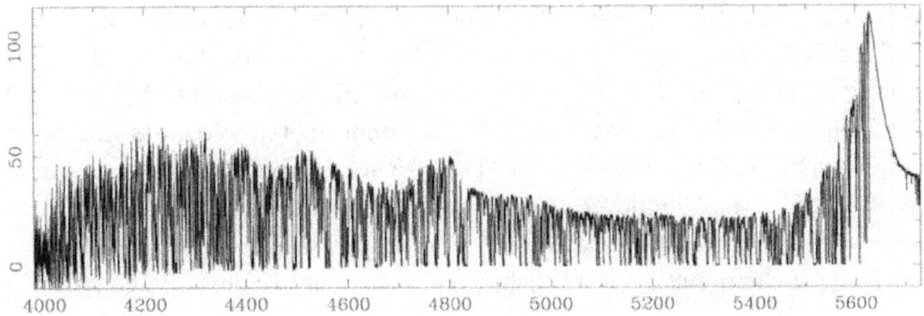

Fig. 5.7. The Lyman α forest in the spectrum of the quasar QSO 1422+2309 with redshift $z = 3.62$, obtained with the Keck telescope. The peak near 5620 Å is the redshifted λ1215 Å Lyman α line in the quasar. The large number of absorption lines at wavelengths below this are produced by nearer structures lying along the line of sight between earth and the quasar. (*Source*: Rauch M. (1998), *ARA&A*, 36, 267, "The Lyman Alpha Forest in the spectra of QSOs," Fig. 1, ©1998 Annual Reviews, www.annualreviews.org, with permission.)

and it is an additional method for studying the large-scale structure, metal abundances at large redshifts (Wolfe, Gawiser, & Prochaska, 2005). Cosmological studies include Hoell & Priester (1991) and Rauch (1998).

5.6. Photometric Redshifts

Although many objects can be observed spectroscopically using MOS instruments with slits or apertures producing accurate redshifts, extending measurements, even with these powerful instruments, to millions of faint objects at larger distances is still time consuming and impossible. So the objective is to obtain redshifts by reducing the spectral resolution and observing fainter objects with widened passbands.

Some studies measure low-dispersion objective prism spectra, which are particularly useful for emission line redshifts, e.g., KISS, the KPNO International Spectroscopic Survey (Salzer *et al.*, 2000, 2001, 2002) and PRIMUS, the PRIsm Multi-object Survey (Cool *et al.*, 2013).

Employing photometric bands lowers the resolution even further. Several programs have been designed to measure redshifts. Two types of codes are being used. The first method employs template fitting where the main idea is to fit an observed spectral energy distribution to a template with standard χ^2 minimization. Some examples of the many that have been used include Hyperz (Bolzonella, Miralles, & Pelló, 2000), EAZY (Brammer, van Dokkum, & Coppi, 2008), and GOODDZ (Dahlen *et al.*, 2010). The second technique employs empirical codes, ANNz uses artificial neural networks (Collister & Lahav, 2004), polynomial fitting (Hsieh *et al.*, 2005), and ArborZ, boosted decision trees (Gerdes *et al.*, 2010). The MegaZ-LRG survey (Collister *et al.*, 2007) has produced a photometric redshift of $>10^6$ luminous red galaxies with $z = 0.4$–0.7. Photometric redshifts are being employed currently in the extensive Dark Energy Survey (DES) (e.g. Amon *et al.*, 2021). MacDonald *et al.* (2004) employ multi-band CCD imaging to accomplish star–galaxy separation and estimate photometric redshifts.

Both of these methods of redshift measurement are designed to perform well for statistical studies of large numbers of faint galaxies. One major problem with the photometric measurements is eliminating false redshifts produced when the observed spectrum is spuriously fit. Typical redshift velocity accuracies are of order $\Delta cz \sim 0.05cz/(1+z)$, but as discussed in Chapter 12, these are large compared to the peculiar velocities that one expects, for the present, they are not employed for most peculiar motion studies.

Chapter 6

Unanswered Questions About Gravity

The nature of gravity is fundamental to the understanding of astrophysics and particle physics. It is intimately connected to stellar and galactic dynamics and in cosmology it ultimately plays a dominant role in large-scale motions and structure. This is an area of extensive effort and we must pass on only discussing some details relevant to the subject of the peculiar motions, but Adler, Bazin & Schiffer (1965), Misner, Thorne, & Wheeler (1973), Will (1993), Hartle (2003) and others provide excellent introductions. Although much is known about observable effects produced by gravity, e.g., spacecraft orbits and elaborate simulations using general relativity and cosmological models are available, several aspects remain obscure. Laboratory measurements of gravity are difficult due to its weakness on these length scales, but a large number of experiments have been carried out to study gravity in the past two centuries. Gillies (1987, 1997) gives useful compilations of earlier measurements of the Newtonian gravitational constant, G.

6.1. Early Empirical Measurements of Gravity

The publication of Newton's law of universal gravitation in 1687 in his famous *Philosophiae Naturalis Principia Mathematica* was built on the formulation of the laws of mechanics and careful astronomical observations based on the earlier work of notably Galileo, Kepler, Brahe and others. Newton gave no explanation for the physical cause of gravity. He and others including Laplace noted a problem with the Newtonian theory of gravity, which is *action at a distance*. This requires that gravity propagates at infinite speed. Until the 20th century, when General Relativity types of theory became available, mechanical theories of gravity such as those of LeSage (1784) were attempted to explain gravity. Versions of these theories postulate that all space is filled by a sea of randomly moving particles pushing on matter, or a in a sense, a different kind of 'aether'. The shadowing of these particles by nearby bodies mimics Newtonian gravity and produce a modified inverse square law of the form

$$F = \frac{\gamma Mm}{r^2} e^{-\int h\rho dr}. \tag{6.1}$$

where ρ is density and h is an absorbing or shielding term and γ is a version of the gravitational constant. The shielding predicts observable effects. Setting aside severe objections that can be raised about the nature of the sea of particles, some properties of gravity have been studied by different investigators from a strictly empirical standpoint.

For example, some stringent limits can the placed on h. Q. Majorana (1920, 1930) conducted two independent and extensive experiments to look for the shielding effect and believed he had measured the absorption coefficient ($h = 6.7 \times 10^{-12}$ and $h = 2.8 \times 10^{-12}\,\mathrm{cm^2 g^{-1}}$ for the two experiments). More recent laboratory experiments push this down to $h \leq 4.3 \times 10^{-14}\mathrm{cm^2 g^{-1}}$ (Unnikrishnan & Gillies, 2000). Additional limits can be placed using observations of spacecraft motions, e.g., the Cassini spacecraft (Bertotti, Iess, & Tortora, 2003) and changes in g on the ground during total solar eclipses when the moon shields the sun (Bottlinger, 1912). Unnikrishnan Mohapatra, & Gillies (2001) discuss this method, which yields $h \leq 2 \times 10^{-17}\,\mathrm{cm^2 g^{-1}}$. It is expected that empirical experiments of this type will be continued, for example, it is still not known what other phenomena the gravitomagnetic equations might produce, such as with spinning superconductors (e.g., Podkletnov & Nieminen, 1992), but no convincing phenomena have been observed. Onofrio (2012), in light of possible vacuum polarization effects, has rediscussed Majorana's measurements.

6.2. Measurements of the Newtonian Gravitational Constant

From this strictly empirical standpoint, there are still questions about the Newtonian gravitational constant, G. It remains the most inaccurately determined fundamental constant and differences in opinion continue about whether it is truly constant with composition, length scale, and time.

Earliest measurements were equivalent to knowing the earth's mean density, $\bar{\rho}$, since at the earth's surface the acceleration of gravity is

$$g = \frac{4}{3}\pi G R_{\oplus}\bar{\rho}, \tag{6.2}$$

Maskelyne (1775), followed by others, tried to determine $\bar{\rho}$ from the attraction of the mountain Schiehallion in Scotland. However, the precision of this method is limited by several effects including, not least, the mountain's complicated geological structure. Consequently, laboratory measurements have been preferred and the first torsion experiment was devised by Cavendish (1798) using two balls arranged in a barbell hung on a narrow wire or filament (Fig. 6.1). Improved modern versions of this apparatus have been used for many measurements.

Since the 19th century, hundreds of measurements of G have been conducted. Since the beginning of the 20th century there have not been

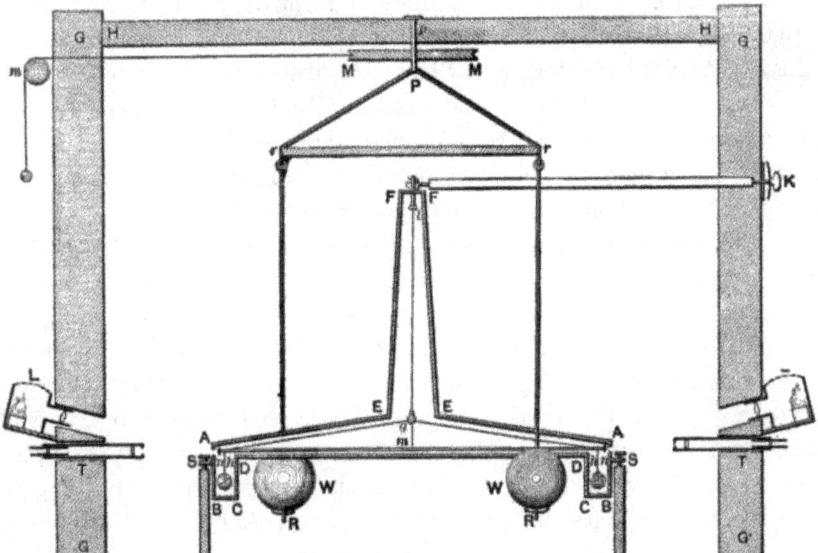

Fig. 6.1. Cavendish's torsion balance inside its protective enclosure, the precursor of many measurements to measure G. Shown are light sources, L, and telescopes, T. The apparatus was about 10 feet tall. The two large lead balls, W, hung on opposite sides of the horizontal rod attached to the two smaller lead spheres near CB. These were suspended by the wire, which rotated from the gravitational attraction of the four spheres and produced a measurement of its force. (*Source*: Cavendish H. (1798), *Phil. Trans. R. Soc. B* 88, 469, "Experiments to determine the density of the earth," Fig. 1, open access.)

enormous improvements. At that time, the best estimate was taken to be $(6.66 \pm 0.01) \times 10^{-11}$ Newton kg^{-2} m^2 (Poynting, 1911). The Committee on Data for Science and Technology (CODATA), that critically examines measurements of physical standards, gives a value of $G = 6.67408(31) \pm 4.7 \times 10^{-11}$ Newton kg^{-2}m^2 (Mohr, Newall, & Taylor, 2016). Recent measurements have individually pushed the experimental errors of G down, but they do not seem to converge on a well-defined value. Schlamminger (2014, 2018) has compared measurements of G measured over the previous 40 years and shows that they span a wide range of 6.671 to 6.676. Qing *et al.* (2018) used two independent apparatuses, the torsion oscillation frequency (TOS) somewhat similar to the Cavendish device and the angular acceleration feedback (AAF) method consisting of two turntables that coaxially rotate the torsion pendulum. Their results are consistent giving $(6.674184 \pm 11.64) \times 10^{-11}$ and $(6.674844 \pm 11.61) \times 10^{-11}$, but they differ from the results of an independent technique using laser-cooled atoms and quantum interferometry. One measurement of G with this technique is $(6.67191(99) \pm 150) \times 10^{-11}$ (Rosi *et al.*, 2014). Therefore, a more exact value of G is still not determined. The more likely explanation of the differences is that all the details of these complex and delicate experiments are not fully evaluated or the other considered less likely is that this disagreement is caused by some unknown modification of gravity. Lu *et al.* (2019) have reviewed current literature on measurements of G.

The change of G with time was discussed as early as 1937 by Dirac (1937). This and the variability of other physical constants has been studied since by several investigators including Garcia-Berro *et al.* (2006) and Solà (2013). Wilczynska *et al.* (2020) measured the fine structure constant, α, in four directions using quasars at redshifts $z = 5.5–7.1$ and found $\Delta\alpha/\alpha \approx -2 \pm 7 \times 10^{-5}$ with no evidence of a time change, but possible directional differences. Mould & Uddin (2014) used data from 580 Type Ia supernovae. They assumed that all used the same physics in burning the ^{56}Ni, which contributes to the regularity of their light curves. They concluded that $\Delta G/G < 10^{-10}$ in the past 9 Gyr covered by the supernova data. Pulsars place constraints on varying G. Wu *et al.* (2015) studied the binary pulsar J1713 + 0747 and obtained $|\dot{G}| = [-0.6 \pm 1.1] \times 10^{-12}\text{yr}^{-1}$. Perrodin & Sesano (2018) discuss using radio pulsars to test alternate gravity theories and place limits on $\dot{G}/G$. Hanimeli *et al.* (2020) consider tests of $\dot{G}$ for low redshift Newtonian cosmologies.

Studies of the fine structure constant also place limits on its variation. Several astrophysical measurements using quasars have been made.

Wilczynska *et al.* (2020) observed α in four directions from quasars at redshifts $z = 5.5$–7.1.

6.3. The Principle of Equivalence

Einstein's geometric theory of General Relativity (GR) assumes the *principle of equivalence.* Loosely speaking, all forms of matter and energy should fall equally in a gravitational field produced by a mass that distorts spacetime near it. The equivalence principle can be broken into two problems (Will, 1993). The first is the *weak* equivalence principle that all bodies irrespective of composition behave the same in gravity and this has been tested to high precision. The second is the *strong* equivalence principle in which all fundamental particles will fall the same in gravity including gravity itself.

The weak case is popularly illustrated by Galileo's legendary Leaning Tower of Pisa experiment showing that different varieties of masses fall at about the same rate, but starting in the early 20th century von Eötvös (von Eötvös, Pekar, & Fekte, 1922) experiments have been conducted with increasing precision. Using rotating torsion pendulum, the difference between the attraction of equal masses of different compositions and the sun or the galactic centre can be determined. The rotation of the device fixes the inertial mass. This is quantified by the Eötvös parameter for two test masses, X and Y,

$$\eta(X, Y) = 2\frac{(m_g)/m_i)_X - (m_g/m_i)_Y}{(m_g)/m_i)_X + (m_g/m_i)_Y}, \tag{6.3}$$

where m_g and m_i are the gravitational and inertial masses of X and Y. In recent work, Schlamminger *et al.* (2008) measured the attraction between the galactic centre and 4.84 g balls of Ti and Be and found $\eta_{\oplus Be-Ti} = (0.3 \pm 1.8) \times 10^{-15}$. Zhu *et al.* (2018) found similar results for chiral masses, using left- and right-handed quartz crystals. Satellite experiments such as the MICROSCOPE mission (Taboul *et al.*, 2017; Begé *et al.*, 2018) also confirm the equivalence principle holds to $\sim 10^{-15}$.

The strong equivalence principle has not been as well tested. Lunar laser ranging has been employed (Williams, Turyshev, & Boggs, 2012) and pulsar-white dwarf triple binary systems set a limit of $\eta < 2.6 \times 10^{-6}$ (Archibald *et al.*, 2018). Binaries containing pulsars can be used to place constraints. Gonzales *et al.* (2011) used 27 pulsar–white dwarf binaries to find $\eta < 4.6 \times 10^{-3}$. Pulsar and black hole binaries will place even stronger constraints on the strong equivalence principle (Wu *et al.*, 2015; Perrodin & Sesano, 2018). Lorimer (2008) gives a general description of binary and millisecond pulsars.

6.4. Tests of the Inverse Square Law

A large number of experiments have been conducted to test the inverse square of Newton's law. One approach assumes that gravity obeys a Yukawa law for the potential,

$$V(r) = -\frac{Gm_1m_2}{r}\left[1 + \alpha e^{-\frac{r}{\lambda}}\right], \tag{6.4}$$

where α and λ are parameters that cause deviations from the inverse square law and are to be fitted. Some early estimates of α used white dwarf redshifts (Wegner, 1989), but measurements using torsion balances, improvements in atomic clocks (Delta *et al.*, 2018; Takamoto *et al.*, 2020), and astronomical data now place limits on these parameters as shown in Fig. 6.2 so that $|\alpha| \ll 1$ for $\lambda \sim 56\mu$m to $\sim 10^{12}$ km (Adelberger, Heckel, & Nelson, 2003; Kepner *et al.*, 2007), which eliminates some theories of extra dimensions.

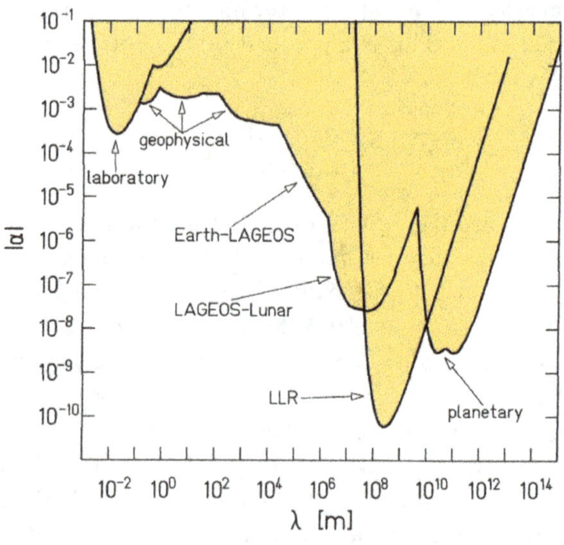

Fig. 6.2. Limits on $|\alpha|$ in the Yukawa formula (6.4) with scale, λ at the 95% confidence level. (*Source*: Adelberger E. G., Heckel B. R., Nelson A. E. (2003), *ARNPS*, 53, 77, "Tests of the gravitational inverse-square Law," Fig. 4, ©2003 Annual Reviews, www.annualreviews.org, with permission.)

6.5. Astronomical Tests of GR

GR is well tested and accurately explains several phenomena that Newtonian gravity cannot. The classical observational tests, known since the early 20th century, include light bending, Mercury's perihelion shift, gravitational redshift, and time delays. Since the 20th century several alternative theories of gravity have been proposed. Some earlier theories and their failures to pass observational tests are described by Misner, Thorne, & Wheeler (1973) and Will (1993, 2014). Experiments on gravitational radiation (Hulse & Taylor, 1975) from binary pulsars and LIGO discussed in Chapter 18 continue to vindicate GR along with the CBR and ΛCDM cosmological models and available observations.

Every physical theory has a range of validity and questions about how gravity behaves on small and large distance scales remain. Tests of gravity show that GR works well on distances of order 10^0–10^{12} cm. Gravity does not enter into the standard theory of particle physics and it is well known that there is no accepted theory on the subatomic size scale and that reaching the Planck length, $\sim 10^{-43}$ cm, a theory of quantum gravity is needed. This has led many researchers to search for new physics. These are included in particle theories such as M-theory, which predict new particles, but none have been found so far with the LHC. On large scales, there is ample evidence for the existence of what has become interpreted as dark matter (DM) and what it is comes into question. One has to accept either the presence of large amounts of DM or a modification of gravity. It has long been known from the dynamics of galaxy clusters (Zwicky, 1933) that estimates of the clusters' total mass, M_{vir} from the virial theorem,

$$M_{\text{vir}} \approx \frac{R_{\text{vir}}\sigma^2}{G}, \tag{6.5}$$

using the measured velocity dispersion, σ when compared to the amount of light, used to estimate the mass of the stars indicates that $M_{\text{vir}}/L \gg 1$, which indicates that the stars only comprise about 5% of the total mass. Further indications for large amounts of dark matter come from the rotation curves of many spiral galaxies (Rubin, 1983). Still, these observations give little information on what DM is.

6.6. Laboratory Experiments for Dark Matter

Astronomical observations give evidence for the existence of dark matter, but do not suggest its nature. The favourite explanation is that it has the form of a particle, or even several different particles. As already discussed they are comprised of cold dark matter. It could represent something wholly different, such as a modification of gravitation on different length scales. So far, laboratory experiments have failed to discover DM particles. Several kinds of DM particles have been proposed. Some assume that the particles are the result of supersymmetry in the early universe. These include Weakly Interacting Massive Particles (WIMPs), perhaps the most extensively explored, but also axions, neutrinos, neutralinos, Majorana fermions, sterile neutrinos, etc.

If the DM consists of WIMPs, then the following properties might be expected: stable on times of order the age of the universe, only acts through gravity and no electro-magnetic interaction or only weak interactions at most occur, large mass in the few to many GeV/c^2 range and cold (CDM) to account for known dynamical effects. Although DM particles do not interact with baryonic matter in any substantial way other than by gravity, one might detect collisions with atomic nuclei that produce observable recoil effects. Two avenues to experimentally detect DM are being explored: direct detection and production in particle accelerators, notably the LHC.

Many direct detection experiments are currently under way. These make use of X, a DM particle's recoil when it collides with an atomic nucleus, N,

$$XN \to XN. \tag{6.6}$$

Possible observable signatures in a DM detection signal would be the modulation produced by the earth's daily rotation, its annual orbital motion of 30 km s^{-1}, and the sun's 220 km s^{-1} motion around the galactic centre (e.g., Caldwell & Kamionkowski, 2009). A maximum in the signal is thereby expected when the earth's motion aligns as much as possible in the direction of the solar apex near the star Vega (first noted by Wm. Herschel) near June 2nd with minimum occurring near December 2nd. The density of WIMPs at the earth can be estimated from galaxy models. ΛCDM models predict that we are within a DM halo and that the density here is $\rho \sim 0.009\ \mathrm{M_{\odot} pc^{-3}}$ or $\sim 0.35\ \mathrm{GeV cm^{-3}}$ (e.g., Kafle *et al.*, 2014). Since the escape velocity from the galaxy at the sun is about 554 km s^{-1}, these place constraints on the velocity distribution of the WIMPs and using nuclear density models, scattering cross-sections can be calculated for an assumed particle mass, m_X, giving the expected nuclear recoil energy spectra and event rates for the different detector materials.

Examples of suitable detector materials range in density from liquid Xe down to liquid Ar and Si. Detectors require large masses, low-energy thresholds, long-time stability, and very low backgrounds. This has led to the construction of laboratories underground, such as WIPP at about 0.5 km depth to Jin-Ping at about 6 km.

The numerous sources of background are a major problem. WIMPs and most matter particles are expected to scatter off atomic nuclei and such different events need to be separated. Atomic electrons will scatter photons and electrons. Gamma rays are produced by reactions induced by muons and produce neutrons. The sun, supernovae, and AGNs can produce neutrinos. The detectors will also produce internal and surface backgrounds due to internal contamination. In liquid detectors, naturally decaying isotopes including ^{85}Kr, Rn, ^{38}Ar, and ^{136}Xe or cosmic rays activating Ge are a problem.

Since about the year 2000, several direct detection experiments have been initiated and the sizes of detectors, such as liquid noble element targets and developments with crystal targets, the oldest, and cryogenic devices have improved the sensitivity.

Detectors employing arrays of scintillating crystals of NaI or CsI are among the simplest and oldest techniques. They run at ambient temperatures and are stable. One ongoing experiment of this type is the DAMA and DAMA/LIBRA experiment at Gran Sasso, Italy, which currently is alone in reporting positive results. This employs a 250 kg array of high purity NaI crystals. Running for 14 years, data show a well-defined yearly modulation very near the expected one year period and phase from galactic motion at high significance for $M_X \sim 10\,\mathrm{GeV}$ and 40–100 GeV with collision cross-sections of $\sigma \sim 10^{-40}$ and $10^{-41}\,\mathrm{cm}^2$, respectively (Bernabei *et al.*, 2018). The COSINE-100 experiment in S. Korea (Adhikari *et al.*, 2018) uses the same type of NaI detector (current devices all have used ~100 kg of detector), but detects no DM signal as does the ANAIS-112 experiment (Amaré *et al.*, 2019).

Other experimental groups have been unable to find detections and have pushed the limits on σ much lower in certain energy ranges. Some of the liquid Xe experiments include XENON100 (Aprile *et al.*, 2017a), Xenon1T (Aprile *et al.*, 2017b), XMASS (Abe *et al.*, 2018) and Bubble chambers, PICO500 and PICO600 (Amole *et al.*, 2018). Cyrogenic bolometers employ crystals at 10–100 mK. CRESST, DAMIC, SuperCDMSSoudan (Agnese *et al.*, 2018) For lower mass WIMPs (1–10 GeV), several other detectors have been in use. Darkside-50 uses liquid Ar as does DEAP.

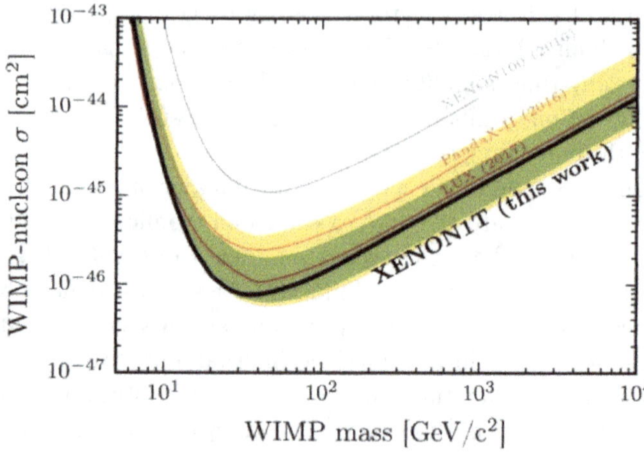

Fig. 6.3. Comparison of some recent limits on the spin-independent WIMP-nucleon cross sections as a function of WIMP mass with the XENON1T experiment (black); 1 and 2 σ limits are in green and yellow colours (*Source*: Aprile E. *et al.* (2017), *PhRvL*, 119, 181301, "First dark matter search results from the XENON1T experiment," Fig. 4, Creative Common 4.0 International.)

In Fig. 6.3, the low mass limits are currently about $\sigma \sim 10^{-42}\,\text{cm}^2$ for 2–5 GeV DM and for higher mass ~60 GeV,it is $\sigma \sim 10^{-47}\,\text{cm}^2$. Detector sensitivity has improved by about five orders of magnitude in the past 20 years and although no DM particles seem to have been detected, the search continues. Schumann (2019) has a good summary of the situation.

Axions are a possible form of DM and they arise from the Peccei–Quinn solution of the strong CP problem of quantum chromodynamics. Typical masses of axions are in the μeV range. Neutrinos would almost work, except they are hot (HDM). Several experiments to detect axions have been conducted using an 'Axion Holoscope'. This consists of a tunable cyrogenically cooled cavity through which a magnetic field ($\sim$ 8 tesla) runs (Boutan *et al.*, 2018). Axions passing through the detector could couple to the magnetic field and be converted to photons, which generate a low amount of power in the cavity ($\sim 10^{-24}$ watt).

6.7. Modified Gravity Theories

Many modified forms of gravity have been proposed. Earlier attempts are listed by, e.g., Will (1987) and reasons given why some do not work. Clifton, Ferreira, & Padilla (2012) give a detailed review of many of the theories of gravity and cosmology. One of the fundamental motivations for understanding gravity stems from particle physics. The literature on this subject is extensive and here we can only describe a few examples.

6.7.1. *Modified Newtonian dynamics*

Modified Newtonian dynamics (MOND) was designed to explain the rotation curves of spiral galaxies (Milgrom, 2015). The acceleration of Newton's second law is modified at large distances,

$$F = m\mu(x)a, \tag{6.7}$$

where m is mass, a is acceleration and $\mu(a/a_0)$ is a function that modifies the acceleration, where a_0 in $x = a/a_0$ is a fundamental constant of the theory and is $a_0 \approx 1.2 \times 10^{-2}$ ms^{-2}, which fits the spiral galaxy rotation curves. Different functions for μ have been proposed, but they all have the property that

$$\mu \to 1 \quad \text{for } x \gg 1 \quad \text{and} \quad \mu \to x \quad \text{for } x \ll 1, \tag{6.8}$$

and the force has the same acceleration as Newton's second law when a is much larger than a_0. Tensor–Vector–Scalar gravity (TeVeS) theory was devised to derive MOND (Bekenstein, 2004).

Famaey & McGaugh (2012), Kroupa (2012) and Hernandez, Jiménez, & Allen (2014) summarize some of the observations that are consistent with MOND. Ogle *et al.* (2019) have obtained new spiral galaxy rotation measurements that cast doubt on MOND. Furthermore, this paradigm runs into difficulties explaining cosmological scale and solar system tests (Hees *et al.*, 2014) and has not received general acceptance.

6.7.2. *f(R) modified gravity theories*

The popular $f(R)$ theories (e.g., Buchdal, 1970; and reviews of Sotiriou & Faraoni, 2010; Clifton *et al.*, 2012) have received extensive attention. One goal of modified gravity theories is to find a mechanism that can predict observable effects and eliminates the need for dark matter and dark energy. This generally requires agreement with GR on observable length scales, but allows modifications on large and small scales where precise tests are difficult.

Peculiar motions have begun to be applied in this area of research. Measuring the value of the growth index, γ (Eq. (13.7)), and possible applications to modified gravity will be discussed in Chapter 20.

Modified $f(R)$ gravity starts with the Einstein–Hilbert action equation, from which Einstein's field equations are derived (Chapter 8), but replaces the Riemann invariant, R, by a function $f(R)$ in the action,

$$S = \frac{c^4}{16\pi G}\int f(R)\sqrt{-g}d^4x + \mathcal{L}_M, \tag{6.9}$$

where G is the Newtonian gravitational constant, c light speed, and $\mathcal{L}_M$ expresses the coupling with matter. Setting $f(R) = R$ and $\mathcal{L}_M = 0$ generates Einstein's field equations (Chapter 8). Using different functions for $f(R)$ and $\mathcal{L}_M$ produces modified gravity theories. Possible forms of $f(R)$ are many, e.g., one or more terms of a power series,

$$f(r) = \cdots + \frac{a_2}{R^2} + \frac{a_1}{R} - 2\Lambda + R + \frac{R^2}{\beta_2} + \frac{R^3}{\beta_3} + \cdots. \tag{6.10}$$

The many forms of $f(R)$ can be constructed to conform to many couplings of fields or particles to gravity and among these are chameleon and galileon theories.

Chameleon $f(R)$ theory uses a broken power law (Hu & Sawicki, 2007)

$$f(R) = -m^2\frac{c_1(R/m^2)^n}{c_2(R/m^2)^n + 1}. \tag{6.11}$$

It predicts that gravity produces a fifth force and behaves differently in dense and low density regions of the universe, where it switches to different interactions. Lombriser *et al.* (2012) and Terukina *et al.* (2014) have employed this to galaxy clusters, such as Coma, but find no significant effects. Arnold, Matteo, & Li (2019), however, have constructed detailed cosmological hydrodynamical simulations using several versions of this $f(R)$ and applied them to galaxies, predicting what future observations might be able to detect.

Galileon theories utilize a more complicated action, which for a minimally coupled covariant model of the form:

$$S = \int d^4x\sqrt{-g}\left[\sum_{i=2}^{5}\mathcal{L}_i(g_{\mu\nu},\phi) + \mathcal{L}_m(g_{\mu\nu},\psi_m)\right], \tag{6.12}$$

where ϕ is the Galileon field, also referred to as the 'fifth force' and ψ_m is the matter field. This produces four-dimensional Newtonian gravity on a 3-brane in a five-dimensional space (Dvali, Gabadadze, & Porrati (DGP), 2000; Nicolis, Rattazzi, & Trincherini, 2009; Brax, Burrage, & Davis, 2011; Renk *et al.*, 2017). Combined with the Vainshtein mechanism

(Vainshtein, 1972; Babichev & Duffayet, 2013), the action is suppressed within the Compton radius for gravity related to the Vainshtein radius,

$$R_v = \left(\frac{2GM}{c^2}\lambda^2\right)^{\frac{1}{3}}, \tag{6.13}$$

where $\lambda \sim 1/H_0$.

Gravitational wave data constrain the Compton radius to $R_c > 1.6 \times 10^{16}$ m (Abbott *et al.*, 2017). Galileon theories give results like GR on small (solar system) scales, but they produce Λ-like effects at large scales, which could eliminate the need for dark energy.

It is currently impossible to exclude $f(R)$ and Galileon theories, but some efforts have been undertaken to observationally constrain them, although so far none of the available results have been stringent. Most analyses have set $f(R) = R + f(R)$. Dossett, Hu, & Parkinson (2014) used the Planck and WiggleZ (Drinkwater *et al.*, 2010) data power spectrum of the large-scale structure to place the limit $\log B_0 < -4.07$ for the Compton radius in Hubble units. Schmidt *et al.* (2009) studied halo statistics; for galaxy clusters in the large-scale regime where the Compton scale $R_c \gtrsim 10$ Mpc/h observable effects could occur and at the small scale regime the effects would be suppressed. Schmidt (2010) considered the effects of $f(R)$ and DGP on derived masses and showed that environmental effects are a major factor that needs to be taken into account. Renk *et al.* (2017) make predictions for observable effects that might be observed in the future.

6.7.3. *Gravitational interactions involving the Higgs boson*

Non-minimal coupling is one possibility for producing modified gravity and there are several possibilities. Among these is that interactions between the Higgs boson and gravity produce observable effects in strong gravity (Onofrio, 2010, 2012). Non-minimal coupling with the curvature of a gravity field and the Higgs field would produce a mass shift in electrons, but hadrons in atomic nuclei get most of their mass from QCD and are little affected. Different gravitational redshifts for atomic lines and molecular lines are produced. Transitions of the electrons in atoms differ from the molecular bands, which arise from the vibration and rotation of atoms. These ideas will only be briefly outlined here.

The magnitude of these effects can be estimated from ϕ^4 theory for a scalar quantum field. The generic Lagrangian density for a scalar field in space-time is

$$\mathcal{L} = \sqrt{-g}\left[\frac{1}{2}g^{\mu\nu}\partial_\mu\phi\partial_\nu\phi - \frac{1}{2}(\mu^2 + \xi R)\phi^2 - \lambda 4\phi^4\right], \quad (6.14)$$

where ξ is the coupling between a scalar field, here the Higgs field, ϕ and gravity, R is the Ricci or scalar curvature invariant, and λ, μ are the Yukawa coefficients. Outside of a spherical mass, R vanishes, but the next highest order non-vanishing invariant is the *Kretschmann invariant.* $K = R^{\mu\nu\rho\sigma}R_{\mu\nu\rho\sigma}$. For a spherical mass,

$$K = 12R_S^2/r^6 \quad (6.15)$$

(e.g., Henry, 2000) where R_S is the Schwarzschild radius. Using the Kretschmann invariant, Eq. (6.14) becomes

$$\mathcal{L} = \sqrt{-g}\left[\frac{1}{2}g^{\mu\nu}\partial_\mu\phi\partial_\nu\phi - \frac{1}{2}(\mu^2 + \xi_K\Lambda_{PL}^2)\phi^2 - \lambda 4\phi^4\right], \quad (6.16)$$

where $\Lambda_P L = (G\hbar/c^3)^{\frac{1}{2}} \approx 10^{-35}$m is the Planck length. For electrons, the curvature produces a mass change,

$$\frac{\delta m_e}{m_e} \approx \frac{1}{2}\xi_K\Lambda_{PL}^2\lambda_\mu^2 K, \quad (6.17)$$

and this equals the atomic line wavelength shift, $\delta\lambda/\lambda$. For iron nuclei the mass change is quite small and works out to be ~0.01 times the electron value. Ideally this effect should be sought in strong gravity fields surrounding black holes, such as in active galactic nuclei and quasars, but their environments are too complicated to be cleanly modelled. However, white dwarfs and neutron stars can be used to provide limits. Searching for modifications to their gravitational redshifts,

$$1 + z = \left(1 - \frac{2GM}{c^2R}\right)^{-\frac{1}{2}}. \quad (6.18)$$

Although the earth's gravitational redshift and that of the sun are accurately known, gravity is relatively weak and much higher density objects are required.

White dwarfs, with about a solar mass and radii about $0.01R_\odot$, have gravitational redshifts of order 60 km s^{-1}or $z \sim 2 \times 10^{-4}$ (e.g., Wegner, 1974, 1979; Wegner, Reid, & McMahan, 1989, 1991). Such stars with known space velocities are required and corrections for pressure shifts of the lines need to be taken into account. Onofrio & Wegner (2014) estimated differential shifts between electronic atomic lines of atoms and the vibrational transitions in the Swan bands of C_2 in the optical spectra of the carbon-rich white dwarfs BPM27606 and Procyon B of 1.3 Å and 0.5 Å respectively and these give corresponding bounds of $\xi_K \leq 5 \times 10^{50}$ and 9×10^{50}.

For neutron stars, also near a solar mass, but radii around 10 km, the gravitational redshift becomes $z \approx 0.2$, but no atomic and molecular features have been found in their spectra. However, in the gamma-ray region an estimate can be made using the e^+e^- annihilation line with laboratory energy 511 keV and atomic nuclear lines, e.g., of ^{56}Fe at 847, 1238, etc. keV and the p^+p^- annihilation feature, presumed to be produced by gamma ray bursts on the surfaces of neutron stars. Although data are sparse in this energy range, these features appear in the Konus catalogues (Mazets *et al.*, 1981a–c) and the Goddard ISEE-3 satellite (Teegarden & Cline, 1980, 1981). Although using these data is more problematic, Wegner & Onofrio (2015) estimate limits of $\xi_K \leq 7 \times 10^{35}$ from differences between the redshifts of these lines in neutron stars and sets limits on Higgs coupling with gravity that are more stringent that those from experiments.

Chapter 7

Magnitude Scale, Photometry, and Reddening

7.1. The Magnitude Scale and Photometry

The intrinsic brightnesses and colours of objects are related to their physical properties. Photometric measurements, corrections, and calibrations constitute a fundamental means for understanding stars, galaxies, and other sources, and are crucial for finding their distances. Historically the apparent brightnesses of stars have been expressed using the *magnitude scale.* In Hipparchus's star catalogue (ca. 129 BC), stars were divided into six magnitude classes with first magnitude being the brightest and sixth magnitude the dimmest visible to the naked eye and this brightness division has been preserved since that time. It is important to note this is an inverse correlation: magnitudes are larger for fainter sources. With the invention of telescopes, fainter stars were brought into view and several definitions of the magnitude scale existed until 1856 when the Pogson scale (Pogson, 1856) was proposed: a ratio of 100 in brightness for two objects, $I_1/I_2 = 100$, corresponds to a difference of 5 magnitudes or

$$m_2 - m_1 = \frac{5}{2}\log(I_1/I_2). \tag{7.1}$$

For example, the sun is $m = -26.7$, the full moon is $m = -12.6$ and so on. Objects as faint as $m = +30$ are regularly detected with the HST.

The colour of a source is measured from the *colour index* defined as the difference in magnitude measured at two wavelengths or through two different pass bands, i.e., $(B - V)$ or the like.

Photometric measurements are on a relative scale. Placing apparent magnitude measurements into a given photometric system is done by comparison with a set of standard stars. Great effort has gone into providing accurate sequences of standard stars (e.g., Hearnshaw, 1996). The earliest efforts used visual photometers of different designs followed by photographic and then photoelectric measurements. The Harvard North Polar sequence is an early set of standard stars extensively used from about 1900 to 1950. It is comprised of 96 stars within 2° of the North Celestial Pole (Leavitt, 1917). This has been superseded by the development of photoelectric photometry, such as in the *UBV* system (Johnson & Morgan, 1953; Cousins, 1973, 1976; Landolt, 1983). A large number of photometric systems designed for stellar and extragalactic measurements (Bessell, 2005) are now available.

Converting magnitude measurements to physical units requires flux standards for which accurate absolute measurements have been made. Two systems are currently often used. The first is the Vega system, with Vega or (α Lyrae) as the fundamental standard. The magnitude at all wavelengths is defined to be 0.0, but as the spectral energy distribution of Vega ($T_{\rm eff} \approx$ 10,000 K) decreases with wavelength, e.g., $F_\nu = 4.06 \times 10^{-20}\,{\rm erg\,cm^{-2}\,s^{-1}\,Hz^{-1}}$ in the B band and $2.416 \times 10^{-20}\,{\rm erg\,cm^{-2}\,s^{-1}\,Hz^{-1}}$ in the I band ($\lambda_{\rm eff} = 0.798\,\mu$), the zero-point changes with wavelength.

The second is the AB system defined such that a source with

$$F_\nu = 3.63 \times 10^{-20} {\rm erg\,cm^{-2}s^{-1}Hz^{-1}} \tag{7.2}$$

has an AB magnitude $m_{AB} = 0$ at every wavelength or filter.

The difference between the apparent (m) and absolute (M) magnitudes is the distance modulus $DM = (m - M)$ and in the absence of any complicating factors such as reddening, it depends purely on the inverse square law and is related to the distance, D, by

$$\log D = \frac{(m - M) + 5}{5} = 1 + \frac{DM}{5}. \tag{7.3}$$

7.2. Photometric Systems

If an object is observed with a detector through a filter, the signal will be

$$F = \int_0^{+\infty} S_\lambda F_\lambda D(\lambda) d\lambda, \tag{7.4}$$

where S_λ is the spectral energy distribution of the source, D_λ is the response of the detector including the transmission of the optics, and F_λ is the transmission curve of the filter. The earliest photometer was the human eye, which peaks in sensitivity at $\lambda \approx 5400\,\text{Å}$ and denoted as visual or m_v. With the development of photography, early plates were sensitive to blue light and denoted m_{pg}. With the development of phototubes and later CCDs, many photometric systems were designed. Photometric systems can be classified by the width of their passbands. Broadband systems are the widest, let through the most photons, and can be used for faint sources. The classical example is the *UBV* system (Johnson & Morgan, 1953), where, e.g., the B filter peaks near $\lambda = 4450\,\text{Å}$ and has a half-width of about 940 Å. An intermediate band system is the Strömgren *ubvy* system, where the corresponding b passband is near 4670 Å and the half width is 180 Å. Additional passbands have been added, such at R and I at longer wavelengths and with infrared detectors, J, H, and $K(2.2\,\mu)$ in the near infrared (NIR). With the development of more detectors and observations from spacecraft the number of filter systems has multiplied, e.g., with the WISE satellite $W1$ through $W4$ bands (3.4–22.2 μ) and GALEX satellite ultraviolet FUV (1452 Å) and NUV (2274 Å) filters. Bessell (2005) gives a compilation of some of the filter systems and their definitions.

For galaxy photometry, narrowband systems have been designed specifically to measure line strengths. The Lick/IDS system (e.g., Worthey *et al.*, 1994; Trager *et al.*, 1998 and references therein) consists of 21 narrow absorption line indices, usually measured directly from the spectroscopic scans, and is one system that has been extensively utilized to study galaxies. Combined with galaxy models, the Lick/IDS indices can be used to estimate ages and different metal abundances. Using this system, Fig. 7.1 gives examples of the positions of galaxies in the (Mg_b, $\langle\text{Fe}\rangle$) diagram, which gives this metal abundances and ages and the ($\langle\text{Fe}\rangle$, $\text{H}\beta$) diagram, which gives their ages and α element abundances.

7.3. Galactic Reddening

Absorption by dust along the line of sight to a source dims it and makes its apparent magnitude fainter. For all measurements involving magnitudes, this is an important effect that requires careful consideration. In general, neglecting redshift effects for the moment, at a given wavelength, λ, the relation between absolute magnitude, M_λ, and apparent magnitude, m_λ, with distance D including absorption A_λ is

$$M_\lambda = m_\lambda + 5\log D - 5 - A_\lambda. \tag{7.5}$$

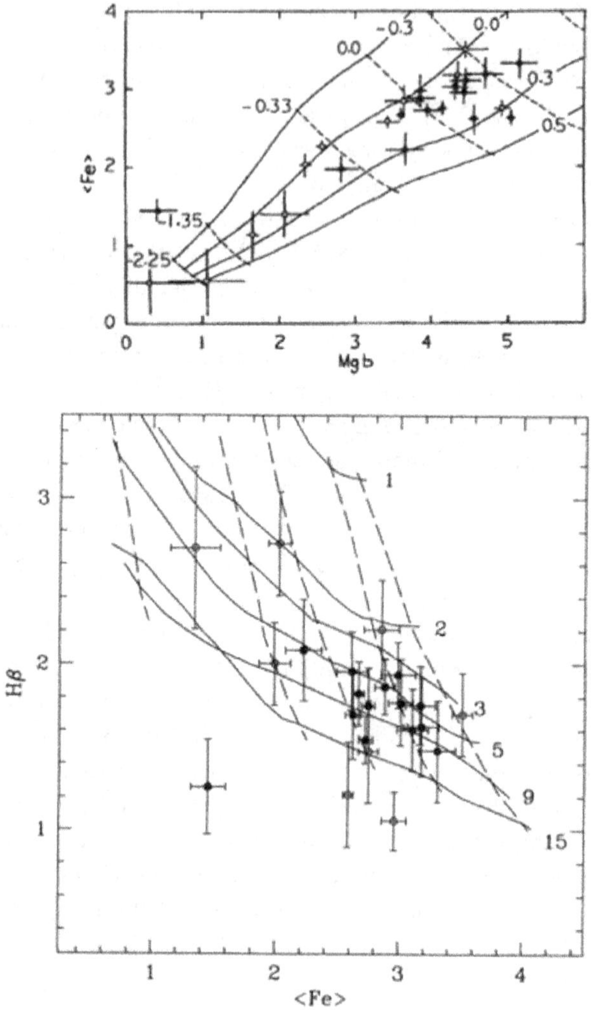

Fig. 7.1. Some early-type galaxies and error bars using Thomas, Maraston, & Bender (2003) for Lick/IDS line indices. Top: the Mg_b index vs. $\langle\mathrm{Fe}\rangle = (\mathrm{Fe5270} + \mathrm{Fe5335})/2$ from the strong iron lines. A 12 Gyr age was assumed. Solid curves are constant $[\alpha/\mathrm{Fe}]$ and dashed curves are [Z/H] constant. Bottom: $\mathrm{H}\beta$ index vs. $\langle\mathrm{Fe}\rangle$ gives age and metal content. Solid curves denote ages in Gyr. Dashed curves give $[\mathrm{Z/H}] = 0.67, 0.35, 0.0, -0.33$, and -1.35 from right to left assuming $[\alpha/\mathrm{H}] = -1.35$. (*Source*: Wegner G., Grogin N. A. (2008), *AJ*, 136, 1, "Ages and metallicities of early-type void galaxies from line strength measurements," Figs. 7 and 8, ©AAS Reproduced with permission.)

This produces an error in the distance if absorption is not accurately taken into account. The wavelength dependence of A_λ is roughly $1/\lambda$ in the visible spectral region, but becomes complicated in the ultraviolet. Several reddening curves, such as the average interstellar extinction curve (Osterbrock, 1989) in Fig. 7.2 are in use for the Milky Way, but large deviations are known, as in the abnormal extinction towards α Sco and in different directions in the galaxy, e.g., Wegner (2002). The absorption in magnitudes can be determined from a colour index using the 'colour excess,' E, if an object's unreddened colour is available. For example for the B magnitude using the $(B-V)$ index the colour excess is $E(B-V) = (B-V)_{\rm obs} - (B-V)_o$ and $A_B = \text{const} \times E(B-V)$ with the constant usually about 3.1.

Absorption by dust in the Milky Way is non-uniform and causes complications for brightness measurements. The distance of a star or galaxy measured using photometry will be affected and will produce a corresponding error in $V_{\rm pec}$. In addition to this local reddening, the observed galaxies themselves can produce different amounts of internal reddening. It was well known in the 19th century that there were dark spaces in the star clouds of the Milky Way, which some investigators believed to be holes or spaces devoid of stars. From the compilations of the Herschels, it was noted that the nebulae are rarer nearer the galactic plane in the 'zone of avoidance'. Early catalogues of the nebulae and Hubble's counts of galaxies revealed this further.

Trumpler (1930) showed that stars in the dark clouds are reddened and established that they are due to absorbing dust. A simple way to picture the role of the dust is if we are near the centre of a plane-parallel layer of absorbing dust that extends infinitely far in azimuthal directions. Then the amount of absorption depends on the angle b above the horizontal in the layer, or the galactic latitude as $\sim \csc b$, or the cosecant law. Several different determinations of this were used in early research, such as the absorption in the blue photographic spectral region of $A_{\rm pg} = 0.25 \csc |b|$ found by Hubble (1934) from galaxy counts.

However, this description neglects that the absorption also depends on galactic longitude, l. It is larger towards the galactic centre in the constellation of Scorpius, and lower in the opposite direction in Perseus. Different authors fit more complicated formulae including de Vaucouleurs & Malik (1969), but these are unable to describe the complicated small-scale reddening effects. Also, there was always the problem of matching the southern and northern hemisphere absorption corrections because the entire celestial sphere is not visible from ground-based observatories.

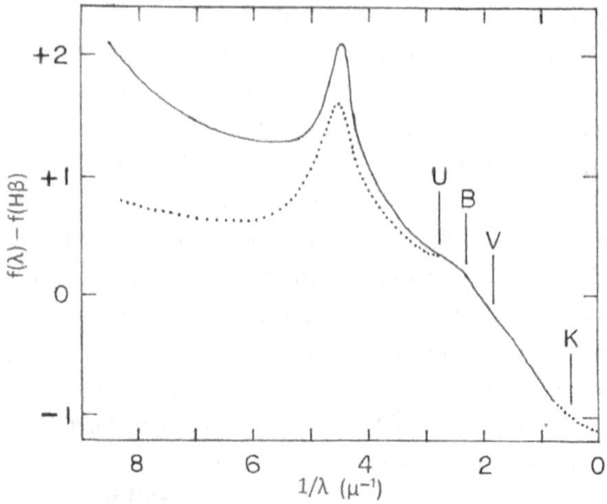

Fig. 7.2. Wavelength dependence of the standard galactic extinction curve (solid curve) and the galactic extinction towards αSco (dotted curve). The wavelengths of some photometric bands are indicated (data from Osterbrock, 1987).

The amount of galactic absorption decreases with wavelength as shown in Fig. 7.2. Some strong peaks occur in the curve including the largest bump near λ2200, thought to result from graphite grains, and molecules including polycyclic aromatic hydrocarbons. At longer wavelengths absorption decreases fairly smoothly and using standard photometric bands, such as the K band in the NIR is a big advantage over the shorter wavelength UBV filters. The main problem is that although a mean extinction curve is often used, strong differences in the extinction curve can occur in different directions.

Angular resolution of the reddening correction is improved by combining the neutral hydrogen H I column density in the radio and galaxy counts. Burstein & Heiles (1978, 1982, 1984) extensively discuss earlier reddening work and used the Shane & Wirtanen (1967) galaxy counts and the H I column density measurements of Heiles (1975) and calibrated the amount of dust along the line of sight. Their reddening map has a resolution of about 0.6°, is valid for b greater than 10°, and they used $A_B = 4E(B-V)$. Burstein & Heiles data were used extensively for absorption corrections until higher angular resolution (about 0.1°) became available by Schlegel, Finkbeiner, & Davis (1998) taking infrared dust emission measured by the COBE/DIRBE and IRAS/ISSA satellites. They employed $A_B = 3.1E(B-V)$. A recalibration in 2011 (Schlafly & Finkbeiner, 2011) has been included in the NASA Extragalactic Database (NED) to provide all-sky extinction calculators that give galactic absorption along the line of sight in different wavelength bands. Systematic errors could still exist in the calibration of the two methods. The zero points of the two reddening laws differ; the Schlegel *et al.* (1998) zero-point in $E(B-V)$ is about 0.02 magnitude larger than Burstein & Heiles. There can be large uncertainties when the extinction is large, particularly near the galactic plane. More recent work with the PLANCK satellite have continued to improve the galactic reddening map.

7.4. Internal Extinction in Galaxies

In addition to nearby galactic extinction, the intrinsic dust absorption in galaxies, especially spirals, requires correction. Although less severe in ellipticals, it is a necessary correction in spirals for employing the Tully–Fisher relation, which requires adequate extinction laws. Otherwise, the incorrect slope, nonlinearity, and biases in inclination and luminosity can be obtained. In some correction procedures, Freeman's Law (Freeman, 1970) is often utilized, where the central disk surface brightness in spirals is roughly constant at $\mu_B(0) = 21.65 \pm 0.30$ mag./arcsec^2 for the zeropoint of the light observed profiles. In spirals, dust is primarily located in the disk, as it is in our galaxy, and a spiral of fixed morphological type viewed pole-on will be less reddened than viewed edge-on. The extinction depends on both the galaxy's inclination and luminosity, e.g., a luminous spiral viewed edge-on is dimmed more than a fainter one.

Figure 7.3 is a schematic of how reddening depends on direction in the Milky Way and an observed galaxy inclined to the line of sight. Both the galactic absorption, which is fairly well under control, and the internal absorption of the observed galaxy, need to be accounted. For the observed galaxy, there are two major problems. The first is the amount of absorption as a function of the galaxy's inclination. The second is the geometrical distribution and amount of the dust. This affects the optical depth in the galaxy, τ. A face-on galaxy may be nearly transparent and most of its light can be seen, whilst at large i, only a fraction of the light reaches the observer.

As in the Milky Way, local deviations in the extinction curve occur and this becomes more severe for extragalactic objects of higher redshift where the stronger ultraviolet portions of the absorption curve are shifted into the longer wavelength bands. Gordon *et al.* (2003) show that the $\lambda 2200$ bump is smaller in the bar of the Small Magellanic Cloud (SMC) and that it is not present in some star forming galaxies.

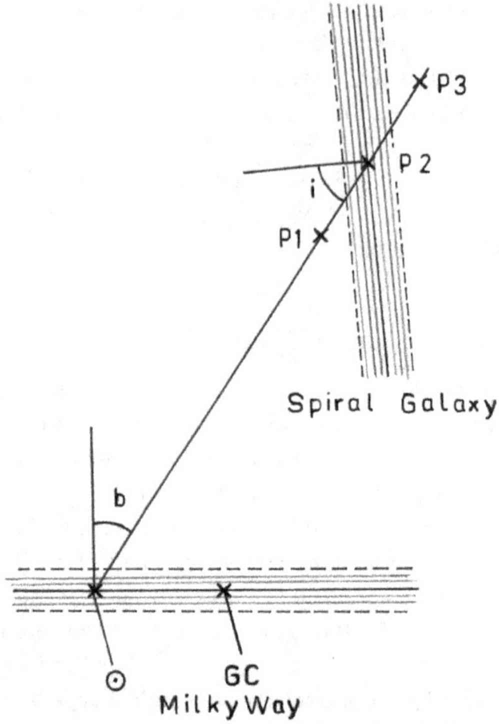

Fig. 7.3. The geometry for extinction correction of observations of a spiral galaxy observed from the Milky Way, both seen edge-on. The absorption depends on the galactic latitude of the observed galaxy, b and the galactic longitude, l, and the inclination of the spiral galaxy, i. The positions p_1, p_2, and p_3 indicate locations where most of the galaxy's light arises depending on the optical thickness of its dust distribution.

For the Tully–Fisher relation, originally constructed using B broadband data ($\lambda = 0.4\,\mu$), the absorption and structure of the galaxy contribute to the scatter in the empirical relation. Furthermore, the differences in brightness and diameter as a function of inclination can produce biases, such as in the derived peculiar velocities. This brings in the problem of knowing the actual geometrical distribution of the dust absorption. In the Milky Way and nearby spirals, this is not smooth and is clumpy. Given these complicated structures for distant spirals, an approximation for the dust distribution is needed that can be employed to correct the observed magnitudes. There are two limiting cases: (1) The disk is completely transparent; the total surface brightness is unaffected by the dust, but the surface brightness depends on inclination. (2) The disk is completely opaque and the disk limits light to the outer regions of the disk. Even more complicated is whether the dust layer is thinner or thicker than the stellar disk. This affects the inclination correction. The dust has been modelled several ways including the 'triplex' model (Disney *et al.*, 1989).

Improvement is made by observations at longer wavelengths, which reduces the reddening because $A_\lambda \sim \lambda^{-1}$. The disk becomes more uniform as the bright H II regions and spiral arms become smoothed out. Thus the internal absorption correction is reduced by making observations at longer wavelengths. Giovanelli *et al.* (1994) used the I band ($\lambda = 0.8\,\mu$) and discussed different absorption models in the SFI survey of peculiar motions based on Sc spiral galaxies. An empirical fit to the change in magnitude with axial ratio, a/b, was found to be

$$\Delta M = M_{\text{tot}} - M = -\gamma \log(a/b), \tag{7.6}$$

where M_{tot} is the magnitude if the galaxy were face-on and M is the observed magnitude. The linear fit to averaged data is shown in Fig. 7.4 with $\gamma = 1.05 \pm 0.08$. Aaronson & Mould (1979) used the near infrared H band ($\lambda = 1.6\,\mu$), and several recent studies utilize $K(\lambda = 2.2\,\mu)$, further reducing the magnitude of the required dust corrections. There is some trade-off when high z galaxies are observed. The shorter wavelengths in the galaxy's rest frame become redshifted into the longer wavelength bands.

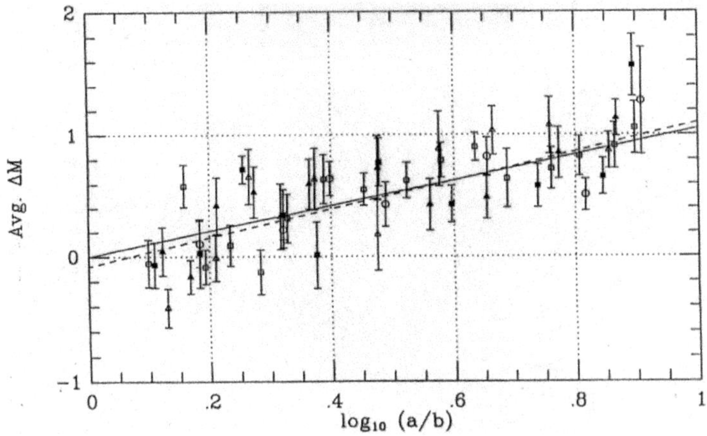

Fig. 7.4. The change in brightness in the I band in magnitudes as a function of axial ratio, a/b, of the projected galaxy from the SFI survey. (*Source*: Giovanelli R., Haynes M. P., Salzer J. J., Wegner G., da Costa L. N., Freudling W. (1994), *AJ*, 107, 2036, "Extinction in SC galaxies," Fig. 12, ©AAS Reproduced with permission.)

Chapter 8

Distance Determinations and Systematic Corrections

Size and brightness data require corrections in addition to the important influence of reddening. This involves both cosmological distortions and instrumental effects. Some of these are described briefly in this chapter. The discussion here is limited to low redshift effects, which apply to current peculiar velocity measurements ($z \lesssim 0.1$). More detailed discussions for high z and different cosmologies can be found elsewhere.

8.1. Basic Types of Distance Determination

The ability to evaluate cosmological parameters, map the galaxy distribution in the universe, and determine peculiar motions depends on how well distances can be measured. These are guided by other considerations of cosmology theory, such as galaxy formation and evolution, the properties of gravitation, dark matter and dark energy, so it must be kept in mind that distances are model dependent.

There are two basic kinds of distance indicators. The first is the *luminosity distance*, D_L. A class of objects of intrinsic luminosity, ℓ_0, often termed a 'standard candle', decreases in apparent luminosity with distance via the inverse square law

$$l \sim \ell_0/D_L^2. \tag{8.1}$$

The second distance indicator is the *angular diameter distance*, D_A. The angle that a class of objects of intrinsic size, d_0, sometimes designated a 'standard ruler', subtends on the sky decreases with increasing distance

$$\Delta\vartheta = d_0/D_A. \tag{8.2}$$

The complication is that not all objects of a given type will have identical ℓ_0 or d_0. Generally, there is a size distribution that introduces scatter and biases. In addition, instrumental distortions can be large and need correction. At small distances corrections will be small. D_L and D_A approach their corresponding Euclidean geometric values and are nearly equal, but with increasing distance they become modified as specified by the cosmological geometry.[1] Hogg (2000) gives a good summary of distance measures.

The commonly held ΛCDM universe model,[2] is most often used as a default to interpret data; it predicts geometrical effects, the expansion redshift $V_{RS} = cz$, D_L, D_A, and other observable quantities from the scale factor, $S(t)$ the expansion with time (e.g., Peebles, 1993; Narlikar, 1993; Linder, 1997; Weinberg, 2008). This is derived starting with the assumption of the 'cosmological principle', that the universe is homogeneous and isotropic. The assumption that we are not at a special location and the observational confirmation of isotropy from the CMB that $\frac{\Delta T}{T} < 10^{-5}$ implies the validity of the simple Robertson–Walker metric,

$$ds^2 = c^2 dt^2 - S^2(t)\left[\frac{dr^2}{1 - kr^2} + r^2(d\theta^2 + \sin^2\theta d\varphi^2)\right]. \tag{8.3}$$

A host of cosmological models and metrics using other assumptions are discussed in the literature (e.g., Clifton *et al.*, 2012), but a review of them is beyond the scope of this book. Einstein's gravitational field equations are derived from the Einstein–Hilbert action (e.g., Landau & Lifshitz, 1975) and are used to describe gravity,

$$R_{ik} - \frac{1}{2}g_{ik}R + \Lambda g_{ik} = -8\pi G T_{ik}, \tag{8.4}$$

and using equation (8.3) to obtain Friedmann's equation for $S(t)$. The observed quantity is the redshift (or blueshift) defined as

$$z = (\lambda_{\rm obs} - \lambda_{\rm emit})/\lambda_{\rm emit}, \tag{8.5}$$

where it is the difference between the observed and emitted wavelength of a photon at rest. In Friedmann models, the relation to the redshift, z, the

[1] When the differences between D_L, D_A, and the Euclidean distances are negligible, the subscripts might be dropped.

[2] A useful source of information on the cosmological parameters, observational survey data, and cosmological simulations and models is: http://lambda.gsfc.nasa.gov.

expansion rate or scale factor is

$$1 + z = \frac{S(t_0)}{S(t_1)}. \tag{8.6}$$

The equations for the scale factor are

$$\left(\frac{\dot{S}}{S}\right) = H_0 E(z), \tag{8.7}$$

where H_0 is the Hubble constant and

$$E(z) = [\Omega(1+z)^3 + \Omega_R(1+z)^2 + \Omega_\Lambda(1+z)^{3(1+w)}]^{1/2}, \tag{8.8}$$

and

$$\left(\frac{\ddot{S}}{S}\right) = H_0^2 \left[\Omega(1+z)^3/2 + \Omega_\Lambda\right]. \tag{8.9}$$

The Hubble parameter is thus defined from this by

$$H(z) = \frac{\dot{S}}{S} = H_0 E(z). \tag{8.10}$$

These other parameters denote $\Omega = \frac{8\pi G\rho_0}{3H_0^2}$, $\Omega_R = \frac{k}{(H_0 S_0)^2}$, and $\Omega_\Lambda = \frac{\Lambda}{3H_0^2}$, with $\Omega + \Omega_R + \Omega_\Lambda = 1$.

The general definition of Ω_x is that it is the ratio of the density of component x to the mean density of the universe,

$$\Omega_x = \rho_x/\rho_0, \tag{8.11}$$

and $w = p/\rho$ is the equation of state of the dark energy. The total matter density is the sum of the baryonic, dark matter, and any other possible constituents,

$$\Omega = \Omega_{DM} + \Omega_b + \cdots. \tag{8.12}$$

Currently as approximations,[3] the Hubble constant, $H_0 \approx 70$ km s^{-1}Mpc^{-1}, $\Omega \approx 0.3$, $k \approx 0$, $w \approx -1$, and $\Omega_\Lambda \approx 0.7$. In general, equation (8.7) can be solved numerically, but for most work on peculiar motions, the redshifts used are $z < 0.1$ in the low redshift universe and low-order approximations suffice.

At redshifts near $z = 0$, the differences between the Euclidean approximations for luminosity and angular size distances, $D_L \sim cz/H_0$ and $D_A \sim d_0/\Delta\vartheta$, where d_0 is the physical size of the object, are small.

[3]More description of estimates for these parameters will be discussed later.

With increasing z, these diverge at large redshifts, $(z > 1)$ as shown, e.g., in Peebles (1993) and elsewhere where cosmological calculator programs are available.

For illustrative purposes, at low redshifts, the expansion parameter in terms of the lookback time $t_0 - t$ is

$$S(t) = S_0 \left[1 - H_0(t_0 - t) - \frac{1}{2} q_0 H_0^2 (t_0 - t)^2 + \cdots \right], \tag{8.13}$$

where

$$q_0 = \frac{\Omega}{2} - \Omega_\Lambda \tag{8.14}$$

is the deceleration parameter. For the currently preferred parameters above, $q_0 \approx -0.55$. Historically, the Einstein–DeSitter model with $\Omega_\Lambda = 0$ was a favourite approximation with $q_0 = \frac{1}{2}$ and was often employed.

The luminosity distance,

$$D_L = (1 + z) r_1 S(t_0), \tag{8.15}$$

for this case can be approximated by

$$D_L \approx \frac{cz}{H_0} \left(1 + \frac{1 - q_0}{2} z\right), \tag{8.16}$$

The accompanying angular diameter distance is:

$$D_A \approx \frac{cz}{H_0} \left(1 - \frac{3 + q_0}{2} z\right), \tag{8.17}$$

and the relation between the two distances is $D_L = (1 + z)^2 D_A$. Mattig (1958) derived useful formulae for Friedmann models with different q_0. It is interesting to note that although D_L decreases with z monotonically, D_A reaches a minimum at $z \approx 1$–2 and increases for large redshifts (e.g., Peebles, 1993 for representative curves).

For $z = 0.1$, D_L is 1.08 times its Newtonian value and D_A is 0.82 times it, so the correction at this low z is relatively small, but non-negligible. For larger redshifts, the difference become quite large.

8.2. Cosmological Dimming

Cosmological dimming of the surface brightness is an effect depending on z. If an object has an absolute brightness ℓ_0 per proper area A, its apparent luminosity will be

$$\ell = \frac{\ell_0}{4\pi D_L^2}, \tag{8.18}$$

or $\ell = (1+z)^{-4}\ell_0/4\pi$). In this case, the correction for $z = 0.1$ is 0.68 of the Euclidean value.

8.3. The K-term

This results from observing objects that are at different redshifts with a finite fixed wavelength passband. The K-correction converts a source's magnitude at redshift z to the rest scale, that is were the object observed at $z = 0$. For galaxies, the corrections depend on morphological type and the passband. After the received radiation is corrected for atmospheric extinction, the incoming flux, $F(\lambda)$, is modified by the wavelength transmission of the optics, $T(\lambda)$ and the wavelength sensitivity of the detector, $\mathcal{D}(\lambda)$ such that the measured radiation is multiplied by

$$S(\lambda) = T(\lambda)\mathcal{D}(\lambda). \tag{8.19}$$

The redshift of the source moves each wavelength of the observed flux redward to $\lambda = (1+z)\lambda_0$ where λ_0 is the rest wavelength. In this case for a passband, p,

$$K_p = m_p(z) - m_p(0). \tag{8.20}$$

The K-term is then:

$$K_p = 2.5\log(1+z) + 2.5\log\frac{\int_0^\infty S(\lambda)F_0(\lambda)d\lambda}{\int_0^\infty S(\lambda)F_z(\lambda)d\lambda}. \tag{8.21}$$

K corrections for low redshifts ($cz < 0.5$) are not large and relatively well determined as the ultraviolet regions of $F(\lambda)$ are shifted into the visible bands and space observations make these wavelengths available. For example in the broadband R, $K_R = 1.00 \times z$ for E galaxies and $0.32 \times z$ for spirals at low redshifts (Saglia *et al.*, 1997), so for relatively near peculiar velocity measurements the K corrections are small, but non-negligible. However with larger z the ultraviolet is redshifted into the visible and grows less well determined, as in the case of the SN Ia with highly jagged $F(\lambda)$, this has been a controversial problem. Some determinations of the K-term with further references are, e.g., Blanton & Roweis (2007), Kinney *et al.* (1996) and Poggianti (1997), which also includes evolutionary effects. Some cosmology calculators that include K-corrections are available (e.g., https://ned.ipac.caltech.edu/help/cosmology_calc.html.)

8.4. Evolutionary Effects

As discussed in Chapter 4, the properties of galaxies change with redshift. In general, star formation in galaxies drops precipitously below $z \lesssim 1.0$ (e.g., Le Floc'h *et al.*, 2005). Starbursts of different ages induced by merging can play a complicated role. Detailed programs for photometric effects in evolutionary calculations have been made by several authors including Bruzual & Charlot (2003). An estimate of the order of magnitude of photometric effects by Tinsley (1972) models the evolution of an E galaxy formed at high redshift with a single initial starburst. These galaxies fade with age because the luminous massive upper main sequence stars in them evolve more rapidly and burn out. Galaxies at high z would be brighter than those nearby from this simple model: If the number of stars, N, of mass M obey a Salpeter initial mass function, $dN/M \sim M^{-(1+x)}$ (where $x \approx 1.35$), using the Einstein–de Sitter approximation, $t \sim (1+z)^{-\frac{3}{2}}$, then the luminosity, ℓ, varies with time as $\ell \sim t^{-(1-0.2x)}$ and

$$\ell \sim 1 + 0.2z, \tag{8.22}$$

which again is not large at small z, but non-negligible. This is only a crude approximation that neglects many factors in galaxy evolution as discussed below.

8.5. Seeing and Resolution Effects

Seeing and telescope resolution affect measurements of galaxy photometry. For peculiar motion studies of ellipticals, measurements using D_n, R_e, SB_e, $\langle SB_e \rangle$ of the $R^{\frac{1}{4}}$ law, the Fundamental Plane, etc., and the required corrections grow with increasing distances. Ground-based astronomical observations are always made through our atmosphere, which is a moving sea of air. This blurring or the 'seeing', whilst it can be reduced by choosing optical observing sites and telescopes, it can never be eliminated. Even in space telescope observations, resolution is limited by the diffraction of light and imperfections in the optics. The 'point spread function' (PSF) results from the combined effects of these factors. Recent investigations obtain these corrections from numerical modelling.

If the observed projected brightness pattern of an object is $I(\mathbf{R})$ in two dimensions on the plane of the sky and $p(\mathbf{R})$ is the point spread function (PSF), where $\mathbf{R}$ denotes the two-dimensional sky position, then the observed intensity and flux through a circular aperture are

$$I^c(\mathbf{R}) = \int_0^\infty \int_0^{2\pi} \mathbf{R}'\mathbf{dR}'\mathbf{d}\theta'\mathbf{I}(\mathbf{R}')\mathbf{p}(\mathbf{R}-\mathbf{R}'), \tag{8.23}$$

$$F^c(\mathbf{R}) = \int_0^R \int_0^{2\pi} \mathbf{R}'\mathbf{dR}'\mathbf{d}\theta'\mathbf{I}^c(\mathbf{R}'), \tag{8.24}$$

and it follows that the resulting convolved magnitude within R is $m^c(R) = -2.5\log_{10} F^c(R)$. The shape of the PSF is often secured by measuring star images and assuming that $I(\mathbf{R})$ is a δ function.

For an ideal circular telescope, the PSF is the well-known Airy disk,

$$p(\theta) = I_0 \left[\frac{2J_1(x)}{x}\right]^2, \tag{8.25}$$

where $x = \frac{\pi q}{\lambda N}$, λ is the wavelength, and J_1 is the first-order Bessel function of the first kind. The Airy function consists of a bright central spot surrounded by faint rings. The first minimum is at $x = 3.8317$ and the resolution is taken to be this width, or $\theta \approx 1.22\lambda/d$. As the intensities of the outer rings are relatively low, the PSF is often approximated by a Gaussian function of the form

$$I'(\theta) = I_0' e^{-\theta^2/2\sigma^2}, \tag{8.26}$$

where 2σ is the 'full width at half maximum' (FWHM).

Including the seeing leads to complications and the effects of seeing were considered, e.g., by Lucey *et al.* (1991), Bower, Lucey, & Ellis (1992) where different analytical formulae were adopted.

Saglia *et al.* (1993b) conducted a detailed study, which was used for the EFAR project. The correction is based on atmospheric turbulence theory (Fried, 1966; Woolf, 1982) which predicts that the PSF is the Fourier transform of the function and $\sim\exp[-(kb)^{5/3}]$ where b is a scaling parameter. Generalizing this function by using γ instead of $\frac{5}{3}$, where a range of γ from 1.3 to 2 was employed where it becomes a Gaussian function,

$$\hat{p}_\gamma(k) = \int_0^\infty 2\pi R J_0(kR) p_\gamma(R) dR = \exp[-(kb)^\gamma], \tag{8.27}$$

where J_0 is the standard Bessel function. Compared to the often used and simpler Moffat (1969) function,

$$p(R) = \frac{\beta - 1}{\pi\alpha^2}\left[1 + \left(\frac{R}{\alpha}\right)^2\right]^{-\beta}, \tag{8.28}$$

p_γ was found to fit the Gaussian core with the extended wings better.

In Saglia *et al.* (1993b) stellar profiles observed with the 1.3 m MDM and 1.0 m JKT telescopes were fitted to determine the optimal γ and the Moffat function parameters α and β. The best values of γ lie near 5/3, which is from Kolmogorov 's turbulence theory, and used for the standard turbulence model, but it can vary in different datasets and needs to be determined for each case.

The results for three values of γ and three circularly symmetric models: (a) *Smo* is a $R^{1/4}$ model smoothed with a $0.04R_e$ Gaussian, (b) the full $R^{1/4}$ model and (c) a $\Phi = 12$ self-consistent model for elliptical galaxies of Bertin & Stiavelli (1984) and Stiavelli & Bertin (1985) are shown in Fig. 8.1. If the $F_H(\gamma = 2)$ has been fitted by a Gaussian, this measure can be converted to $F_H(\gamma)$ using

$$\frac{F_H(\gamma)}{F_H(\gamma = 2)} = 1 - 0.14(2 - \gamma), \tag{8.29}$$

found from fits to stars. The top diagram of Fig. 8.1 uses the ratio of the observed seeing convolved effective radius, $R_e^c/F_H(\gamma)$ to determine the corrections to the observed R_e and surface brightness parameters. The lower diagram in the figure gives the corrected D_n from the observed seeing convolved D_n^c.

The magnitude of the seeing effects, as expected, grow with decreasing $R_e/F_H(\gamma)$ or increasing distance. For nearby intrinsically large galaxies, it is small but for distant intrinsically small galaxies it becomes large. In general, uncorrected D_n and fundamental plane distances will be overestimated and lead to spurious negative v_{pec}. Saglia *et al.* (1993) conclude that seeing effects must be corrected for clusters with cz larger than about 8000 km s^{-1}.

8.6. Aperture Correction

The aperture correction is required for early-type galaxies. It arises because the projected velocity dispersion, σ, decreases with radius and the physical aperture size of the spectrograph on the galaxy changes with distance.

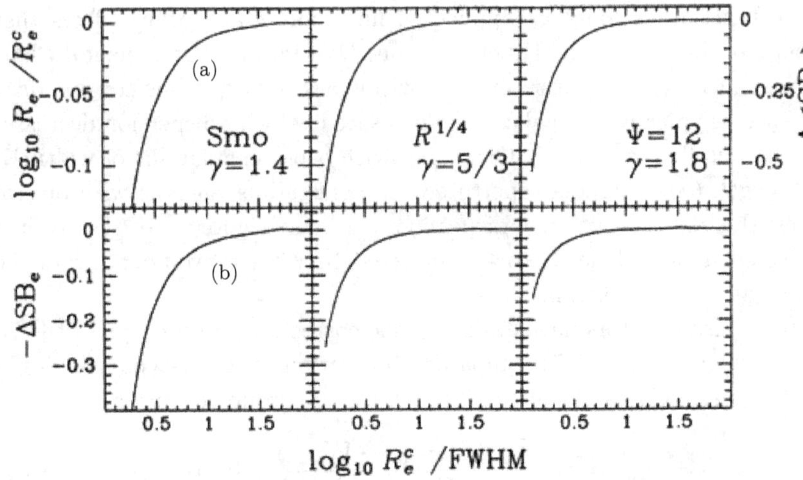

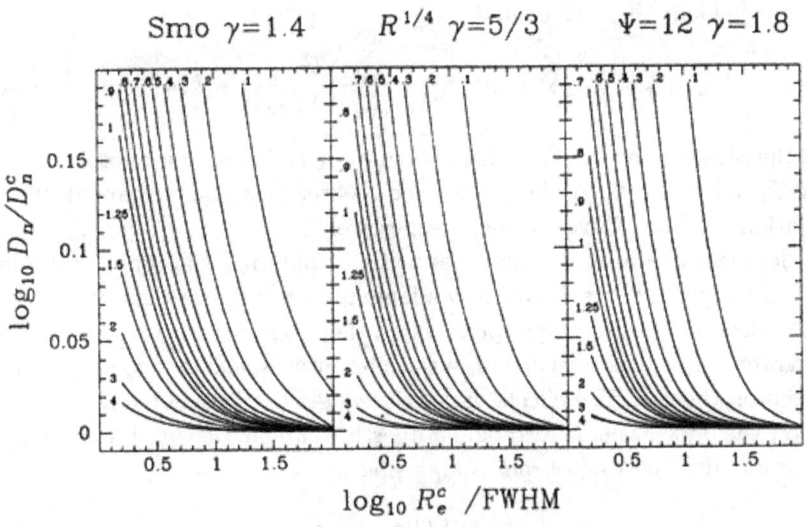

Fig. 8.1. Diagrams that show the sizes of seeing corrections to elliptical galaxy data as a function of the $R_e^c/FWHM$ ratio where R_e^c is the observed seeing convolved effective radius. Three values of the parameter γ in Eq. (8.27) and different circularly symmetric models are given to be used with Eq. (8.29). Upper panel: Corrections observed R_e^c effective radius, $\langle \mathrm{SB}_e^c\rangle$ effective surface brightness, and SB_e the effective surface rightness. Lower panel: Corrections to the observed D_n^c parameters. The labels on the curves represent constant D_n^c/R_e^c. (*Source*: Saglia R. P., Bertschinger E., Baggley G., Burstein D., Colless M., Davies R. L., McMahan R. K. Jr., Wegner G. (1993), *MNRAS*, 264, 961, "The effects of seeing on the photometric properties of elliptical galaxies," Figs. 8 and 9, Reproduced by permission of Oxford University Press on behalf of the Royal Astronomical Society.)

Corrections also should be applied to line indices and other effects that depend on radius. In the Faber–Jackson, $D_n - \sigma$, and Fundamental Plane methods, the value of σ is taken to be within a standard radius at a standard distance. Typically in a galaxy, the measured velocity dispersion decreases with radius like, $\sigma \sim R^{-x}$. Observing with a rectangular slit of width W and length ℓ or a circular aperture of a fixed radius, such as with optical fibres, the observed velocity dispersion, $\sigma_{\rm obs}$, is weighted over the covered projected area and needs to be corrected to what it would be were the galaxy at a chosen distance, $\sigma_{\rm cor}$.

Some investigators have evaluated the correction empirically. In the 7S study, Davies *et al.* (1987) compared observations of early-type galaxies in the Virgo and Coma clusters and found for the group velocity averages,

$$\sigma_{\rm cor} = \sigma_{\rm obs}\left[0.05\frac{(V_{\rm Group} - V_{\rm Virgo})}{(V_{\rm Coma} - V_{\rm Virgo})} + 0.95\right], \tag{8.30}$$

for the velocity dispersions and

$$(Mg_2)_{\rm cor} = (Mg_2)_{\rm obs}\left[0.06\frac{(V_{\rm Group} - V_{\rm Virgo})}{(V_{\rm Coma)} - V_{\rm Virgo})} + 0.94\right], \tag{8.31}$$

for the Mg_2 line index. Thus, the aperture correction for σ is approximately linear and amounts to about 5% over the velocity range of about 1000–7000 km s^{-1} from Virgo to Coma respectively.

Jørgensen, Franx, & Kjaergaard (1995) obtain an accurate aperture correction procedure subsequently adopted in several investigations. Using published data for 51 early-type galaxies, they constructed two-dimensional diagrams for velocity dispersion, surface brightness, and rotation angle projected on the sky. The effects of position angle can be ignored as they are below the 0.5% level. A rectangular aperture can be converted to an equivalent circular aperture of radius $r_{\rm ap}$ using

$$r_{\rm ap} = 1.025\sqrt{w\ell/\pi}, \tag{8.32}$$

where w and ℓ denote the slit's width and length. Then the aperture correction is given by

$$\sigma_{\rm cor} = \sigma_{\rm obs}\left[\left(\frac{r_{\rm ap}}{r_0}\right)\left(\frac{cz}{cz_0}\right)\right]^{0.04}, \tag{8.33}$$

where $r_0 = 0.54h^{-1}$ kpc or 1.7 arcsec and cz_0 are taken at Coma. An alternative approximation that is accurate to within 1% in the range of apertures

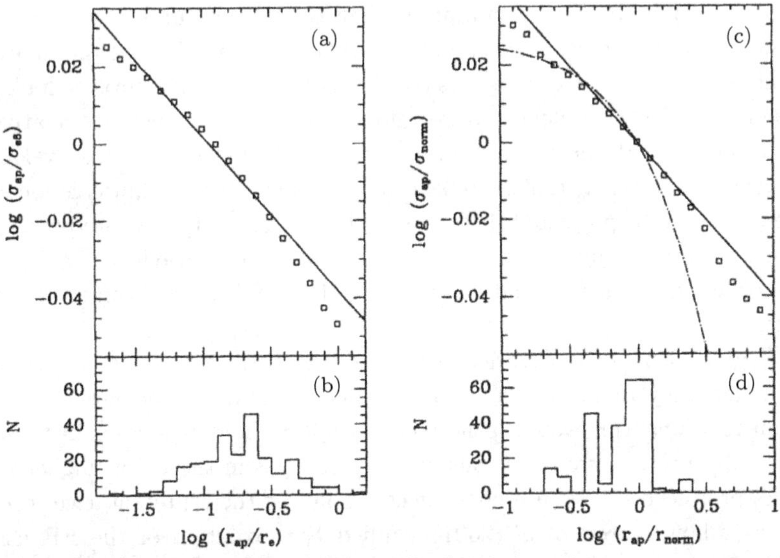

Fig. 8.2. Showing the aperture correction derived by Jørgensen *et al.* (1995). The median values from empirical galaxy models are the open squares. (a) The line shows the fit to Eq. (8.33) using an $(r_e/8)$ aperture. Panel (c) shows the fit using an aperture of metric diameter ($2r_{\rm norm} = 1.19h^{-1}$ kpc and compares with the aperture correction formula (8.30), the dot-dashed curve. Panels (b) and (d) are the distributions of the data. (*Source*: Jørgensen I., Franx M., Kjaergaard P. (1995), *MNRAS*, 276, 1341, "Spectroscopy for E and S0 galaxies in nine clusters," Fig. 4, Reproduced by permission of Oxford University Press on behalf of the Royal Astronomical Society.)

is to use a relative aperture defined in terms of the effective radius, $R_e/8$,

$$\sigma_{\rm cor} = \sigma_{\rm obs}\left(\frac{r_{\rm ap}}{R_e/8}\right). \tag{8.34}$$

The fits to the data by Jørgensen *et al.* (1995) in Fig. 8.2 give a good approximation over a wide range of aperture sizes. This approximation for the aperture correction has been used for several studies including EFAR (Wegner *et al.*, 1999), ENEAR (Wegner *et al.*, 2003) and the 6dF survey (Campbell *et al.*, 2014).

8.7. Environmental Effects

8.7.1. *Early-type (elliptical) galaxies*

Using the D_n–σ and FP relations for early-type galaxies and the Tully–Fisher relation for spirals to estimate distances for peculiar motion studies makes the fundamental assumption that they follow universal relations. However, star formation in galaxies is well-known to depend on environment, especially at large redshifts (see Section 8.4). Early efforts to look for the effects of environment and evolution indicated that they are relatively weak, at least at the redshift limits ($z < 0.3$) of many peculiar velocity surveys. Lucey *et al.* (1991) looked at 51 Coma cluster ellipticals located from the core to projected distances within $\sim 4h^{-1}$ Mpc, where projected density decreases by ~150, to check for environmental effects on D_n–σ, and concluded that the relation is independent of distance from the cluster centre.

Saglia, Bender, & Dressler (1993) analyzed 51 Coma cluster galaxies. Although they found a weak trend by disky galaxies to deviate from the mean relations, this would insignificantly affect the derived peculiar velocities. More recent investigations indicate changes in the FP at higher redshifts produced by evolutionary effects. Some of the representative results are as follows. Treu *et al.* (2001) studied the evolution of the FP using 25 field galaxies with $z \approx 0.1$–0.5. They found that the scatter in the relation does not change with redshift, but for a ΛCDM model, the intercept changes by $\Delta \log R_e = \tau z$ with $0.44 < \tau < 0.56$. Blakeslee *et al.* (2002) compared distances for the FP and SBF methods and looked for systematic trends. Although the overall agreement is good, there is some indication that galaxies with low Mg_2 indices can deviate due to recent star formation activity. At even higher redshift, van Dokkum & Stanford (2003) compared the FP from three galaxies in a $z = 1.27$ cluster with the Coma FP and found offsets in $\log R_e$ of ~0.5 in B and ~0.3 in R, which they conclude is consistent with passive evolution. Moran *et al.* (2007) studied the buildup of S0 galaxies from spirals in the different environments of two galaxy clusters ($z = 0.39$ and 0.54). More recent investigations using NIR data and extensive datasets are discussed in Chapter 11 below.

8.7.2. *Tully–Fisher relation*

For the TF relation, many investigations have considered whether the coefficients of the linear relation change with redshift; these earlier studies do not seem to have reached agreement on whether it changes significantly. The interpretation of the TF relation with z will depend on the passband going from UV to IR as different types of stars are represented, e.g., the blue stars dominating the B band and red disk stars in the NIR. In addition to its application as a distance indicator, it has been used to study galaxy properties (e.g., M/L) and star formation history. Earlier investigations utilized 21-cm line profiles and visual wavelength photometric bands, but more recent applications use Hα and NIR K-band photometry.

Peuch *et al.* (2008) concluded that some redshift-dependent effects could be due to fitting methods. Koen & Lombard (2009) found that the K-band TF relation shows no statistically significant difference with distance. Fernandez *et al.* (2010) studied TF relations using B, V, R, I, and K_s for $0.1 < z < 0.3$, $1.1 < z < 1.4$ and although they found some evidence for colour evolution at the short wavelengths, it is unimportant in the NIR K_s-band. Sorce *et al.* (2013) have also considered calibration of the TF and found that in the 3.6μ band, a small colour correction is needed. Tiley *et al.* (2016) studied the K-band TF relation at redshift $z \sim 1$. Evolution effects in galaxies will change the zero-point of the relation at this distance.

Chapter 9

The Malmquist and Other Biases

9.1. Order-of-Magnitude Limitations

The accuracy of all distance measurements have ranges of validity. For peculiar velocity determinations, the order of magnitude of a single galaxy's peculiar velocity, V_{pec} is typically below 600 km s^{-1}. Comparing this with the expansion redshift or Hubble flow velocity, cz, V_{pec} and expansion become equal when $V_{\text{RS}} = cz \approx 70D$ km s^{-1}, so V_{pec} is lower than V_{RS} at $D \gtrsim 8$ Mpc. If a given distance indicator is accurate to $\Delta D \sim \pm 10\%$, the error becomes comparable to the expansion at ~80 Mpc, measuring peculiar velocities becomes a problem of extracting a weak signal from noisy data, which can distort the final answer. This is also an important factor in some derivations of H_0.

In addition to employing a more accurate distance indicator to probe greater distances, one can average a number of galaxies, N, e.g., in clusters at the same distance to reduce the statistical error by $\sqrt{N}$, and this is done to calibrate the TF and FP relations. As an example, the nearby Coma cluster has $V_{\text{RS}} \sim 7000$ km s^{-1}or $D \sim 100$ Mpc. If the ΔD is $\pm 20\%$ for the error of one cluster member, $\Delta D \sim \pm 20$ Mpc. The 7S observed about 30 galaxies in Coma, so its distance should be known to $\sim \pm 3.6$ Mpc. However before this can be determined, further corrections must be made to the raw observed data in addition to those already discussed. In a given direction, the observations do not have uniform depth coverage, and further systematic errors arise making the data analysis increasingly tricky with distance. These are sources of bias in the data.

9.2. Malmquist and Malmquist-Like Bias

Most astronomical data involving counts of objects at different distances are in some way affected by biases, due to one's inability to detect all sources. This is an important consideration in all branches of astronomy and cosmology. G. Malmquist (b. 1893–d. 1982; Wegner, 2000b), after whom this general problem is named, first encountered it from considering counts of stars, but it is even more serious and complex for extra-galactic astronomy. Willick (1994), Strauss & Willick (1995), Teerikorpi (2015), Binney & Merrifield (1998), and others cover many of the details of these biases. In general, biases depend on several factors including the spread in intrinsic properties such as diameter or brightness, spatial distribution, and the method of selection as well as the chosen scaling relation. Instrumental limitations including sensitivity and colour response are additional factors. In some situations, environmental and evolutionary effects may be present and differences in cosmological corrections can become important at high z. If these distortions are uncorrected, the final answer may be greatly in error. This is particularly true where biases in the distances measured can lead to fallacious peculiar velocities and in deriving H_0. For the near universe, Malmquist and Malmquist-like effects resulting from selection and incompleteness are major contributors to bias.[1]

Biases affect the derived spatial distribution of sources measured in magnitude limited surveys. This arises when objects of a fixed given type with an intrinsic scatter such as in brightness or diameter are observed at increasing distance with a limit imposed by the observing instrument or selection process. As the depth of observation grows, the numbers of objects in a shell increases as r^2, but more intrinsically faint sources below the magnitude limit of the survey will be lost, i.e., undetected with distance. This leads to an overestimate of the sample's mean intrinsic brightness with distance.

In general, one knows neither the intrinsic brightness distribution nor the spatial distribution of sources along the line of sight. The observed number of objects brighter than magnitude $m_{\rm lim}$ in solid angle Ω can be written schematically as an integral equation along the line of sight:

$$N(m_{\rm lim}) = \Omega \int_0^{\infty} \int_{-\infty}^{m_{\rm lim}} S(M + 5\log r - 5)P(M)dMD(r)r^2dr, \qquad (9.1)$$

[1]There are different definitions of Malmquist bias in the literature. For extragalactic sources Lynden-Bell *et al.* (1988) referred to Malmquist bias as related to effects produced along the line of sight.

where $D(r)$ is the density function, $S(m) \equiv S(M+5\log r-5)$ is the selection function, and $P(M)$ is the luminosity function or brightness distribution of the sources. The bias in the counts is proportional to $N(M)^{\rm obs} - N(M)$. Reddening and environment produce additional complications. In general, this cannot be solved without some additional information about these functions, such as the luminosity function, $P(M)$, which could be a Gaussian, a Schechter, or some other function. The density, $D(r)$, might be determined from redshifts or distances employing the variety of methods to be described.

Spaenhauer (1978) discussed obtaining solutions to equation (9.1) if $P(M)$ is a Gaussian function. Figure 9.1 shows the number of sources in a homogeneous distribution with distance. The curves represent different magnitude limits. Objects fainter than the limit are above the curves and unobserved so one sees that the average magnitude of the observed sources becomes brighter than the true average.

9.2.1. *Homogeneous Malmquist correction*

Malmquist (1920) developed a bias correction to star counts with a magnitude cutoff and simplifying equation (9.1) with the assumptions that the

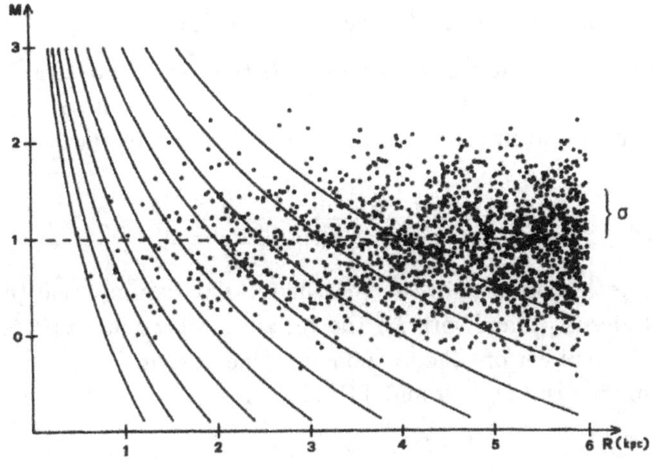

Fig. 9.1. The magnitude distribution M of stars with distance from a homogeneous Gaussian distribution with a dispersion σ and a mean magnitude of 1.0 (dashed line). The solid curves are $M = m_{lim} - 5 - 5\log R + 5$ where $m_{lim} = 9.5 + 0.5n(n = 9.5 + 0.5n = 0, 1, 2, 3, \ldots)$ and represent different magnitude limits with distance R. Faint stars above each curve are lost for the survey with that magnitude limit. (*Credit*: Spaenhauer A. M. (1978), *A&A*, 65, 313, "A systematic comparison of four methods to derive stellar space densities," Fig. 2, Reproduced by permission ©ESO.)

spatial density of the sources is homogeneous and their brightness probability distribution is Gaussian,

$$P(M) = \frac{1}{\sqrt{2\pi}\sigma} \exp[-(M - M_o)^2/2\sigma^2], \tag{9.2}$$

where σ is the dispersion and M_o is the peak of the distribution. Since $m = M + 5\log r - 5$, all objects will be observed within the cutoff $r = 10^{[5+(m-M)]/5}$. This is the case illustrated in Fig. 9.1 where curves of constant cutoff magnitude are shown. The number of sources for a homogeneous distribution increases as r^3, so the average absolute magnitude in the magnitude limited sample will be

$$\bar{M} = \frac{\int P(M) M r^3 dM}{\int P(M) r^2 dM}. \tag{9.3}$$

It is easy to find the bias of the mean magnitude $\bar{M}$

$$\bar{M} = M - 1.38\sigma^2. \tag{9.4}$$

This illustrates the important feature of the Malmquist bias: as the dispersion, σ, of the distance indicator (assumed here to be Gaussian), is reduced, the bias decreases.

9.2.2. *Malmquist biases for scaling relations*

Applied to scaling relations of galaxies, the situation is more complex. For a scaling relation, such as the TF, of the form $\log d_{\text{raw}} = 0.2(m_I - a_{\text{TF}} \log w - b_{\text{TF}})$, the general formula for correcting the observed or raw distance is

$$d_{\text{cor}} = \frac{\int_0^\infty r^3 n(r) \exp(-\frac{[\ln(r/d_{\text{raw}})]^2}{2\Delta^2}) dr}{\int_0^\infty r^2 n(r) \exp(-\frac{[\ln(r/d_{\text{raw}})]^2}{2\Delta^2}) dr}, \tag{9.5}$$

where d_{cor} is the corrected distance estimate. This assumes that the scatter in magnitudes is normal with Δ^2 the variance in $\log d$ and $n(r)$ is proportional to the number of galaxies along the line of sight.

Determining the $D_n - \sigma$ and TF relations,

$$\log D = a + b \log \sigma, \tag{9.6}$$

$$M = a_{\text{TF}} \log W + b_{\text{TF}}, \tag{9.7}$$

requires evaluating the constants in the relations and determining unbiased distances. Following Burstein & Faber (1988) and Strauss & Willick (1995), there are two approaches:

- Method I: Use distance indicators to predict distance and v_{pec}. Calibrate the relations from galaxy clusters.
- Method II: Use redshifts to estimate distances in redshift space. Determine the constants from a global model, the velocity field, and minimize the residuals.

The biases differ for the two methods. In Method I, there will be a magnitude limit and the sample will have a Malmquist bias. In Method II, Malmquist bias is less or eliminated, but there will be selection bias. In using clusters, the assumption is made that all galaxies are at the same distance and the scatter about the relation, σ, changes due to the missing faint galaxies. Two alternate methods of analysis can be applied for Methods I and II using the relations in equations (9.6) and (9.7) can be employed. The first is the 'direct' relation in which the 'direct' formula

$$M = a + b \log W_0, \tag{9.8}$$

is used. The second is the 'inverse' relation

$$\log W_0 = a' + b'M. \tag{9.9}$$

Line widths and magnitudes for the Tully–Fisher relation can be replaced with D_n and σ, etc., for other distance estimators.

Schechter (1980) pointed out that in the Faber–Jackson and Tully–Fisher relations, no selection effects are produced by measuring velocity dispersions, σ, or the line-widths, W, which should eliminate the Malmquist effect. However, this does not necessarily eliminate the other kinds of biases. Aaronson *et al.* (1982b) were among the first to apply the inverse method. Several investigators (e.g., Willick, 1994; Strauss & Willick, 1995; Teerikorpi, 1993; Gould, 1993 and references therein) considered both the direct and inverse relations in detail and found that neither approach is completely free of biasing.

9.2.3. D_n–σ homogeneous correction

The simplest approximation to the density distribution is to assume that it is homogeneous. Lynden-Bell *et al.* (1988) used the direct D_n–σ relation

$$\log R_e = 1.20 \log \sigma - \log D_n + \log\left(\frac{1 + \frac{7}{4}z}{1 + \frac{7}{4}z_{\text{Coma}}}\right) + 1.411, \tag{9.10}$$

where R_e is the estimated distance and the cosmological corrections are included.

Following Lynden-Bell *et al.* (1988) and Strauss & Willick (1995) assuming Gaussian distributions, the distance R corrected for forward Malmquist bias using equation (9.5) can be put in the form relative to the estimated distance, R_e,

$$R = R_e \frac{\int_0^\infty x^3 n(r) e^{\frac{-(\ln x)^2}{2\Delta^2}} dx}{\int_0^\infty x^2 n(r) e^{\frac{-(\ln x)^2}{2\Delta^2}} dx}, \tag{9.11}$$

where $x = r/R_e$ and $\Delta = (\frac{\ln 10}{5})\sigma \approx 0.4605\sigma$ and becomes

$$R = R_e \exp\left[1 + \left(\alpha + \frac{1}{2}\right)\Delta^2\right]. \tag{9.12}$$

For a uniform distribution $\alpha = 3$ and for groups and clusters of N galaxies, Δ^2 is replaced by Δ^2/N and this simple formula illustrates how the distances for elliptical galaxies are affected by bias.

$$R = R_e \exp(3.5\Delta^2), \tag{9.13}$$

which is the homogeneous correction applied by several investigators.

As an example, Lynden-Bell *et al.* (1988) adopted $\Delta \approx 0.21 \pm 0.02$ as the dispersion in the $\ln D_n$, taken to be a Gaussian relation and the correction can be written relative to the Coma cluster ($cz = 7200\,\mathrm{km\,s^{-1}}$ was used) rather than in absolute units.

9.2.4. *Corrections to the TF relation*

Corrections to the Tully–Fisher relation have been discussed extensively as it has been used to determine the Hubble constant in addition to peculiar velocities (e.g., Teerikorpi, 1984; 1987; 1993; Willick, 1994; Bottinelli *et al.*, 1988).

Landy & Szalay (1992) discussed how to improve this homogeneous correction by including the radial distribution of the galaxies. They found an approximation,

$$R_{\mathrm{cor}} = R_e \exp\left(\frac{\Delta^2}{2} \frac{h[\ln R_{\mathrm{raw}} + \Delta^2]}{h[\ln R_{\mathrm{raw}}]}\right), \tag{9.14}$$

where $h(R)$ is the galaxy distribution observed along the line of sight. Although relations such as these can be easily applied to a homogeneous distribution, along the line of sight, they are difficult to apply globally when there are smaller scale inhomogeneities.

9.2.5. *Non-homogeneous Malmquist corrections*

Assuming a uniform galaxy distribution is not necessarily a good approximation, for example, in the structure of the large-scale structure the presence of filaments and voids is well known. Returning to equation (9.12), if α is replaced by a power series expansion, $\alpha + \frac{d \ln n(r)}{d \ln r} + \cdots$, and the galaxy distribution has a slowly changing density gradient, at distance R_e,

$$\mathfrak{g}(R_e) = \left. \frac{d \ln n(r)}{d \ln r} \right|_{R_e},$$

expanding this to first order gives

$$R = R_e \left[1 + (3.5 + \mathfrak{g}(R_e))\, \Delta^2\right]. \tag{9.15}$$

If $(g) = 0$ this is the homogeneous formula, but $\mathfrak{g} \neq 0$ is an example of a non-homogeneous Malmquist correction. Hudson (1994a) has considered the inhomogeneous Malmquist bias for the D_n–σ relation. Figure 9.2 illustrates how a non-homogeneous density distribution produces a spurious peculiar velocity. A proper evaluation of this in general requires a numerical calculation and involves considerably more effort.

Freudling *et al.* (1995) demonstrate using Monte Carlo simulations to estimate the non-homogeneous Malmquist bias and apply it to peculiar velocity maps. This numerical approach has been followed by subsequent investigations. Both the direct and inverse relations can be utilized with this method and applied to the TF relation and show that it can be applied to the D_n–σ relation. Spatial distributions of galaxies found by redshift surveys are corrected using

$$\beta = \Omega / b^{0.6}. \tag{9.16}$$

Additional model galaxies are made, which fill in the density field. A simulated sample based on the selection criteria of the observed sample, e.g., diameter or magnitude incompleteness, or reddening, is constructed and distances are calculated using the assumed distance estimator. This allows one to include all known bias sources. Subtracting the observed distance from the estimated "true" distance yields the bias of the observed sample in redshift space. Peculiar velocities can be computed from the bias corrected distances.

Freudling *et al.* (1995) used the final density field constructed in the Hydra-Centaurus direction (Wilmer *et al.*, 1995) that covers approximately 50% of the sky based on an optical-IRAS catalogue (Freudling, da Costa, & Pellegrini, 1994) using the 1.9 Jy IRAS sample of galaxies (Wilmer *et al.*, 1995; Strauss *et al.*, 1992). This method was used to derive bias corrections

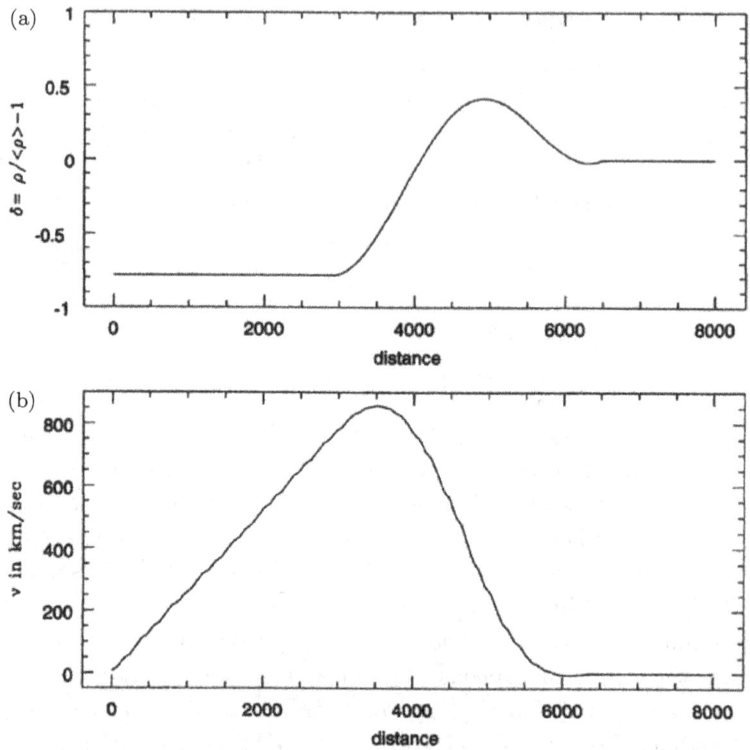

Fig. 9.2. Density and peculiar velocity assumed for the toy model used in Fig. 9.3. This shows a non-homogeneous density distribution with a non-uniform change, roughly as a step, near 4000 km s^{-1} and the associated velocity distribution, appearing as the bump. (*Source*: Freudling W., da Costa L. N., Wegner G., Giovanelli R., Haynes M. P., Salzer J. J. (1995), *AJ*, 110, 920, "Determination of Malmquist bias and selection effects from Monte Carlo simulations," Fig. 3, ©AAS Reproduced with permission.)

for the SFI TF data (Haynes *et al.*, 1999a,b), but it was applied as a check to other samples, e.g., the 7S D_n–σ sample (Lynden-Bell *et al.*, 1988), which was found to be free of bias. A similar approach is described by Masters *et al.* (2008).

Figure 9.3 illustrates results of the direct and inverse relations for the Malmquist bias determination employing this method using the arbitrary or 'toy' model, shown in Fig. 9.2 for which the density and $v_{\rm pec}$ change.

Other statistical methods have been proposed for dealing with the Malmquist bias problem. Watkins & Feldman (2015b) have pointed out that the derived peculiar velocities depend on the measures having a Gaussian distribution of errors. However, these distributions can have skewed

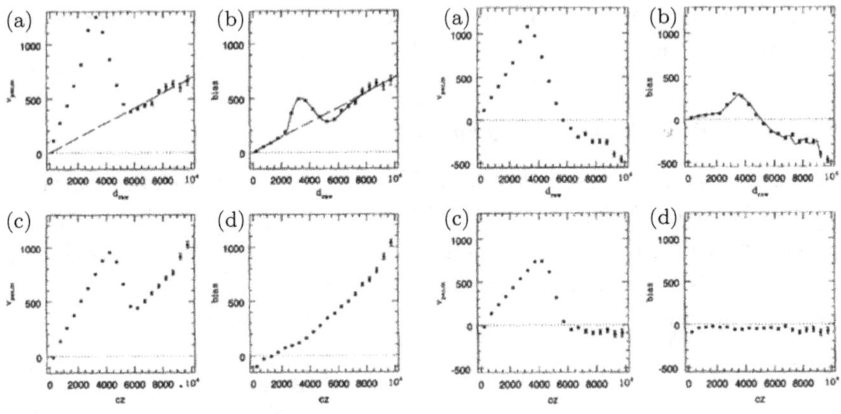

Fig. 9.3. Observed peculiar velocities and biases for the direct and inverse relations. Left four panels: Direct TF relation: (a) and (c) are observed $v_{\rm pec}$ with binning using $d_{\rm raw}$ and cz, respectively; (b) and (d) show the biases. Solid lines in (a) and (b) are homogeneous Malmquist corrections. The solid curve in (b) is the bias computed by Eq. (9.5). Right four panels: Indirect TF relation: panels have the same relation. Note that the bias in (d) is largely reduced by the inverse relation and demonstrates its superiority. (*Source*: Freudling W., da Costa L. N., Wegner G., Giovanelli R., Haynes M. P., Salzer J. J. (1995), *AJ*, 110, 920, "Determination of Malmquist bias and selection effects from Monte Carlo simulations," Fig. 4, ©AAS Reproduced with permission.)

tails and from objects contributing to these tails, large errors are introduced as often the measurement errors are larger than the estimated $v_{\rm pec}$. Starting with $v = cz - H_0 r$, a new estimator, which does not use a bias correction is given by

$$v_e = cz \log(cz/H_0 r_e), \tag{9.17}$$

and z is replaced by $z_{\rm mod}$ where

$$z_{\rm mod} = z\left[1 + 0.5(1 - q_0)z - \frac{1}{6}(1 - q_0 - 3q_0^2 + 1)z^2\right]. \tag{9.18}$$

Since

$$v = \frac{z_{\rm mod} - H_0 r}{1 + H_0 r/c} \approx \frac{z_{\rm mod} - H_0 r}{1 + z_{\rm mod}}, \tag{9.19}$$

this gives

$$v_e = \frac{cz_{\rm mod}}{1 + z_{\rm mod}} \log(cz_{\rm mod}/H_0 r_e), \tag{9.20}$$

and the error is

$$\delta v_e = \frac{z_{\rm mod}\delta\ell_e}{1 + z_{\rm mod}}. \tag{9.21}$$

Watkins & Feldman (2015b) apply this on the SFI++ and CF2 catalogues and conclude that this estimator produces unbiased Gaussian errors.

9.3. Sample Selection Functions

Usually the faintest or smallest limit of a survey using a selection variable, X, the selection function $S(X)$, will not be a sharp cutoff. One might expect this to be a smooth function of the form $S(X) = 1 - \mathrm{erf}(X, a, b, \ldots)$. Dividing the counts into k bins, $X_k - \frac{1}{2}\Delta X \leq X_k < X_k + \frac{1}{2}\Delta X$, yields the selection function

$$S(X_k) = N_k^{\mathrm{obs}}(X_k)/N_k^{\mathrm{all}}(X_k). \tag{9.22}$$

This requires a method of finding $N_k^{\mathrm{all}}(X_k)$, which is not always easy but can sometimes be obtained using lower precision data that goes fainter than the studied sample. The EFAR study compared CCD diameters with those measured off enlargements prints of the POSS-1 glass photographic plates, D_W (Wegner *et al.*, 1996). Examples of selection functions are shown in Fig. 9.4 for the 84 clusters in the study.

One useful approximation for $S(X)$, which models a cutoff is a fitting formula,

$$S(\log D_W) = \frac{1}{2}\left\{1 + \mathrm{erf}[(\log D_W - \log D_W^o)/\delta_W]\right\}, \tag{9.23}$$

where δ_W is the width and $\log D_W^o$ is the midpoint of the cutoff. This can be fitted using the linear regression using the inverse error function,

$$\mathrm{erf}^{-1}[2S(\log D_W) - 1] = (\log D_W - \log D_W^o)/\delta_W. \tag{9.24}$$

Saglia *et al.* (2001) employed this formula for an analysis of the EFAR data peculiar velocities for a maximum-likelihood fitting of the Fundamental Plane and analysis of peculiar velocities.

Many other approximations are possible. Masters *et al.* (2006) have employed another fitting formula depending on absolute magnitude to correct for the incompleteness bias of Tully–Fisher measurements in galaxy clusters for the SFI++ project

$$C(y) = \frac{1}{e^{(y-y_F)/\eta} + 1}, \tag{9.25}$$

where $y = M - 5\log h$.

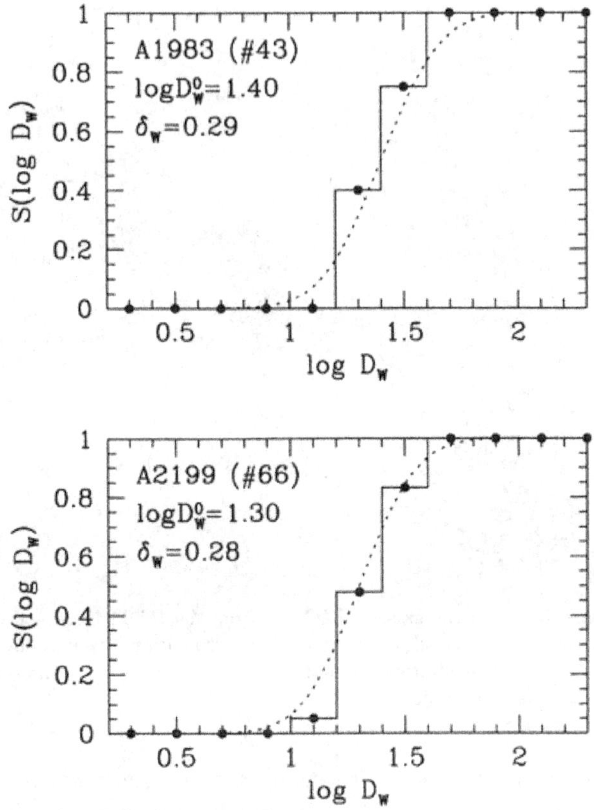

Fig. 9.4. Selection functions for two galaxy clusters. The dots and histograms are the measured selection functions. The dashed curve is the fit using Eq. (9.23). (*Source*: Wegner G., Colless M., Baggley G., Davies R. L., Bertschinger E., Burstein D., McMahan R. K., Jr., Saglia R. P. (1996), *ApJS*, 106, 1, "The peculiar motions of early-type galaxies in two distant regions. I. Cluster and galaxy selection," Fig. 6, ©AAS Reproduced with permission.)

9.4. Correction for Fore- and Background Sources

Contamination by intervening sources along the line of sight, e.g., foreground stars, background galaxies, etc., must be corrected. Several programs are now available to correct CCD images. Many of these stem from the GRASP programs of Cawson (1987). Figure 9.5 is an example of the correction for the *I*-band image of the spiral galaxy NGC 259. The faint sources on the galaxy image have been blocked out and elliptical isophotes have been fitted to the galaxy's light, which are used to interpolate for the

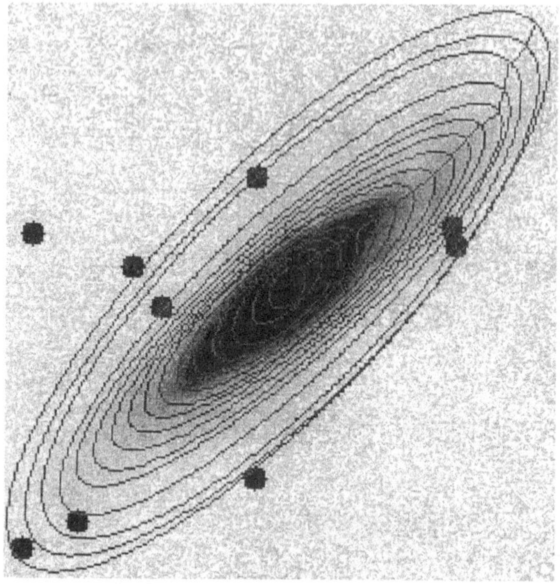

Fig. 9.5. I-band image of the galaxy NGC 259 from the MDM Observatory 1.3 m telescope utilized for photometry showing the faint foreground and background objects that have been eliminated. The elliptical isophotes fitted to the data were employed to replace the missing data where the objects were.

missing light. In this instance, the correction is relatively small and usually galaxies close to the galactic equator where the number of foreground sources becomes large are not used. Here the correction is relatively small, but in some instances, bright neighbours require larger corrections and this adds to the uncertainty of the measurement.

9.5. Cosmic Variance

Cosmic variance, σ_v, is a source of uncertainty in cosmological and large-scale investigations. Only portions of the universe can be observed from the earth. When averaging a measured quantity from different survey regions, statistical errors result. The problem is to estimate the magnitude of this uncertainty. Measurements from the earth are affected by the large-scale structure. Observations as simple as galaxy counts from one part of the sky compared to the same measurement in a different field can give differences exceeding the sample variance. Cosmic variance is important for a number of observations including the CMB and has received numerous attention.

If N is the number of objects counted in a volume V, σ_v is defined by its mean and variance by

$$\sigma_v = \frac{\langle N^2 \rangle - \langle N \rangle^2 - \langle N \rangle}{\langle N \rangle}. \tag{9.26}$$

Peebles (1980) shows that σ_v for a set of objects in a volume V with a known 2-point correlation function, $\xi(r)$ is

$$\sigma_v^2 = \frac{1}{V^2} \int_0^R \xi(|\mathbf{r}_1 - \mathbf{r}_2|) dV_1 dV_2. \tag{9.27}$$

If $\xi(r) = (r_o/r)^\gamma$, for spherical shells, this is

$$\sigma_v^2 = \{72/[(3-\gamma)(4-\gamma)(6-\gamma)2^\gamma]\}(r_o/r)^\gamma. \tag{9.28}$$

Thus σ_v drops with increasing survey volume, V. Somerville *et al.* (2004) and Moster *et al.* (2011) model it for CDM. Driver & Robotham (2010) have derived an empirical formula for some galaxy surveys based on the SDSS data release DR7. For a volume V, the basic result is

$$\sigma_v \times 100 = 219.7 - 52.4 \log V + 3.21 (\log V)^2. \tag{9.29}$$

This gives a rough estimate that $\sigma_v \lesssim 10\%$ when $V \sim 10^7 h_{0.7}^3\,\mathrm{Mpc}^3$.

Most recent investigations involved with peculiar motions have estimated the cosmic variance using mock catalogues.

Chapter 10

The Extragalactic Distance Ladder

Distance determinations have always been a major problem in the development of astronomy and the lack of accurate distances produced several controversies including the famous Great Debate of Shapley and Curtis (Gingerich, 1999) over the distances of the nebulae and later the Hubble constant. In the early 20th century, many astronomers believed that the sun was near the centre of the Milky Way, based on the 'Kapteyn universe', derived from a massive photographic compilation of the distribution of stars across the sky to measure the structure of the Milky Way, similar to Wm. Herschel's survey a hundred years earlier. This did not take into account interstellar absorption, which was generally unknown in his time and in 1900 was not yet understood.

The zero points for determining distances are in the calibration of ℓ_0 and d_0 in Eqs. (8.1) and (8.2). For the different classes of extragalactic objects, this depends on observing representative members of a sample of identical sources with known distances and making the proper corrections. As observations of them are limited by intrinsic properties, e.g., size, luminosity, spatial density, etc., a hierarchy for distance determinations must be employed, to build the overlapping steps or rungs of the 'extragalactic distance ladder'. Nearer objects of one type that permit their distances to be determined with highest precision are used to calibrate other classes of objects at increasing remoteness. These in turn are employed to calibrate remoter objects.

The distance scale begins with solar system distances that can be determined directly with high precision. *Primary distance indicators* are sources in nearby galaxies that can be calibrated directly using examples in our galaxy. *Secondary distance indicators* are objects not found in our galaxy and are generally more remote requiring a second step in their calibration via the primary distance indicators. These may in turn be used to calibrate even more remote and less accurate *tertiary distance indicators.* As distance increases, estimates become increasingly insecure.

Many distance indicators have been devised and detailed lists and descriptions can be found, e.g., Rowan-Robinson (1985) and Jacoby *et al.* (1992). Freedman & Madore (2010) discuss calibration of distance indicators used for determining the Hubble constant, and these are also important for peculiar motion studies. Some methods were important historically and have been heavily investigated; others have appeared promising, but have never been fully developed. New distance indicators continue to be invented. In the following, an account of distance methods that are currently important for peculiar motion measurements and large-scale structure will be given. Some of the more promising new methods, including the SZ effect, gravitational lensing delays, baryonic acoustic oscillation, and standard sirens are described in later chapters.

10.1. Primary Distance Indicators

In the Milky Way, solar system distances calibrate the astronomical unit and secure trigonometric parallaxes to the nearest stars as the first rung of the ladder. The baseline of the trigonometric triangle for measuring the parallactic angle, π, is the astronomical unit (1 au), the mean distance of the earth to the sun, which has now been well determined using radar telemetry and space probes. This has also been made possible by the high precision measurement of the speed of light in vacuum ($299,792,458\,\mathrm{ms}^{-1}$) by definition (Mohr *et al.*, 2016) and the invention of atomic time standards such as those kept at the US Naval Observatory and the Royal Greenwich Observatory. This gives an astronomical unit, au $= 149,597,870,700\,\mathrm{m}$, or about 93 million miles (Capitaine, Klioner, & McCarthy, 2012).

The geometry for the determination of the astronomical parallaxes is well known (e.g., Smart, 1962). This is approximated very accurately as a skinny triangle with a base of 1 au and the star at the vertex. Measurement of the star's position relative to distant background sources over the year as the earth follows it orbit about the sun yields the parallax in the parallactic ellipse,

$$\pi = 206,265/D\,\mathrm{au}, \tag{10.1}$$

where D is the distance and π is measured in arcsec. A unit of measurement, the parsec (pc), is introduced where

$$D = \frac{1}{\pi}\,\text{pc} \tag{10.2}$$

(one parsec is 3.086×10^{13} km or about 3.26 light years). Positional measurements over several years additionally reveal a star's proper motion, μ. F. W. Bessell (1838) measured the first reliable parallax to the star 61 Cygni in 1838. This has been followed by increasing numbers of ground-based measurements, e.g., in *The General Catalogue of Trigonometric Parallaxes* (van Altena, Lee, & Hoffleit 1995) which contains data for 8112 stars with an average standard error (se) of $\Delta\pi \approx \pm0\overset{''}{.}004$.

It is clear that the most distant objects with the smallest parallaxes have the largest distance errors as $\Delta D/D = \Delta\pi/\pi$. Thus, ground-based parallax measurements with these observational errors have limited reliable distance determinations to the nearest stars with $D <$ 25–50 pc.

A revolution in parallax determinations began with the Hipparcos satellite (Prusti *et al.*, 2016), which took data in 1989–1993, followed by the Gaia satellite in 2013. Both devices are operated by the European Space Agency (ESA). Hipparcos covered the entire sky and produced positional data, parallaxes with proper motions, and photometry for 118,200 stars brighter than 12.5 mag. with a typical accuracy of $\Delta\pi \approx \pm0\overset{''}{.}001$. The accuracy was substantially improved after a second data reduction (van Leeuwen, 2007). This was combined with scans of 22,000 glass plates from the Astrographic survey series begun near the end of the 19th century into the Tycho-2 catalogue which contains positional data for about 2.5 billion stars. These data have considerably expanded understanding in stellar astronomy as well as the calibration of the primary distance indicators.

The Gaia satellite by ESA in 2013 December followed Hipparcos, which was intended to operate for about five years. Gaia's goal is to secure parallaxes, proper motions, photometry, and radial velocities for about one billion stars and to create a three-dimensional map of the Milky Way. The parallaxes are expected to have accuracies of better than $\pm20\,\mu$arcsecs and give distances to 20 million stars good to better than $\pm1\%$ and to 200 million stars to better than $\pm10\%$. Both of these satellites have begun to have far reaching impacts on all branches of astronomy and cosmology. The data are publically available on the internet.

High precision parallaxes have been measured for a handful of stars outside the galaxy in the Local Group using the HST, but due to this instrument's narrow field of view, large numbers of parallaxes such as those already mentioned, are not to be expected.

An additional quantity needed for sorting out the observed velocities is the motion of the sun relative to the galactic centre. This can be determined using VLBI observations of the source SgrA* and background quasars, which gives $R_o = 7.9^{+0.8}_{-0.7}$ kpc (Reid & Brunthaler, 2004; Reid *et al.*, 2009) and $R_o = 7.86 \pm 0.14 \pm 0.04$ from the astrometry of faint stars near the galactic centre (Boehle *et al.*, 2016).

Even with these great expansions in the quantity and precision of trigonometric parallaxes, providing improved calibration of primary distance indicators, plus data on stars and stellar dynamics in the Milky Way and Local Group, it does not reach even the nearer of the more distant galaxies and other distance methods have to be employed. The nearest of these contain the primary distance indicators.

10.1.1. *Variable stars and Cepheid variables as distance indicators*

Many, if not all, stars throughout the HR diagram vary in light output at some level. Even the sun is variable and has 5-minute oscillations. Generations of effort have gone into studying these objects. Since the end of the 20th century precision measurement of stellar light curves and stellar modelling that parallel the advances in extragalactic astronomy have opened vast fields for research in astroseismology (Aerts, Christensen-Dalsgaard, & Kurtz, 2010) and helioseismology (e.g., Christian-Dalsgaard *et al.*, 1996; Gough *et al.*, 1996), enabling the study of stellar interiors and physics, outside the subject of this book. A by-product of this active field has been to place the understanding of variable stars as distance indicators on a firmer foundation. Many variable stars are easily recognized observationally and they cover a fascinatingly wide range of stellar evolutionary states. However using some of them as extragalactic distance indicators requires that an object is bright enough to be seen in other galaxies, the variability can be calibrated to yield a luminosity, and there are enough of them to be easily found. Although there are still many kinds of object that can be utilized as distance indicators, for some types the luminosity is too low or the irregular light variations and associated physics are too complicated.

10.1.1.1. *The different kinds of variable stars*

Diverse objects come under the heading of variable star, but not all varieties are useful for determining distances. The less useful types are the impulsive or extrinsic variables where the star's variability does not arise from the internal physics of the star itself, but from external effects. Cataclysmic variables are an example; mass flows between members of a close binary system onto a degenerate white dwarf, neutron star, or black hole from which, nuclear reactions are built up, causing an outburst. Recurrent novae such as RS Oph and dwarf novae like U Gem that have 20–100 day outbursts are further examples. Flare stars are produced by faint M dwarfs that combine rotation with surface spots and solar-like flare outbursts. Rotating non-spherical stars and highly magnetic stars can also be light variable. Eclipsing systems from the well-known naked eye variable Algol binary system down to stars with transiting exoplanets are also variable.

More interesting for distance indicators are intrinsic variables of which, the most important are the pulsating stars and cover the Cepheids and Cepheid-like variables, which include the classical or Type I Cepheids, Type II Cepheids, RR Lyraes, δ Scuti, rapidly varying Ap stars (rvoAp), and variable white dwarfs. These lie in the 'instability strip' in the HR diagram which extends roughly from spectral type F to G. Although for different reasons, most stars in this strip will oscillate. Outside this instability strip are other variables. The β CMa stars are one of the hottest type of spectral class B1–B3 as are the variable DB white dwarfs.

On the red colour side of the HR diagram among the M supergiants are the irregular and semiregular variables. These include the Mira (o Ceti) variables that show large light amplitudes in the visual and have long periods ($P \sim 300$–600 days) and may have asymmetrical pulsations based on optical telescope array imaging interferometer observations (Ragland *et al.*, 2006). Among these objects is the bright star αOri or Betelguese (Guinan *et al.*, 2020). Miras are bright enough to be located in Local Group galaxies, have a period–luminosity relation, and have been utilized as extragalactic distance indicators, but are not as well understood as the classical Cepheids (Robertson & Feast, 1981; Glass & Evans, 1981; Yuan *et al.*, 2017). Considerably more data are becoming available on these and other stars, e.g., the Radial Velocity Experiment (RAVE) which covers 457 thousand stars (Kunder *et al.*, 2017).

The Cepheid variables divide into two types. The classical or Type I Cepheids have been intensively studied and used as distance indicators, have a period–luminosity relation with $P \sim 1$–100 days, absolute magnitudes $M_V \sim -3$ to -5 mag., and mostly lie in the disks of spiral galaxies with solar-like metal abundances. The Type II Cepheids are relatively metal poor and their period–luminosity relation is ~2 mag. fainter. Baade (1953) showed that their existence required a major revision in the distance scale and consequently H_0. The Type II Cepheids have several subclasses and often complicated light curves (Wallerstein, 2002): These include RV Tauri stars; they are found in globular clusters and their light curves ($P \sim 70$–140 days) alternate between deep and shallow minima. The W Virginis stars arise in the most metal poor globular clusters and show emission lines, possible due to shocks in their atmospheres. The BL Herc stars show the shortest periods of this class ($P \sim 1.3$ days). Even shorter dwarf cepheids are known.

The RR Lyrae stars are related to the Type II Cepheids, although fainter with $M_V \sim +1.0$ mag. and shorter periods ($P \sim 0.25$–0.75 days). They are numerous and found in globular clusters and the galactic halo in large numbers (Smith, 2003). They have been employed as distance indicators, but at limited distances. Thackeray & Wesselink (1953) used RR Lyraes in the Magellanic clouds and further confirmed the major revision of the extragalactic distance scale indicated by Baade (1953). Three subtypes are recognized based on light curve shape and sometimes complicated light curves are found according to the Blazhko effect, which is a long period irregular variation in the period and amplitude of the light curve. The RR Lyraes lie on the horizontal branches of globular clusters where they have been extensively studied, e.g., in the globular cluster M22 (Kunder *et al.*, 2013).

Other variables in the instability strip include δ Scuti stars, variable Ap stars (Kurtz, 1978; Kurtz & Wegner, 1979; Mkrtichian *et al.*, 2008), and variable DA or ZZ Ceti stars. These have short periods ($P \sim 10$ mins) and undergo non-radial or f-mode oscillations, where a wave moves around the star's surface, and will not be considered further.

Type I and II Cepheids and RR Lyraes are radial pulsators, also known as the g-mode. These variables and the M giants are evolved stars, initially in the mass range $M \sim 4$–$20M_\odot$. After exhausting hydrogen in their cores, they leave the main sequence and rapidly move horizontally at nearly constant luminosity to the cooler side the HR diagram. They cross the instability strip the first time after which they ignite the shell burning phase, and become red giants. Subsequently their helium cores ignite and they follow a 'blue loop' that takes them twice, or possibly more times, across the instability strip before reaching the asymptotic giant branch (AGB). Miras and other red giants represent more advanced states of evolution and lie on the AGB. The RR Lyraes are lower mass stars ($M \sim 0.8M_\odot$). They undergo a helium flash in their red giant phase and quickly move to the horizontal branch.

Early explanations for Cepheid variability in the 19th century assumed they are a type of binary star. Shapley (1914) concluded that they pulsate and this was corroborated by the researches of Eddington (1917, 1959). Figure 10.1 presents the light and radial velocity curves of the of the dwarf Cepheid RS Gru, which is also a member of a binary system. This shows the radial pulsation curve derived from these data, which is ~10% of the radius.

For each type of star, there is a period–density relation

$$P = \frac{Q}{\sqrt{\bar{\rho}}}, \tag{10.3}$$

where Q depends on the type of star and can be understood from the internal structure and radial pulsations. The radial pulsators are taken to be spherically symmetric and the radii vary periodically from a maximum to a minimum as opposed to the non-radial pulsations.

Here, it is important to digress from peculiar motions and briefly mention the many steps necessary to understand the basic mechanisms needed to understand these stars. The perfection of modelling them has required several generations of effort in both stellar structure and stellar atmospheres. The period is an eigenfrequency of the star. By analogy to a vibrating string with fixed ends, there are a fundamental and overtones. In the case of a spherical star, one considers it to be in equilibrium and distorts it radially. The vibration frequencies will be described employing Euler's equations. With finite boundary conditions, the star oscillates unless a damping mechanism operates. Neglecting viscosity, the equations of motion are

$$\rho\frac{dv}{dt} = -\nabla p - \rho g, \tag{10.4}$$

$$\frac{\partial\rho}{\partial t} = -\nabla\rho v, \tag{10.5}$$

where v is velocity, ρ is density, p is pressure, and the gravity $g = GM/r^2$.

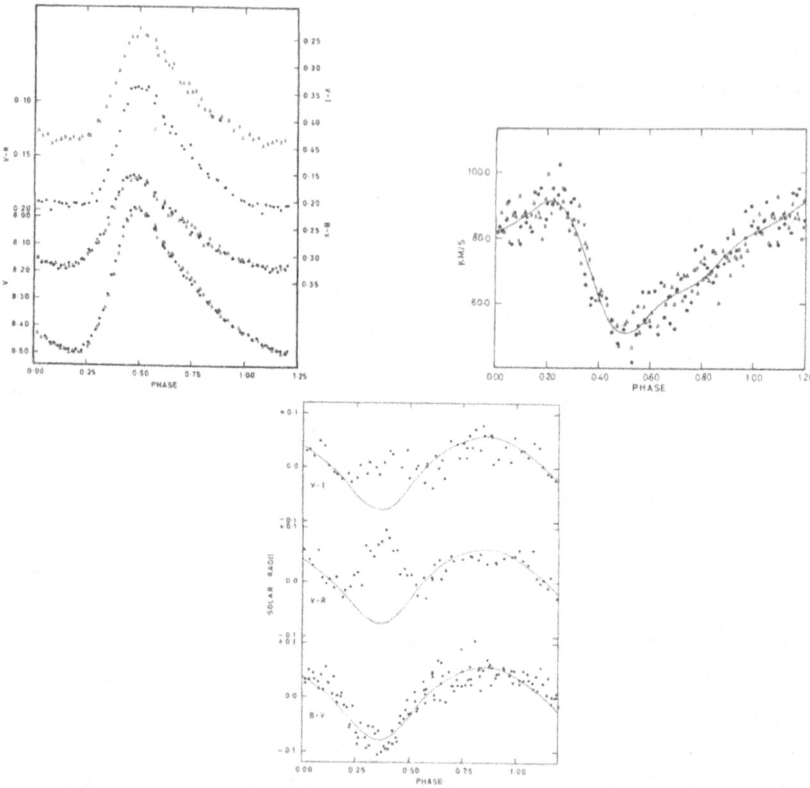

Fig. 10.1. Examples of data for the short-period Cepheid variable RS Gru ($P \approx 0.147$ day). (a) Upper left: photometric data from bottom to top are the V-band light curve, $(B-V), (V-R)$, and $(V-I)$ colour indices. (b) Upper right: the corresponding radial velocity measurements and fitted curve. (c) Bottom: change of radius during the pulsation using the above data and a version of the Baade–Wesselink method with an adopted radius of 2.9 $R_\odot$. (*Source*: Balona L. A., Martin W. L. (1978), *MNRAS*, 184, 1, "RS Gruis: a dwarf cepheid in a binary system," Figs. 1–3, Reproduced by permission of Oxford University Press on behalf of the Royal Astronomical Society.)

In spherical coordinates, $\nabla = \partial/\partial r + 2/r$ and this leads to the equation of motion

$$\frac{d^2v}{dt^2} = \frac{1}{\rho^2}\frac{d\rho}{dt}\frac{\partial p}{\partial r} - \frac{1}{\rho}\frac{d}{dt}\left(\frac{\partial p}{\partial r}\right) - \frac{dg}{dt}, \tag{10.6}$$

where the change of the pressure gradient with time, $\frac{d}{dt}(\frac{\partial p}{\partial r})$, and the change in pressure in time, $\frac{dP}{dt}$ are needed

$$\frac{d}{dt}\left(\frac{\partial p}{\partial r}\right) = \frac{\partial}{\partial r}\left(\frac{dp}{dt}\right) - \frac{\partial p}{\partial r}\frac{\partial v}{\partial r}. \tag{10.7}$$

Assuming an ideal gas,

$$\frac{dp}{dt} = \frac{\gamma P}{\rho}\frac{d\rho}{dt} + (\gamma - 1)(\rho\varepsilon - \nabla F), \tag{10.8}$$

where $\gamma = c_p/c_v$ is the ratio of the specific heats, F is the heat flux, and ε is heat liberated irreversibly by subatomic processes.

Putting these relations into equation (10.6) gives

$$\frac{d^2v}{dt^2} = \frac{1}{\rho}\frac{\partial}{\partial r}(\gamma p \nabla v) + \frac{4gv}{r} + \frac{2v}{r}\frac{dv}{dt} - \frac{1}{\rho}\frac{\partial}{\partial r}[(\gamma - 1)(\rho\varepsilon - \nabla F)], \tag{10.9}$$

the equation of motion for radial pulsations.

The solution of this equation, including the full physics is difficult, involves non-adiabatic pulsations, and has been the subject of many numerical calculations (e.g., Cox, 1980; Buchler, 2010). However, to gain insight, assume adiabatic oscillations as an approximation, setting $(\rho\varepsilon - \nabla F) = 0$, and eliminate the last term. Starting from an equilibrium solution, linearize the equation taking small perturbations from the equilibrium, $p = p_0 + p'; \rho = \rho_0 + \rho'; g = g_0 + g'$, etc. corresponding to the small displacement ζ where $r = r + \zeta$ and

$$\zeta = \zeta(r)\exp(i\sigma t). \tag{10.10}$$

Inserting these into Eq. (10.9) and eliminating the last term yields the linear adiabatic equation,

$$\frac{\partial}{\partial r}(\gamma p \nabla \zeta(r)) + \left(\sigma^2 + \frac{4g}{r}\right)\rho\zeta(r) = 0. \tag{10.11}$$

This is further simplified by defining $\zeta = \nabla\varphi$, which yields a wave equation,

$$c_s^2\nabla^2\varphi + \left(\sigma^2 + \frac{4g}{r}\right)\varphi = 0, \tag{10.12}$$

where $c_s^2 = \gamma p/\rho$ is sound speed.

Rosseland (1964) considers a few analytical solutions. The simplest case is the homogeneous spherical star, $\rho(r) = \bar{\rho} = \text{const.}$ and $\gamma = \text{const.}$, which Ritter (1879) first solved. For this

$$p = \frac{1}{2}\rho\frac{g_0}{R}\left(R^2 - r^2\right), \tag{10.13}$$

with $g = \frac{g_0}{R}r$ and $c_s^2 = \gamma\frac{1}{2}g_0R(1-x^2)$, where $x = r/R$. Then, the equation of motion becomes

$$(1-x^2)\nabla^2\varphi + J\varphi = 0, \tag{10.14}$$

where $J = \frac{2}{\gamma}(\frac{R\sigma^2}{g_0} + 4)$.

Employing a power series solution $\varphi = \Sigma_n a_n x^n$ gives

$$a_{n+2} = a_n \frac{n(n+1) - J}{(n+2)(n+3)}. \tag{10.15}$$

The boundary conditions, $\zeta = 0$ at the centre and finite at the boundary indicate that this can only happen if $J = m(m+1)$ and that even values of n provide a convergent series, whilst odd powers do not. Combining these conditions,

$$J = \frac{2}{\gamma}\left(\frac{R\sigma^2}{g_0} + 4\right) = m(m+1), \tag{10.16}$$

with $m = 2, 4, \ldots$ permitted. Thus in order to pulsate,

$$\sigma^2 = \frac{g_0}{R}\left[\frac{1}{2}\gamma m(m+1) - 4\right] \geq 0. \tag{10.17}$$

Since $P = 2\pi/\sigma$, for the fundamental $m = 2$ mode,

$$P\sqrt{\bar{\rho}} = \frac{2\pi}{\sqrt{G\frac{4}{3}\pi(3\gamma - 4)}} = Q, \tag{10.18}$$

the period–density relation. Higher frequency overtones ($m \geq 4$) are observed in some stars. Note that this depends on the sphere's mean density and not its mass. For $\gamma = 5/3$, $P\sqrt{\bar{\rho}} = 0.14$. Analytical solutions can be found for some non-homogeneous density distributions such as $\rho \sim x^{-2}$, the point source or Roche model, and some polytropes (e.g., Eddington, 1930, 1932; Rosseland, 1964). The polytrope of index 3 is found to fit Cepheids as an approximation with $Q \sim 0.29$.

The excitation of the Cepheid variables and many other types of pulsating stars occurs in the instability strip in the HR diagram is due to the κ-mechanism, first described by Eddington (1917, 1930). The radiative opacity, κ, at every layer in a star depends on the local physical conditions, e.g., temperature, pressure, and composition. In a layer, where H, He, or another element is nearly ionized, κ becomes large and acts like a blanket that slows the outward heat flow, so the temperature builds until the ionization is so high that the opacity drops. This produces a cyclic process and the layer periodically moves back and forth from larger to smaller radii in the star. Helium ionization is considered to cause Cepheid variability and hydrogen ionization is thought to drive variability in Miras, variable Ap stars (roAp) and ZZ Ceti variable white dwarfs.

10.1.1.2. *Discovery of the period–luminosity relation*

Cepheid variables are one of the primary distance indicators. They are found in the Milky Way and the Magellanic Clouds and calibrate the zero point of the period–luminosity relation. The naked-eye star δ Cephei discovered in 1784 by John Goodricke is the archtype of this class of variable. At a distance of 273 parsecs (890 light years), it varies regularly between V = 3.4 and 4.4 magnitude with a period of 5.4 days. It has an absolute magnitude, $M_V = -3.5$. η-Aquilae of similar apparent brightness was discovered visually also in 1784 by E. Pigott. The brightest Cepheids with periods of about 200 days have $M_V \sim -8$, but even at this brightness, they are restricted to the nearest galaxies. There are several effects that need to be calibrated in order to obtain accurate distances from the period–luminosity relation as they may differ in element composition and extinction along the line of sight.

The logjam in the extragalactic distances was broken by Leavitt's (Leavitt & Pickering, 1912, 1917) discovery of the period–luminosity relation for Cepheid variables from studies of variable stars in the Large and Small Magellanic Clouds from measurements of photographic plates. This distance indicator became one of the most important discoveries of astronomy and Leavitt's contribution cannot be overstated.

The period–luminosity relation was utilized by Shapley (1918) to determine the diameters of globular clusters that are unaffected by absorption and found that the sun is 26,000 to 39,000 light years from the centre of a much larger galaxy (current value is about 26,000 LY). Part of the argument against the nebulae being outside of the Milky Way came from van Maanen's (1916) claim, later disproven, that he had measured the rotation of the spiral M 101, which would exceed the speed of light if it were not inside the Milky Way. In 1925, Hubble (1925) observing with the Mount Wilson 100-inch (2.5 m) telescope found Cepheid variables in the Andromeda galaxy (M31) and other galaxies in the local group and used Leavitt's period–luminosity relation to estimate that it is at a distance of about 930,000 light years (current estimate is about 2.5 million), well outside the Milky Way.

Hubble (1929) had estimated that $H_0 = 500\,\mathrm{km\,s^{-1}\,Mpc^{-1}}$ (in $V = H_0 D$), nearly ten times too large using Cepheids and some other incorrect distance indicators such as the diameters of H II regions The idea of the expansion had also been discussed by, e.g., Wirtz (1924), Lundmark (1925), Lemaître (1927). This made the expansion age of the universe to be only about 2×10^9 years, creating a problem as this is less than the age of the earth (4.6×10^9 years) found from radioactive isotopes in terrestrial rocks. For this reason, at the time, there were diverse other cosmological theories discussed. However in 1952 as already noted, it was realized that there are the two types of Cepheids depending on their metal abundances (Baade, 1953). Type I Cepheids are in galactic disks and the ones found near the sun. Type II Cepheids have metals decreased from the solar by ten or more and are found in galactic halos. A Type II is about two magnitudes fainter than a Type I. Taking this into account, $H_0 = 75\,\mathrm{km\,s^{-1}\,Mpc^{-1}}$ (Sandage, 1958).

The Hubble key project and the Carnegie Hubble Program (CHP) to measure the period–luminosity relation in nearby galaxies extends the distance scale to approximately the Coma cluster (Freedman *et al.*, 2001; Freedman & Madore, 2010). Figure 10.2 shows examples of the derived light curves for Cepheids in the galaxy NGC 4725. In all, distances to approximately 545 galaxies in 35 clusters are being derived using this method.

The period–luminosity law can be fit by linear relations between the absolute magnitude and the period as seen in Fig. 10.3. For the near infrared magnitude, M_I, and the period in days P,

$$M_I = -2.962 \log P - 1.66, \tag{10.19}$$

as found by Udalski *et al.* (1999). The zero point from the distance to the LMC is required and was assumed to be $DM = 18.22$ mag. The corrections to the data outlined previously apply. Effects that seriously affect the calibrations are reddening, metallicity variations that change the colours, and the interiors, and the distances to the calibrator stars as well. Large efforts to calibrate these in order to improve the accuracy and by moving measurements farther into the infrared where reddening and metallicity line-blanketing effects are reduced.

10.1.2. *Tip of the Red Giant Branch (TRGB) method*

Low mass ($M \leq \sim 1.8 M_\odot$) post-main sequence stars develop degenerate helium cores and become red giants. They rise in luminosity as their cores contract until helium ignites in their cores and ends their luminosity at the

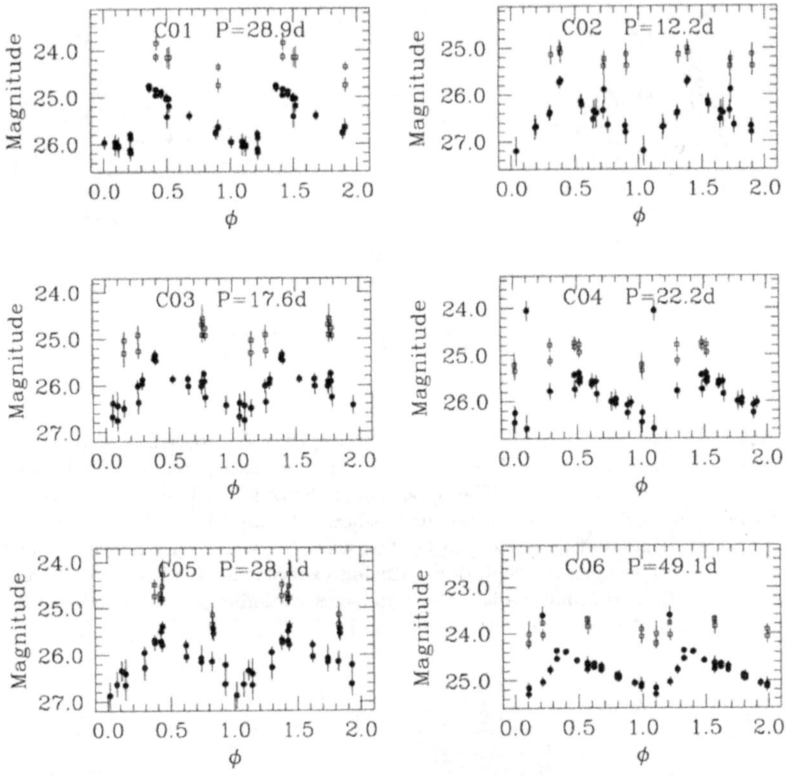

Fig. 10.2. Light curves for some of the Cepheids observed in the nearby galaxy NGC 4725 (type SAB(r)ab pec, $cz = 1206\,\mathrm{km\,s^{-1}}$) by the HST in two colours with phase ϕ. The derived periods are shown. The upper open squares are observations in the I-band; the lower filled dots are V-band data. (*Source*: Gibson B. K. *et al.* (1999), *ApJ*, 512, 48, "The hubble space telescope key project on the extragalactic distance scale. XVII. The Cepheid distance to NGC 4725," Fig. 4, ©AAS reproduced with permission.)

red giant tip at a nearly unique brightness. Although some early investigators (e.g., Humphreys & Davidson, 1979) proposed using the brightest red supergiants as distance indicators, the method required development of advanced imaging instruments.

These brightest red giants can be seen directly in nearby galaxies and the Milky Way, thus the TRGB method is a primary distance indicator. Using the near-infrared bands, it is less sensitive to metallicity and age (see Fig. 10.4). It is a Pop II indicator and unlike Cepheids, can be used in all galaxy types. The method is further described in Sakai (1999), Makarov *et al.* (2006), Rizzi *et al.* (2007), and Mould & Sakai (2009), where

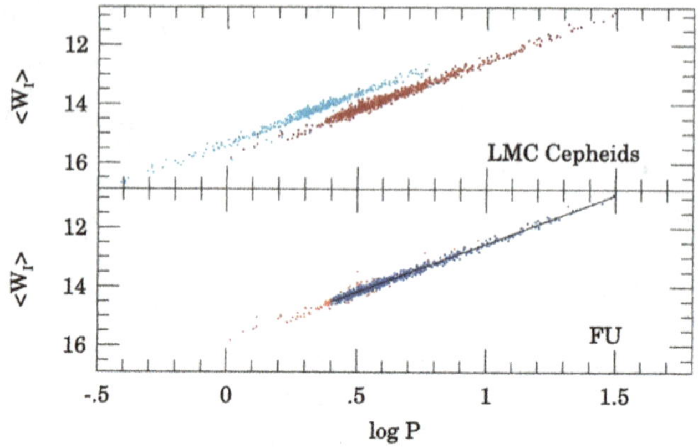

Fig. 10.3. The period–luminosity relations for Cepheid variables in the Large Magellanic Cloud from the OGLE-II survey using the reddening free I-band Wesenheit index, $\langle W_I \rangle = I - 1.55(V - I)$. The upper panel shows FU Cepheids (blue dots) and FO Cepheids (red dots). The bottom panel shows the final fit. (*Credit*: Udalski A., Szymanski M., Kubiak M., Pietrzynski G., Soszynski I., Wozniak P., Zebrun K. (1999) *AcA*, 49, 201, "The optical gravitational lensing experiment. Cepheids in the magellanic clouds. III. Period-luminosity-colour and period-luminosity relations of classical Cepheids," Fig. 4, Reproduced with permission.)

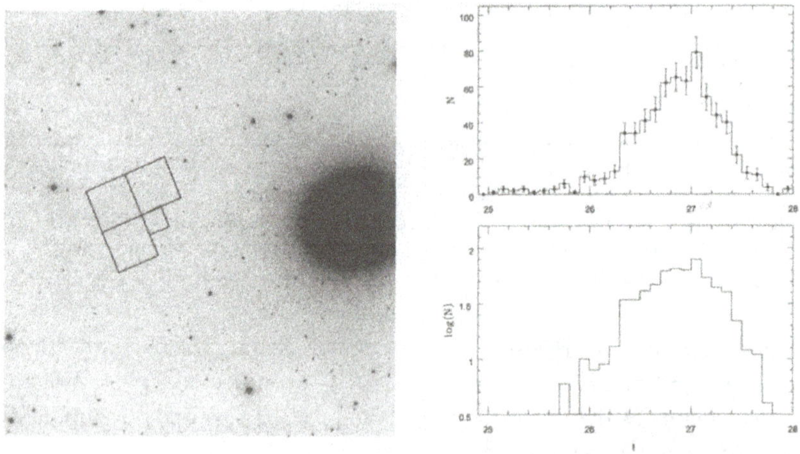

Fig. 10.4. Illustrating the TRGB method used to determine the distance to NGC 3889. Left panel: ground-based I-band CCD image of the galaxy showing the location of the placement of the HST WFPC2 camera in its halo, 6 arcmin West of the nucleus. Right panel: the distribution of observed sources showing the peak corresponding to the fitted distance modulus of 30.30 mag. (*Source*: Sakai S., Madore B. F., Freedman W. L., Lauer T. R., Ajhar E. A., Baum W. A. (1997) *ApJ*, 478, 49, "Detection of the tip of the red giant branch in NGC 3379 (M105) in the Leo I Group using the hubble space telescope," Figs. 1 and 3 combined, ©AAS Reproduced with permission.)

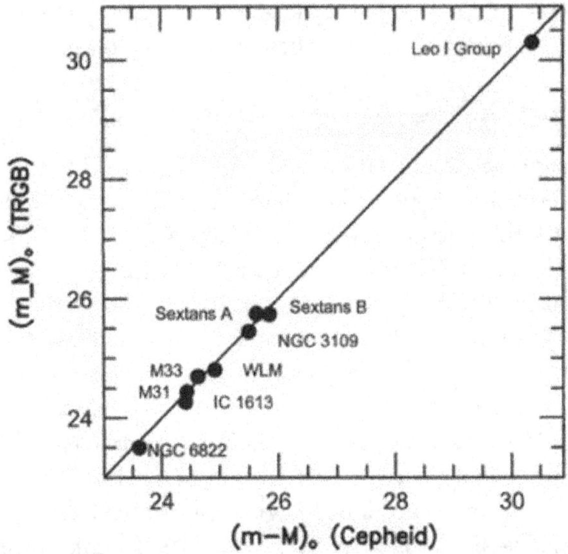

Fig. 10.5. Cepheid period–luminosity distance compared to those from the TRGB method. The unit slope was not fit to the data. (*Source*: Sakai S., Madore B. F., Freedman W. L., Lauer T. R., Ajhar E. A., Baum W. A. (1997), *ApJ*, 478, 49, "Detection of the tip of the red giant branch in NGC 3379 (M105) in the Leo I Group using the hubble space telescope," Fig. 7, ©AAS Reproduced with permission.)

$M_I \approx -4.0 \pm 0.1$ mag. This method has been extended to the IR where the effects of absorption are further reduced. Approximately 250 galaxies have TRGB distances. It has the advantage over Cepheid distances since following a light curve is not needed and only single observations are required, but can deliver similar precision as shown in the comparison in Fig. 10.5.

10.2. Secondary Distance Indicators

10.2.1. *Supernovae as distance indicators*

Different classes of supernovae have been known for a long time (e.g., Zwicky, 1965; Turatto, 2003). The two basic classes are divided by the visibility of spectral lines of H I in their spectra and their light curves are different. There are many subtypes of both classes. The subclasses of Type I do not generally show hydrogen lines in their spectra and are thought to represent explosive events involving the end states of lower mass evolved stars that have produced white dwarfs with masses very close to the Chandrasekhar limit of $M_{CH} = 1.4\ \mathrm{M}_\odot$ or in some cases, core collapse.

The subclasses of Type II supernovae generally show strong lines of hydrogen and are interpreted to involve the core collapse of massive stars.

10.2.1.1. *Type Ia supernovae*

The Type I supernovae are subdivided into Types Ia, Ib, and Ic (Turatto, 2003). The spectra of Type Ia are characterized by the strong broad line of Si II near 6150 Å but including other lines of O I, Mg II, Si II, Si III, S II, Ca II, Fe II, and Fe III superimposed on a continuum (Parrent, Friesen, & Parthasarathy, 2014). Types Ib and Ic do not show the Si II feature, but Type Ib has strong helium lines and Type Ic shows little He. These latter two types are not as well understood, may be formed by core collapse related to Wolf–Rayet stars, and are not as useful as distance indicators.

The Type Ia supernovae have been employed extensively as standard candles. They are sufficiently bright to be observed at large distances; at maximum, they have an absolute magnitude of about $M_B \approx -19.5\,\text{mag}$. Large surveys have been constructed to study the Hubble constant and the expansion at high redshifts from which dark energy was found (Perlmutter *et al.*, 1999; Schmidt *et al.*, 1998). A sufficient number are known over the sky to be useful for studying large-scale flows.

In addition to their spectra, the Type Ia supernovae (SNe Ia) are recognized by their characteristic light curves and spectra. As early as the 1970s, it was realized that these objects showed promise as standard candles at large distances due to their uniform light curves (e.g., Pskovskii, 1971; Barbon, Ciatti, & Rosino, 1973). They occur in all types of galaxies and although one single model of the explosive event cannot be agreed upon (e.g., Hillebrandt *et al.*, 2013). It is thought that SNe Ia's are produced by white dwarfs in binaries with $M_{WD} \approx M_{Ch} \approx 1.4 M_\odot$. In a typical SNe Ia, the light curve rises rapidly to a maximum brightness of $M_B \approx -19.5\,\text{mag}$. in about 10–20 days. This is followed by an exponential decline of the supernova's brightness with a half-life, $HL = 110$ days, due to the combined radioactive decay of $^{56}\text{N}i \rightarrow ^{56}\text{Co} \rightarrow ^{56}\text{Fe}$ produced in the outburst from the original CNO core of the white dwarf.

Using the SNe Ia as standard candles therefore requires empirical calibrations to obtain accurate distances from them as they are not perfectly uniform. The absolute magnitude depends on the decay rate. A faster Ia is fainter than a slower one with a spread in the maximum brightness of order two magnitudes, as shown in Fig. 10.6, which also shows they have differences in colour. The decay rate parameter $\Delta M_{15}(B)$ follows the Phillips

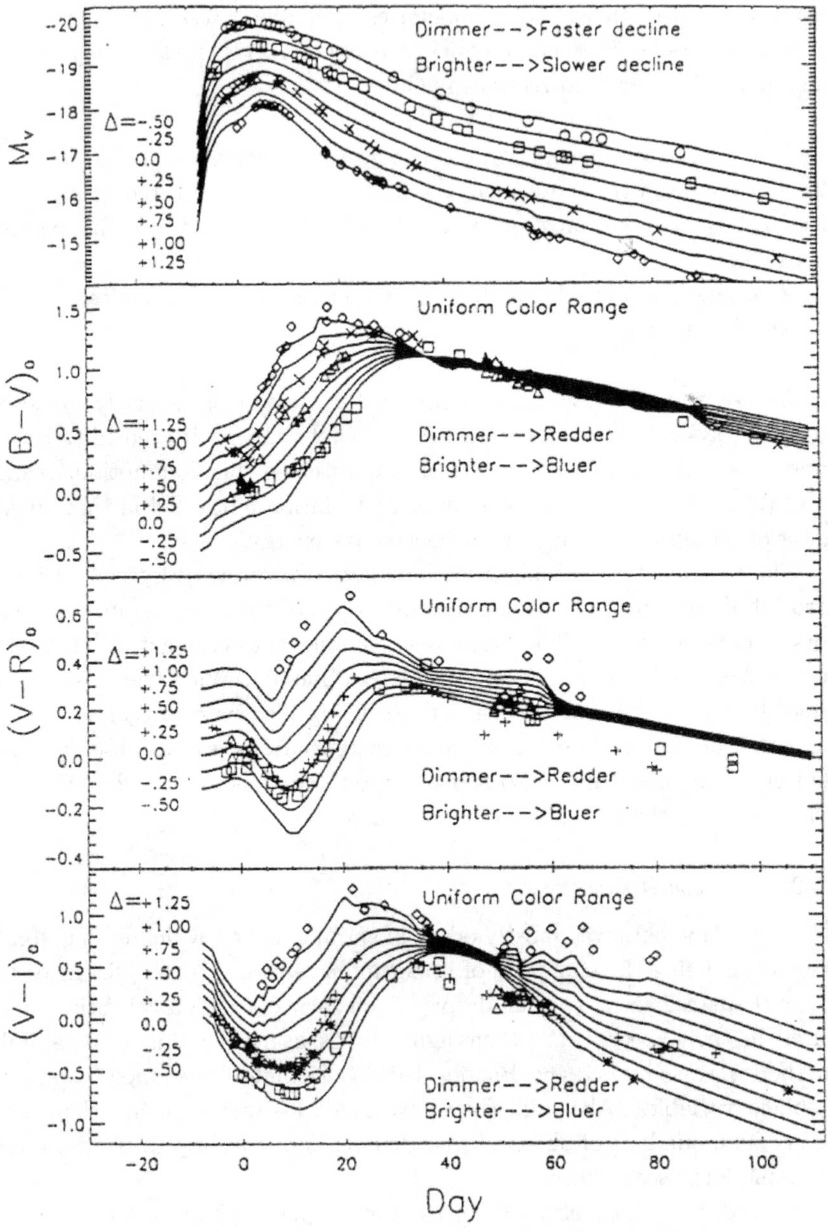

Fig. 10.6. Light and colour curves of fast and slow Type Ia supernovae as a function of the decay rate parameter Δ. (*Source*: Branch D. (1998), *ARA&A*, 36, 17, "Type IA supernovae and the hubble constant," Fig. 8, ©1998 Annual Reviews www.annualreviews.org, with permission.)

relation (Phillips, 1993) and parametrizes these differences. This is defined as how much the B band magnitude declines in the first 15 days after maximum. The main corrections include:

(1) The decay rate depends on colour and this depends on $\Delta M_{15}(B)$.
(2) The colour of the parent galaxy, $\Delta M_{15}(B)$ is larger for red galaxies.
(3) Dust corrections must be made for the Milky Way and the parent galaxy.
(4) K-corrections must be made. This increases for large redshifts, $z > 1.0$ and is a problem.

An additional important assumption that has been generally made is that the physical properties of the SN Ia do not change with redshift. However, recent observations of high z have questioned this assumption. Kang *et al.* (2019) find that there is luminosity evolution in the SN Ia that could lower or eliminate the need for Λ hence dark energy.

The distance of a well-observed Ia supernova can be estimated with an error of about $\pm 5\%$. Consequently, this is one of the best distance indicators at large distances. The major disadvantage is that to get the distance to a galaxy, a Ia supernova needs to be observed. With the passage of time, larger numbers of supernovae are being discovered by many ongoing searches that can be found on the internet from several observatories and including amateur astronomers. Tripp (1998) has suggested additional refinements to the distance modulus determinations.

10.2.1.2. *Type II supernovae*

The Type II supernovae mostly occur in spiral galaxies, which is consistent with them being the end state of upper main sequence stars. Distances to Type II supernovae have been derived from their expansion velocities measured in their spectra and their light curves using the Baade–Wesselink method (Wesselink, 1946; Baade, 1953), originally developed to study Cepheid variables. Although they have given estimates of H_0, there are not enough numbers of objects using this method currently to be useful for studying large-scale flows.

Consider the geometry of the expanding photosphere of the Type II supernova in Fig. 10.7 (Kirshner & Kwan, 1974; Schmidt *et al.*, 1994). The leading edge is a mean free path λ from R_p, the radius of the photosphere. Here, we will approximate the energy distribution of the supernova by a

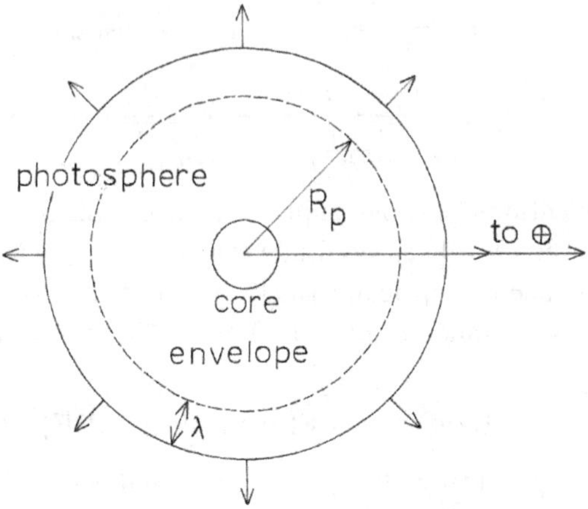

Fig. 10.7. Sketch of the geometry of the expanding photosphere of a Type II supernova. The leading edge is a distance λ beyond R_p, the radius of the photosphere.

black body. Then, the observed flux, obtained from photometry is

$$f_\nu = \frac{\pi R_p^2}{D^2} B_\nu(T_p), \tag{10.20}$$

where $B_\nu(T_p)$ is the Planck function for temperature T_p, which is found using spectra or colours. During the early initial stages of the expansion, the envelope's density compared to the interstellar medium $\rho_{\rm env} \gg \rho_{\rm ism}$, so the motion of the envelope essentially expands freely. Then, in terms of the photospheric expansion velocity, $v_p(t)$,

$$R_p(t) \approx R_p(t_0) + \int_{t_0}^{t} v_p(t')dt'. \tag{10.21}$$

Generally t_0 is unknown, but $v_p(t)$ is measured from the Doppler shifts and it can be eliminated using two observations separated by two times using:

$$\begin{aligned} Rp(t_1) &= R_p(t_0) + v_p(t_1 - t_0), \\ Rp(t_2) &= R_p(t_0) + v_p(t_2 - t_0), \end{aligned} \tag{10.22}$$

which yields

$$R_p(t_2) - R_p(t_1) = v_p(t_2 - t_1). \tag{10.23}$$

From the above, $R_p(t) = \sqrt{\frac{f_\nu(t)}{\pi B_\nu(t)}} D$ and the distance is

$$D = \frac{v_p(t_2 - t_1)}{\sqrt{\frac{f_\nu(t_2)}{\pi B_\nu(t_2)}} - \sqrt{\frac{f_\nu(t_1)}{\pi B_\nu(t_1)}}}. \tag{10.24}$$

With modifications taking supernova atmosphere structures into account, this method has been used in early stages when T_p is below about 6000 K when the shell is optically thick. These distances agree well with Cepheid distances (Schmidt *et al.*, 1994) $H_0 = 73 \pm 6 \pm 7\,\mathrm{km\,s^{-1}\,Mpc^{-1}}$.

10.3. The Surface Brightness Fluctuation (SBF) Method

This method is specifically designed to be used with CCDs developed by Tonry & Schneider (1988). It is a distant relation of the brightest star method. It requires some knowledge of the stellar luminosity function. Tonry *et al.* (1997, 2000, 2001), Blakeslee *et al.* (1999), and Jensen *et al.* (2003) have applied this method to the Coma cluster and extensively elsewhere. Distance errors of ±5% are attainable.

Consider two identical galaxies at different distances that are observed with a CCD. The resolution of details on one CCD pixel will change with distance. On a nearby galaxy, the resolution scale per pixel is higher than on the more distant twin, so more stars and structure fall on the pixel of the distant galaxy. This produces a more uniform distribution, i.e., the 'roughness' changes with distance. Stars in a galaxy are divided into luminosity classes where all stars of class i have the same intrinsic luminosity L_i. Observing with a detector such as a CCD, a pixel will receive energy ℓ from the stars,

$$\ell = \frac{\sum_i N_i L_i}{4\pi D^2}, \tag{10.25}$$

where N_i is the number of stars of class i on the pixel and D is the distance. Stars are not directly seen, but differences in ℓ between separate parts of the galaxy on other pixels can be measured. If the fluctuations obey Poisson statistics,

$$\langle (N_i - \langle N_i \rangle)(N_j - \langle N_j \rangle) \rangle = \delta_{ij} \langle N_i \rangle, \tag{10.26}$$

where brackets denote averages. The extraction of the variance requires sophisticated analysis of the galaxy images. Then

$$\langle (\ell - \langle \ell \rangle)^2 \rangle = \frac{\bar{L}}{4\pi D^2}, \tag{10.27}$$

where $\bar{L}$ is the mean luminosity weighted stellar luminosity of the galaxy. If as an approximation, it is assumed that $\bar{L}$ does not vary between galaxies, this can be calibrated using Cepheid distances in terms of the $(V - I)$ colours of the galaxies. Tonry *et al.* (2001) compiled a catalogue of 300 early-type galaxies. Blakeslee (2012) has further described the application of this method.

Chapter 11

Scaling Relations as Distance Indicators

The scaling relations for spirals and ellipticals are secondary distance indicators and for individual objects do not produce the accuracy of the Cepheids, SN Ia, and SBF distance indicators (~15–20% compared to ~5%), but they have two advantages. Firstly, they can be used for galaxies at larger distances than Cepheids and SBF, and secondly being ubiquitous one does not have to await a supernova event in the desired object. Consequently, they have been employed in many large-scale flow studies.

Scaling relations for spirals and ellipticals are determined by observing a number of galaxies in clusters. These galaxies can be assumed to be at identical distances because usually the thickness of the cluster can be neglected. An additional assumption is that these objects follow the same relation, usually taken to be linear, and after the corrections outlined above, differences in their apparent brightness or angular size are due to their distance. Determining the scaling relation then becomes the problem of fitting for the relative distances and the zeropoint is obtained from clusters of known distance.

11.1. Spiral Galaxies and the Tully–Fisher (TF) Relation

In the early 20th century, spectroscopic observations of some of the brightest nebulae, or galaxies as they are now called, began to reveal that they have large redshifts, but also that they rotate. Slipher (1914) detected the rotation of M31, followed by measurements of other galaxies, such as NGC

4594 by Pease (1916).[1] The Milky Way's galactic rotation was studied by B. Lindblad and confirmed by Oort (1927), who proposed the presence of dark matter in the galaxy's disk (Oort, 1932) independently of Zwicky (1933) who used the velocity dispersion of the Coma galaxy cluster. Until the 1980s, it was generally held that spiral galaxy rotation profiles rose with radius to a maximum until reaching the edge of the galaxy and then fell off with a Keplerian $R^{-0.5}$ curve. However, larger samples of spiral galaxy rotation curves reaching larger radii became available. It was realized that they mostly have extended flat rotation curves (e.g., Rubin, Ford & Thonnard, 1980; Rubin, 1983; Burstein & Rubin, 1985). These flattened rotation curves are illustrated in Fig. 11.1 that also shows some of the trends with different luminosities and rotation rates. Persic, Salucci, & Stel (1996) compiled optical and radio rotation curves for about 1100 spiral galaxies. They fitted the rotation curves by a 'universal rotation curve' (URC) where the URC surface is shown in Fig. 11.1; this implies a universal DM/LM, dark matter to luminous matter scaling relation, that scales inversely to the luminosity as well as other properties of the spirals. Ostriker & Peebles (1973) shows that an extended DM spherical halo will stabilize the rotating flattened disk. The stability of such disks, spiral structure, and warps are covered in detail by Binney & Tremaine (2008).

11.1.1. *Measurement of the Tully–Fisher (TF) relation*

The rotation curves of spiral galaxies display the notable property that the maximum rotation velocities correlate closely with the total luminosities and this relationship can be utilized as an important distance indicator. Tully & Fisher (1977) discovered one of the most important secondary distance indicators, which relates the luminosity or absolute magnitude, M_o, with the rotation rate:

$$M_o = 3.5 - 6.25 \log W_o + m_{pg}(o), \tag{11.1}$$

where the inclination corrected width of the lines in the spectra of spirals is W_o and the magnitudes are face-on values.

Originally this relation was between the H I 21 cm line width and the absolute total m_{pg}, or B magnitude, but other lines and infrared magnitudes can be used. The sizes of spiral galaxies are often specified using

[1]These were heroic efforts. With the photographic detectors and instruments available, it took Pease 80 hours of exposure with the Mount Wilson 60-inch (1.5 m) telescope to acquire the data on this one bright galaxy.

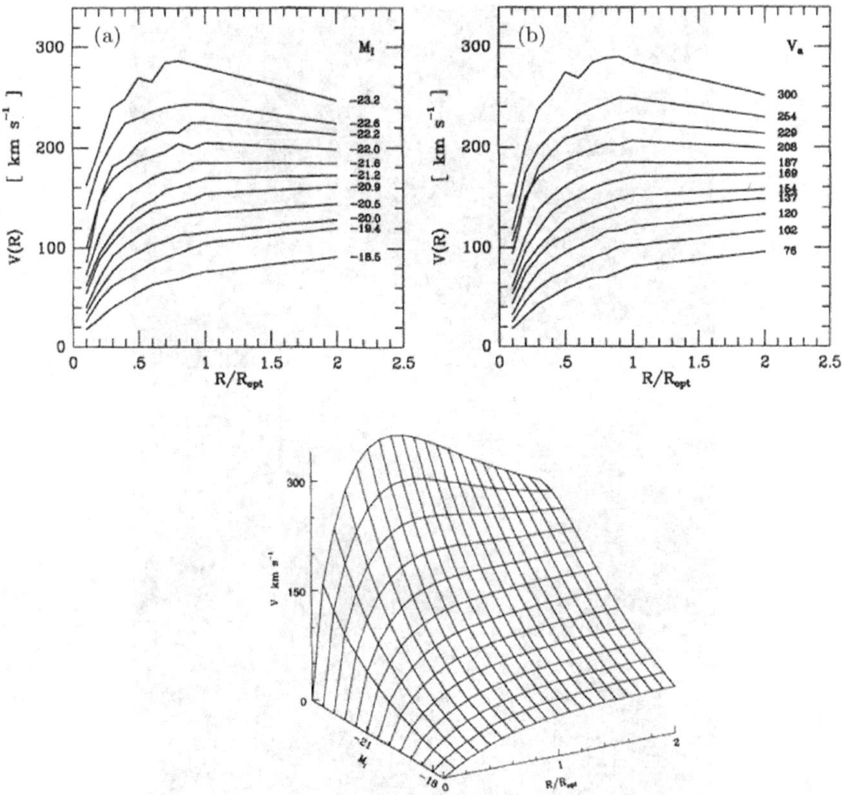

Fig. 11.1. Persic *et al.* (1996) universal rotation curves for spiral galaxies in optical I band constructed from about 1100 optical and radio rotation curves. (Top) Velocity with radius in terms of absolute magnitude (M_I) and (V_a), the velocity amplitude (panels (a) and (b), respectively). (Bottom) The universal rotation curve surface. Radii are in units of $R_{\rm opt}$, the radius containing 83% of the light. (For an exponential disk a useful approximation is that $I(r) \propto \exp(-r/R_D)$ where R_D is the exponential scale-length, $R_{\rm opt} = 3.2R_D$.) (*Source*: Persic M., Salucci P., Stel F. (1996), *MNRAS*, 281, 27, "The universal rotation curve of spiral galaxies I. The dark matter connection," Figs. 7 and 10, Reproduced by permission of Oxford University Press on behalf of the Royal Astronomical Society.)

the 'Holmberg radius'. This is defined by the B-band isophote where the surface brightness reaches $\mu_H = 26.5\,\text{mag.arcsec}^{-2}$ (Holmberg, 1958). The TF relation has been extended using global Hα and other lines and infrared magnitudes. Sc galaxies give the best result. A radio image of the nearby Sc galaxy M33 is in Fig. 11.2, which illustrates the rotating neutral hydrogen in its disk.

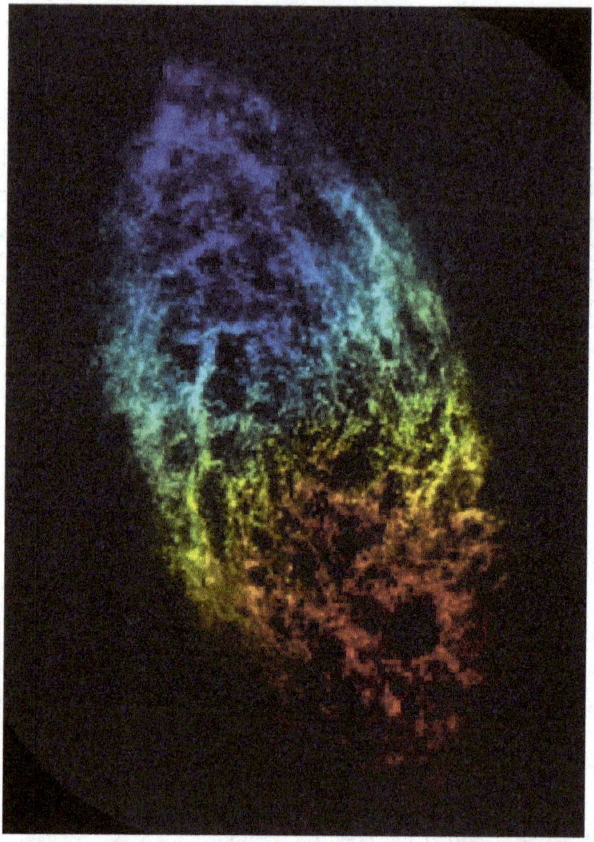

Fig. 11.2. The 21 cm radio image of the nearby type Sc spiral triangulum galaxy, M33 (distance 2–3 million light years). This illustrates its rotation. The false colour represents the Doppler shift due to rotation, blue is approaching and red receding. This was produced from data from the VLA in New Mexico and the WSRT in the Netherlands. (Image courtesy of NRAO/AUI and NSF © nrao.edu/pr/2001/m33gas/.)

The line-width is measured from the resulting 'horned' profile of the rotating disk. Some examples containing the observational noise are shown in Fig. 11.3. This profile arises as the hydrogen in the spiral galaxies is confined to a ring which is an inclined rotating disk. Early utilization of the infrared Tully–Fisher relation (ITFR) was pioneered by Aaronson, Huchra, & Mould (1979) and Aaronson *et al.* (1981, 1982, 1986) using the near infrared H-band (1.6 μ).

Figure 11.4 is an example of the I-band (0.81 μ) TF relation for Sc spirals constructed by combining data for objects in 24 galaxy clusters. In

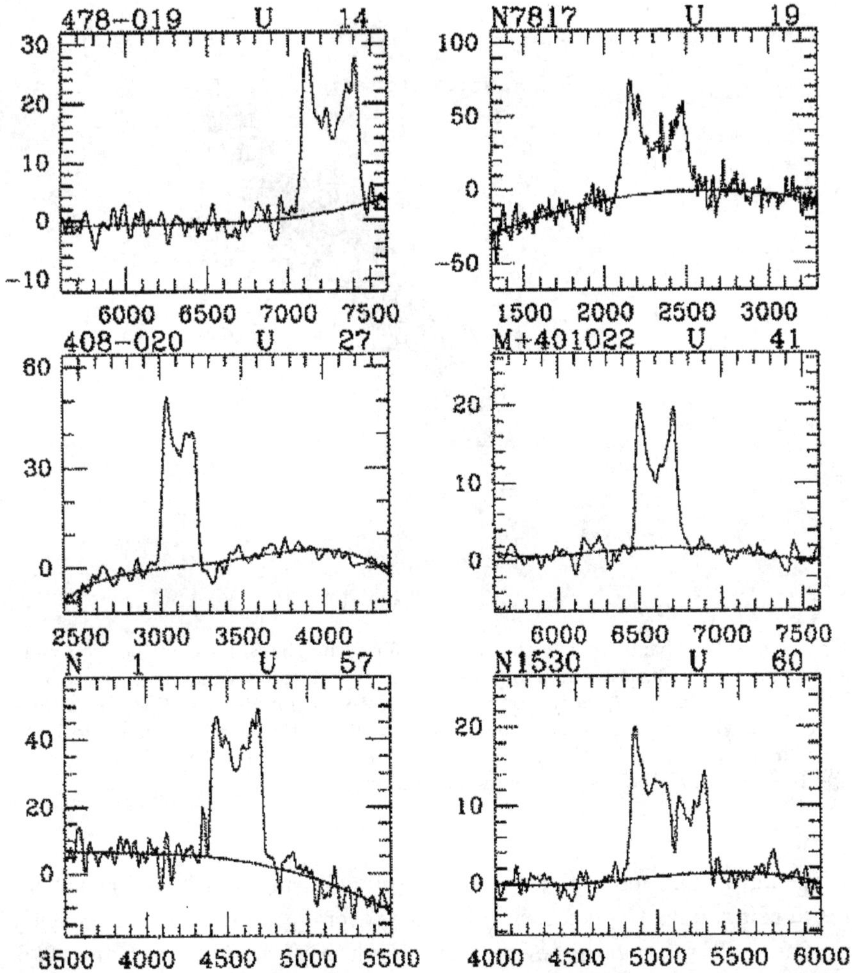

Fig. 11.3. Some observed 21 cm line profiles for spiral galaxies in the SFI study showing typical 'horned' profiles. The horizontal axis is the radial velocity in km s^{-1}and the vertical axis is the flux in mJy. (*Source*: Haynes M. P., Giovanelli R., Chamaraux P., da Costa L. N., Freudling W., Salzer J. J., Wegner G. (1999), *AJ*, 117, 2039, "The *I*-band Tully–Fisher relation for SC galaxies: 21 centimeter H I line data," Fig. 1, © AAS Reproduced with permission.)

general linear relation is fitted,

$$M_X = -a \log W_o + zp,$$

where M_X is the absolute magnitude in a photometric band, X, and W_o is the inclination corrected line width. The slope of the relation, a, varies with

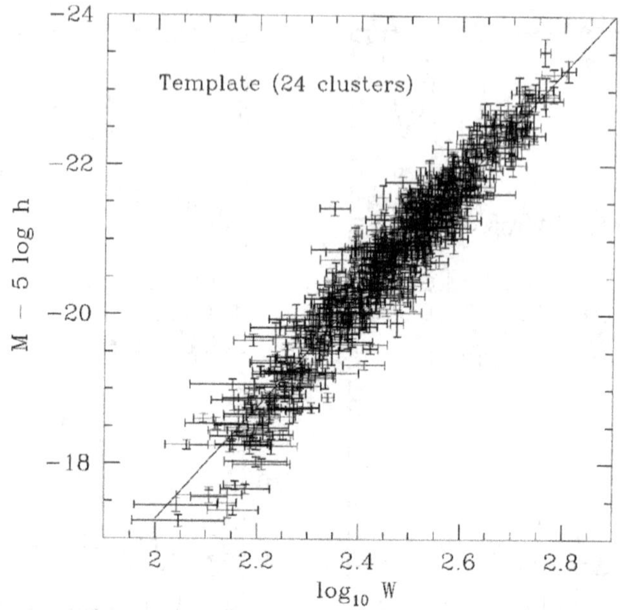

Fig. 11.4. The I-band Tully–Fisher relation found by the SFI study. The top panel shows the data points without error bars and the bottom panel contains the same data, but with error bars. M_I is the corrected absolute magnitude and W is the inclination corrected 21 cm line width. (*Source*: Giovanelli R., Haynes M. P., da Costa L. N., Freudling W., Salzer J. J., Wegner G. (1997), *ApJL*, 477, L1, "The Tully–Fisher Relation and H_0," Fig. 1, ©AAS Reproduced with permission.)

wavelength. The zero point, zp, is determined using sources of known distance using, e.g., Cepheids, SNe Ia, etc. Several groups have subsequently produced TF relations on several wavelengths as (e.g., Masters *et al.*, 2006; Tully *et al.*, 2016) and many of the rotation widths utilized optical lines rather than radio measures. Corrections to the data outlined in Chapter 9 are required to obtain the template relations used for determining peculiar motions as outlined, e.g., by Giovanelli *et al.* (1997a,b) and many varieties of fitting and measurement have been used. Table 11.1 presents a small sample of published determinations showing the slope of the relation. It is clear that the slope increases with wavelength, approaching $a \approx -10$ in the infrared. A morphological correction is also observed. Sa galaxies have the lowest slope, followed by Sb and Sc with the highest slope. These effects are related to the structure of the galaxy, such as gas content, star formation rate, etc. (e.g., Mould, 2020).

Table 11.1. Sample slopes measured for the Tully–Fisher relation.

Band	a	References
B	-6.80 ± 0.60	Yasuda, Fukugita, & Okamura (1997)
R	-7.64 ± 0.14	Pierce & Tully (1988)
I	-7.68 ± 0.13^a	Giovanelli *et al.* (1997b)
J	-10.61 ± 0.12	Masters, Springob, & Huchra (2018)
H	-9.36 ± 00.16^b	Aaronson *et al.* (1979)
K	-10.74 ± 0.10	Masters, Springob, & Huchra (2018)
[3.6]	-9.16 ± 0.17	Tully *et al.* (2016)

[a]Later new value -7.85 ± 0.1 from SFI++ (Masters *et al.*, 2006).
[b]From 16 UMa galaxies.

11.1.2. *Origin of the horns*

The horns arise from a rotating ring in the galaxy's disk illustrated in Fig. 11.5, which shows the ring in the infrared image of the spiral galaxy M104. Similar rings and emission line profiles occur, not only in spiral galaxies, but in other objects, e.g., stars and black holes with rotating disks or rings. Models of the horned emission line profiles have been discussed by many authors including Huang (1972), Roberts (1978), Smak (1981), Schulz & Mücke (1994). The geometry of an axially symmetric rotating disk with inner radius R_1 and outer radius R_2 as shown in Fig. 11.6. Here, several simplifications are made. The disk is thin compared to its radii and optically thin. All orbits in the disk are circular and effects such as spiral structure, turbulence and instrumental resolution are neglected. For

Fig. 11.5. Composite infrared image of the Sa type spiral galaxy NGC 4594 (M104, the Sombrero galaxy, D = 11 Mpc) in the 3.6, 4.5, 5.8, and 8.0 μm bands showing its rotating ring. (Image: NASA/JPL-Caltech and The Hubble Heritage Team (STScI/AURA).)

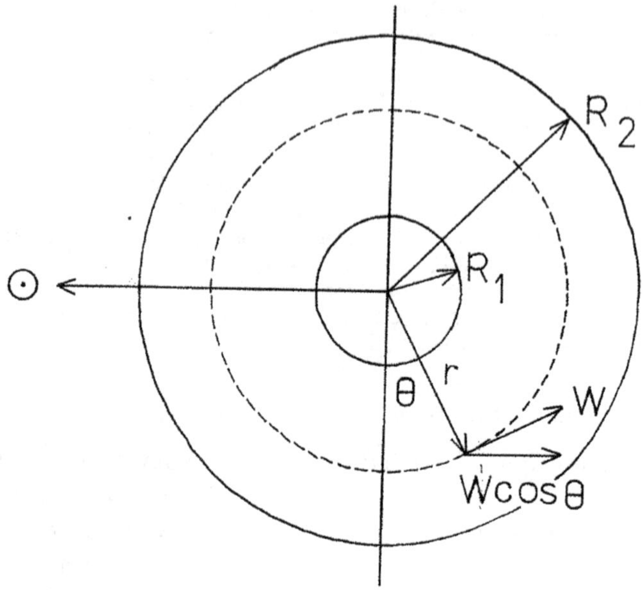

Fig. 11.6. Top view of a rotating ring of inner radius R_1 and outer radius R_2. A point in the ring at radius r and angle θ rotates at speed W. The projected velocity observed is $V = W \cos\theta$.

a mass point in a circular orbit along the line of sight in the disk r is the radius of the orbit, $W(r)$ is the velocity, which has been corrected for the inclination of the galaxy, ζ. This is obtained from fitting elliptical isophotes to the outer regions of the disk as in Fig. 9.5, where $\sin\zeta = \frac{a}{b}$ with a, b as the measured axes of the isophotes. Then $W = \frac{W_{obs}}{\sin\zeta}$. The photometry of the total magnitudes must be corrected for extinction in our galaxy and the internal extinction in the observed galaxy, which depends on ζ. The component of the velocity observed along the line of sight is $V = W\cos\theta$, where θ is the angle around the disk plane. In general, $W = \omega(r)$, the rotation rate in the ring. The shape of the horned profile depends on the forms of $W(r)$ and $j(r)$, the emissivity across the ring. As a simplification, assume $W \sim r^a$. Limiting cases include $W \sim r^{-0.5}$ (Keplerian), $W \sim r$ (solid body), and $W \sim r^0$ (constant). Real galaxies will be some complex mix of varying a such as sketched in Fig. 11.7(a). These produce curves of constant V on the galaxy called 'spider diagrams' shown in Fig. 11.7(b).

The simplified model of the ring constructed using dimensionless variables has the boundary conditions:

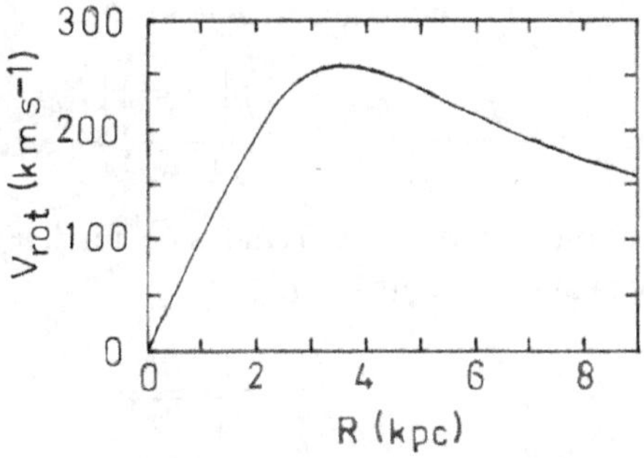

(a)

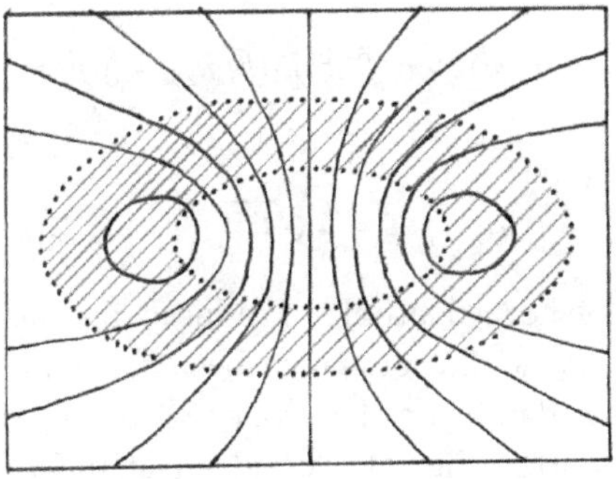

(b)

Fig. 11.7. Sketch of rotation curve and resulting spider diagram for a typical spiral galaxy. (a) Rotation curve $V_{\rm rot}(R)$ for a typical flat spiral galaxy. (b) The spider diagram; this is projected onto the inclined disk (here $i = 30°$) seen as a shaded elliptical ring and produces the constant velocity contours of the spider diagram. In the inner linear or solid body portion of the rotation curve, velocity contours are evenly spaced; near the peak, the turnover produces the closed oval curves; at large radii, the nearly flat portion of the curve leads to radial contours.

inner: R_1; $v_1 = R_1^a$,
outer: R_2; $v_2 = 1$.

Setting $u = v\cos\theta$, the observed frequency function, $F(u)$ along the radius is

$$F(u) = \int_{R_1}^{R_z} j(r)\frac{d\theta}{du}dr \approx \int_{R_1}^{R_z} \frac{j(r)r^{-a}}{\sqrt{1-u^2r}}dr, \tag{11.2}$$

where the limit of integration, R_z, is $R_z = 1$ when $u \leq 1$ and $R_z = u^{-2}$ when $u > 1$.

Changing the variable of integration to $x = ur^{-a}$,

$$F(u) = \frac{-u^{\frac{2}{a}-1}}{a} \int_{x_1}^{x_z} \frac{x^{-\frac{2}{a}} j\left[\left(\frac{x}{a}\right)^{-\frac{1}{a}}\right] dx}{\sqrt{1-x^2}}. \tag{11.3}$$

If the Keplerian rotation law, $a = -0.5$ and as a further simplification a radial emissivity law is assumed, $j(r) \sim r^b$,

$$F(u) \sim u^{-5-2b} \int_{x1}^{x_z} \frac{x^{4+2b} dx}{\sqrt{1-x^2}}. \tag{11.4}$$

These can be integrated numerically, but if $4 + 2b = m$, is an integer, there are analytic formulae,

$$F(u) \sim u^{-5-2b}[\mathfrak{J}(b, x_z) - \mathfrak{J}(b, x_1)], \tag{11.5}$$

e.g., when $b = -1$,

$$\mathfrak{J}(b, x) = -\frac{x}{2}\sqrt{1-x^2} + \sin^{-1} x. \tag{11.6}$$

Some examples of horned profiles produced by this simple model are in Fig. 11.8. The peaks of the horns have been normalized to 1.0. Smak (1969, 1981) gives further details on the properties of this type of ring model.

It can be seen from Fig. 11.8 besides instrumental effects such as resolution and signal-to-noise, the shape and width of the horned profile is

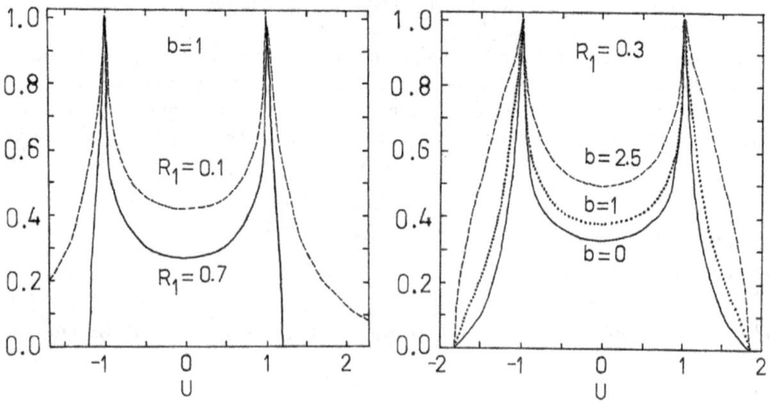

Fig. 11.8. Examples of horned emission line profiles produced by simple rotating ring models following the Keplerian law $W \sim r^{-0.5}$ in Eq. (11.6). Left: rings of different width, R_1 and fixed emissivity law. Right: profiles for rings of width R_1 fixed and different emissivity laws. The abscissae U are related to the wavelength, λ, and the rotation speed V_2 at the outer edge of the ring by $\Delta\lambda = \lambda\frac{V_2}{c}U$.

governed by $j(r)$, $\omega(r)$, and R_1 of the rotating disk in the spiral galaxies. Thus in order to obtain the highest precision in the TF relation, the measurement of the width, W must be considered. Different authors have considered the best methods to estimate W. Haynes & Giovanelli (1987), Haynes *et al.* (1999) discuss, e.g., using 50% of the mean flux or 20% of the peak flux in the line and the extent of the errors produced. Stewart, Blyth, & Blok (2014) have discussed more sophisticated modelling and fitting of these profiles.

Methods for improving the Tully–Fisher relation have been discussed by several investigators. This was discussed by Weiner *et al.* (2006) studying 1089 galaxies using optical emission line widths and Kassin *et al.* (2007) who investigated the TF relation to $z = 1.2$. Replacing the width, W, with $S_{0.5}$ reduces the scatter,

$$S_{0.5} = \sqrt{0.5W^2 + \sigma_t^2}, \tag{11.7}$$

where σ_t is a velocity dispersion from random velocities in the disk, and using a fit to the rotation curve using a function of the following form explained by those authors:

$$\begin{aligned} I(x,v) &= G(x)\exp\left[-\frac{(v-V(x))^2}{2\sigma_{2d}^2}\right], \\ G(x) &= \frac{I_{\text{tot}}}{2\pi r_1}\exp\left[-\frac{(x-x_0)^2}{2r_1^2}\right], \\ V(x) &= V_{\text{rot}}\frac{2}{\pi}\tan^{-1}(x/r_v). \end{aligned} \tag{11.8}$$

The infrared Tully–Fisher relation can be considered to be more representative of the disk. Keeping in mind that the exponents a and b and the TF relation in Eq. (11.1) do not remain constant with passband, in the blue, the spiral arms the relatively young blue stars can dominate the light, while the infrared light arises mostly from an older stellar population that is spread around the disk and is a better tracer of the mass. Although there is no agreement for a detailed derivation of the Tully–Fisher relation, a crude derivation giving the right order of magnitude of the infrared relation can be made using three simple assumptions. First, if the gravitational potential drives the rotation velocity v_{rot},

$$v_{\text{rot}}^2 \sim \frac{M}{R}, \tag{11.9}$$

where M is the galactic mass and R is the radial distance. Second, if the spirals have the same mass-to-light ratio (M/L), $v_{\rm rot}^2 \sim L/R$, and third if all spirals have a constant mean surface brightness, $\bar{I}$, the luminosity is

$$L \sim \bar{I}R^2, \tag{11.10}$$

which yields

$$L \sim v_{\rm rot}^4. \tag{11.11}$$

Each of these three assumptions can be criticized. For example, the simple assumption of centripetal support may be too simple, the constancy of (M/L) and $\bar{I}$ have been called into question (see, e.g., Strauss & Willick, 1995).

The exponent in the luminosity – rotation velocity relation $L \sim v^x \sim v^{-0.4a}$. In Table 11.1 for the shorter visible wavelengths such as blue B $(0.4\,\mu)$, $x \approx 2.7$, while for the longer wavelength near infrared bands H $(1.6\,\mu)$, K $(2.2\,\mu)$, x is closer to 4. Covington *et al.* (2010) considered merger simulations and their relation to stellar mass in the TF relation.

11.1.3. *The baryonic Tully–Fisher relation*

An important version of the TF relation is the baryonic Tully–Fisher relation (BTFR), which relates a galaxy's total baryonic mass, M_b, from stars and gas and its flat rotation speed, v_f.

$$M_b = M_* + M_g. \tag{11.12}$$

The stellar mass is obtained from an assumed mass-to-light ratio and the measured luminosity, L,

$$M_*/L = \Upsilon_*, \tag{11.13}$$

which is usually derived from colours and population models; M_g comes from radio measurements, but requires inclusion of all types of gas. This gives the relation

$$M_b = Av_f^x, \tag{11.14}$$

where $A \approx 47\ \mathrm{M}_\odot\ \mathrm{km}^{-4}\mathrm{s}^4$ and $x \approx 4.0$ (McGaugh, 2012 and references therein). The BTFR has less scatter than the TF relation and extends validity from about 10^8 to $10^{11}\ \mathrm{M}_\odot$. It is clear that the BTFR is produced by the physical processes of galaxy formation and that it is important for testing galaxy formation theories as well as gravity theories, particularly MOND, which makes definite predictions for the BTFR. Namely, $x \equiv 4$ and that there is no scatter in the relation. Early CDM simulations were unable to match the observed BTFR relation and several investigations concluded that this is a good test for MOND (e.g., McGaugh, 2012; Lelli, McGaugh, & Schombert, 2016; Lelli *et al.*, 2019). Newer ΛCDM hydrodynamical simulations (Sales *et al.*, 2017) and observations (Papastergis, Adams, & van der Hulst, 2016) seem to indicate that the BTFR is reproduced well in ΛCDM and that the case of no intrinsic scatter is unsupported. Desmond & Katz (2019) have considered sample size effects theoretically and conclude that larger sample sizes will be required to settle these questions.

11.2. Scaling Relations for Elliptical or Early-Type Galaxies

It was realized that the elliptical galaxies have a correlation between absolute magnitude and colour (e.g., Sandage, 1972; Sandage & Visvanathan, 1978). This can be seen in Fig. 11.9. As well, the Faber–Jackson (1976) relation, $L \propto \sigma^4$ was known. However, even more exact relations can be derived that have reduced scatter. Terlevich *et al.* (1981) conducted an analysis of velocity dispersions, σ, line strengths of the Mg_2 index, and absolute luminosity, L of 24 elliptical galaxies. They found that combining the three measurements considerably reduced the scatter in the (L, σ) and (L, Mg_2) relations, which proved a precursor to using ellipticals as distance indicators. Subsequently, elliptical galaxies have been employed as distance

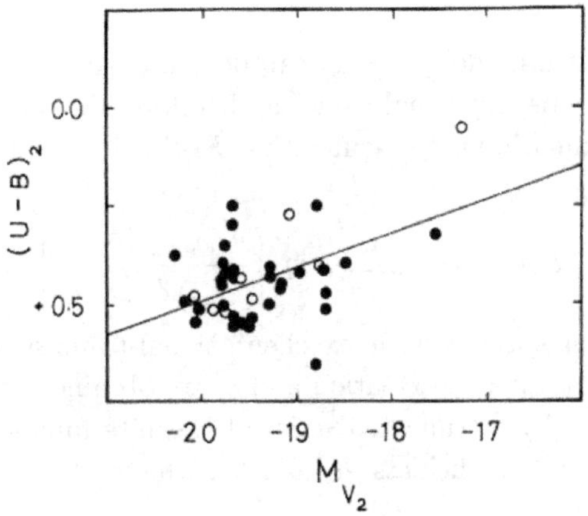

Fig. 11.9. The relation between $(U - B)$ colour and absolute V magnitude for elliptical and spheroidal galaxies. The solid line is Sandage's (1972) relation. (*Source*: Wegner G. (1979), *Ap&SS*, 60, 15, "UBV photometry of 262 southern galaxies," Fig. 6(b), © 1979, Reprinted by permission from Springer Nature.)

indicators using the following related approximations: the Faber–Jackson, the D_n–σ, and the Fundamental Plane.

11.2.1. *Velocity dispersion measurement*

Most early-type galaxies have smaller rotations and less detectable gas compared to spirals, so generally, their dynamics are studied using their optical absorption line spectra. There is an extensive literature on these techniques in general, a good introduction being Binney & Merrifield (1998). If a galaxy's redshift, v along the line of sight is $v \ll c$, the spectral lines all shift by the Doppler effect, $\Delta\lambda = \lambda v/c$. In addition, the net sum of the light and orbital motions of the stars along the line-of-sight broaden the lines, which is the line-of-sight velocity dispersion $\sigma_{\rm los}$. The $\sigma_{\rm los}$ contains information about the light distribution and orbital properties of the galaxy. It can be measured at points on the projected surface of the galaxy to construct its overall dynamics as mentioned in Chapter 4, but in many investigations, only the bright central $\sigma_{\rm los}$ is measured in the galaxy. For a distribution of stellar velocities in the galaxy along the line of sight, $F(v_{\rm los})$, one writes the usual definitions of the mean velocity, $\bar{v}_{\rm los}$, and the velocity variance which is the velocity dispersion, $\sigma_{\rm los}$:

$$\bar{v}_{\rm los} = \int dv_{\rm los} v_{\rm los} F(v_{\rm los}), \tag{11.15}$$

$$\sigma_{\rm los}^2 = \int dv_{\rm los} (v_{\rm los} - \bar{v}_{\rm los})^2 F(v_{\rm los}). \tag{11.16}$$

For some analytical models, $F(v_{\rm los})$ can be computed. For moderate spectroscopic resolutions and signal-to-noise, therefore, the simplest and reasonable first approximation is to assume that $F(v_{\rm los})$ is a Gaussian function of the form

$$F(v_{\rm los}) = \frac{\gamma}{\sqrt{2\pi\sigma^2}} \exp\left[-\frac{(v_{\rm los} - \bar{v}_{\rm los})^2}{2\sigma_{\rm los}^2}\right]. \tag{11.17}$$

Higher resolution spectra with excellent signal-to-noise yield additional information. Typically, the rotation and σ are obtained, but the line profiles are described by a truncated series of Hermite functions, H_3 and H_4, and measuring their coefficients, h_3 and h_4, where

$$F(v_{\rm los}) = \frac{\gamma}{\sqrt{2\pi\sigma^2}} \exp\left[-\frac{(v_{\rm los} - \bar{v}_{\rm los})^2}{2\sigma_{\rm los}^2}\right] [1 + h_3 H_3(v_{\rm los} - \bar{v}) + h_4 H_4(v_{\rm los} - \bar{v})]. \tag{11.18}$$

Old-fashioned measurements used microscopes to measure the shifts and widths of lines in photographic spectra, but current methods utilize all the information contained in the digital spectra. There are two methods that have been used for determining velocity dispersions and redshifts from digital spectra.

The first is the *Fourier Quotient Method,* to which an extensive literature is devoted (e.g., Simkin, 1974; Sargent *et al.*, 1977; Bender, 1990). This uses the well-known convolution theorem and the FFT or Fast Fourier Transform (e.g., Brault & White, 1971 for references and applications to astronomy). The measurement employs $\ln\lambda$ rather than wavelength λ as the independent variable; this way the Doppler shift slides the spectra rather than stretching them. The observed galaxy spectrum, $G(\ln\lambda)$ is the convolution of the true galaxy spectrum, $S(\ln\lambda)$ and the broadening function, $F(\ln\lambda)$, in Eq. (11.18),

$$G(\ln\lambda) = S(\ln\lambda) \star F(\ln\lambda), \tag{11.19}$$

where '$\star$' denotes the convolution of the two functions. The convolution theorem yields,

$$\tilde{F}(k) = \frac{\tilde{G}(k)}{\tilde{S}(k)}, \tag{11.20}$$

where tilde denotes the Fourier transform, and

$$\tilde{F}(k) = \frac{\gamma}{\sqrt{2\sigma_{\text{los}}}^{2}} \exp\left[-2\pi\sigma_{\text{los}}^{2}k^{2} + i2\pi\bar{v}_{\text{los}}k\right]. \tag{11.21}$$

A least squares fit to $\tilde{F}$ provides $\bar{v}_{\text{los}}, \sigma_{\text{los}}$, and γ. An estimate of the unbroadened galaxy spectrum, S is required and this is obtained from the spectra of comparison stars usually taken with the same instrument and time as the galaxy spectrum. From the colours and spectra of early-type galaxies, it is known that the light is dominated by luminous red giants, so the comparison stars are chosen from bright stars of spectral classes in the range near K0 III and K2 III. Examples of the spectra of a comparison star and a galaxy spectrum, along with the fitted transform of the spectrum, are in Fig. 11.10.

The application of the Fourier Quotient Method requires several steps to the data. These include normalizing the spectrum to its continuum, treating the ends of the finite scan, and rebinning the data to 2^N pixels as required by the FFT method. Further details can be found in the reference given here and elsewhere. Figure 11.11 demonstrate how the spectra of

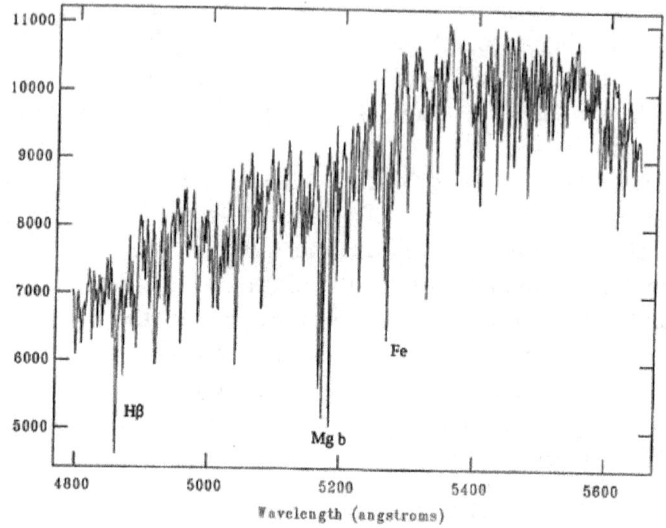

(a)

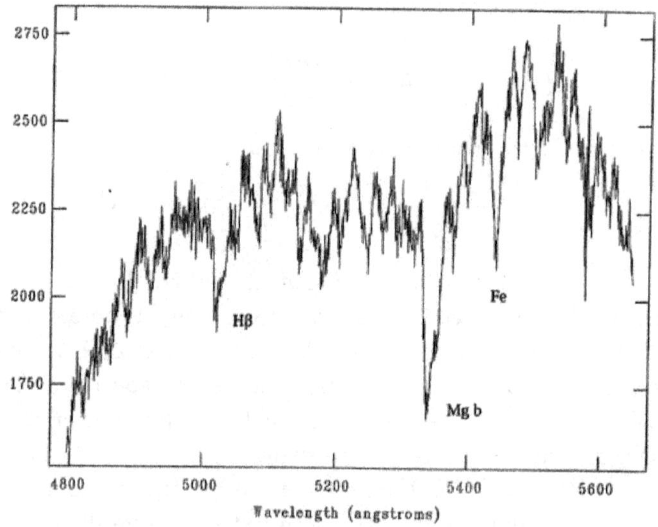

(b)

Fig. 11.10. The spectra of a red giant star, typical of the light from stars in an early-type galaxy, and the spectrum of the galaxy showing the broadened lines due to the velocity dispersion of its stars and the redshift. Some of the same strong lines in the spectra are labelled. Spectra were obtained with the Hiltner 2.4 m telescope of the MDM Observatory on Kitt Peak, Arizona. (a) Optical spectrum of a K0 III comparison star. (b) The same spectral region of an elliptical galaxy.

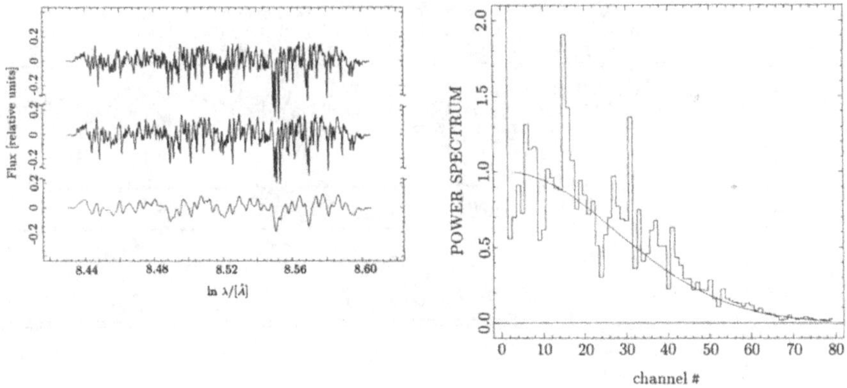

Fig. 11.11. Example of using the Fourier quotient method for velocity dispersion. The top two spectra in the left panel show the input data for K0 III and K2 III stars after they have been normalized to their continua in the wavelength region $\lambda\lambda 4600\text{Å} - 5400\text{Å}$. The bottom scan is the spectrum of the K2 III broadened by a Gaussian of $\sigma = 224$ km s^{-1} with an instrumental resolution of 70 km s^{-1}. The right panel is the Fourier transform of the spectra and shows the fit to the data. (*Credit*: Bender R. (1990), *A&A*, 229, 441, Figs. 1 and 2, ©ESO Reproduced with permission.)

standard star and a broadened galaxy spectrum are combined to obtain the Fourier transform.

The second method widely used for measuring galaxy spectra to obtain redshifts and velocity dispersions by many observing projects is the *Cross-Correlation Method* described by Simkin (1974) and Tonry & Davis (1979). The cross-correlation function is computed from

$$v_{\text{los}} = S \star G = \int_{-\infty}^{+\infty} dx S(x - v_{\text{los}}) G(x), \tag{11.22}$$

where again G and S denote the galaxy and a bright comparison star of known radial velocity as in Fig. 11.10. A number of programs utilizing the Fourier cross-correlation are available, e.g., FXCOR in IRAF and several papers, which provide computational details (e.g., Wegner *et al.*, 1999). The steps carried out in the application of the cross-correlation method are described in the references. Figure 11.12 shows the implementation of the cross-correlation method for the measurement of v_{los} and σ_{los} in a galaxy. The position of the peak gives v_{los}. The width of the peak provides a measurement of σ_{los}. This is calibrated by computing cross-correlation functions of the star with spectra of the star broadened, e.g., a Gaussian function of known velocity dispersion. One source of error is 'template mismatch'. This arises because the light of a galaxy is from a mix of stars and

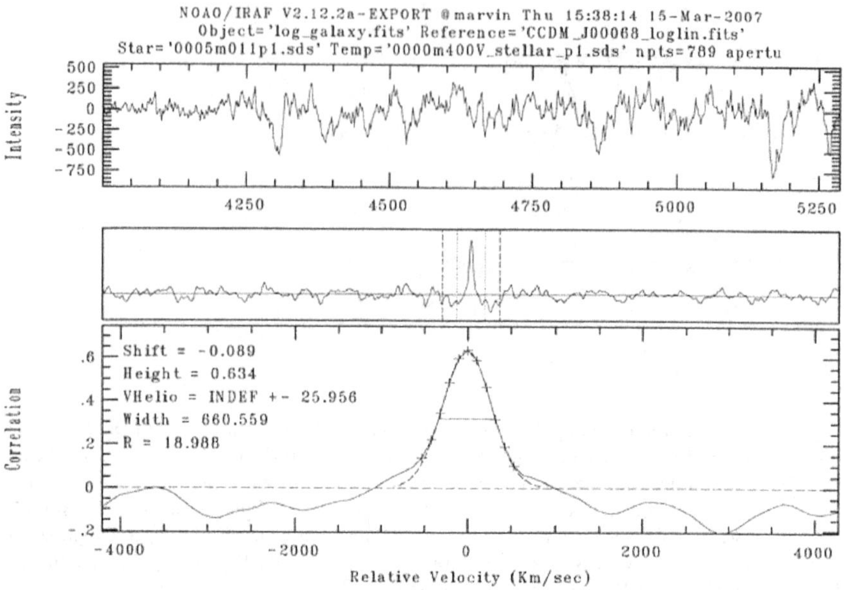

Fig. 11.12. Showing the measurement of the redshift and the velocity dispersion using the cross-correlation method implemented in IRAF. The top panel shows the galaxy spectrum. The middle panel shows the cross-correlation and the lowest panel shows an enlargement of the cross-correlation peak. (*Credit*: Campbell L., "The 6dF peculiar velocity survey," PhD Thesis, The Australian National University, Fig. 2.3, Reproduced by permission.)

the template is from only one star (e.g., Bender, 1990; Binney & Merrifield, 1998).

11.2.2. *Early-type surface brightness measurements*

Compared to most spirals, ellipticals generally have little or no H I and dust, if they are not examined in great detail, but actually their structures are quite complex. Setting these complications aside for the present, their scaling relations provide distance accuracy comparable to the TF relation.

From the earliest measurements of the light profiles of ellipticals, Reynolds (1913) and Hubble (1930) found that in their central regions, the projection of light on the sky can be fit with,

$$I(R) = \frac{I_0}{\left(1 + \frac{R}{a}\right)^2}, \tag{11.23}$$

with $1/a \approx 0.1R_e$, where R_e is the 'effective radius', the distance from the centre which contains half of the emitted light of the galaxy. Although this describes the inner regions of the galaxy profile, the total luminosity does not converge to a finite value of for large R (equation (11.28)).

The de Vaucouleurs $R^{\frac{1}{4}}$ relation (de Vaucouleurs, 1948) gives a remarkably good fit to the surface brightnesses of many ellipticals as seen in Fig. 11.13. It has the form:

$$I(r) = I_e \exp\{-7.67[(r/R_e)^{\frac{1}{4}} - 1]\}. \tag{11.24}$$

This has only two adjustable parameters, where R_e is the effective half light radius and I_e is the surface brightness at R_e. The constant 7.66925 provides half the light inside R_e. The mean surface brightness inside R_e is

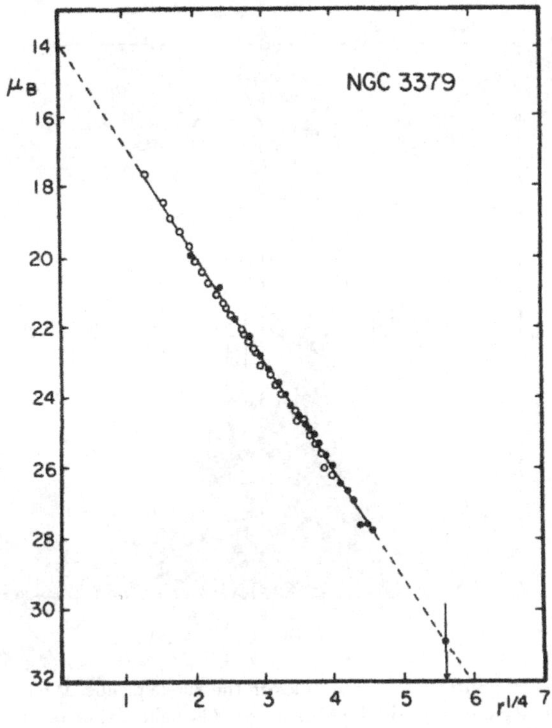

Fig. 11.13. The intensity profile of NGC 3379 (E1) from photoelectric measurements plotted against $r^{\frac{1}{4}}$. The abscissa is in units of arcsec$^{\frac{1}{4}}$ and the ordinate is in magnitudes per arcsec. (*Source*: de Vaucouleurs G., Capaccioli M. (1979), *ApJS*, 40, 699, "Luminosity distribution in galaxies. I. The elliptical galaxy NGC 3379 as a luminosity distribution standard," Fig. 2, ©AAS Reproduced with permission.)

$I = 3.61 I_e$ and the peak central brightness is $I(0) = 10^{3.33} \approx 2143 I_e$. For galaxies with non-circular elliptical isophotes, of major and minor axes, a and b, the effective radius is approximated by $R_e = \sqrt{ab}$.

Replacing the $\frac{1}{4}$ by $\frac{1}{n}$ improves fits to many galaxies' profiles and is known as the Sersic formula (Sersic, 1968). The choice of $n = 4$ is generally a good approximation, but it is not a universal constant. The distribution of n among 558 early-type galaxies in the EFAR sample was found by de Jong *et al.* (2004) examining correlations with structural parameters, such as the bulge to total light (B/T). Most early-type galaxies have n between 3 and 5 as shown in Fig. 11.14 with the median value $n \approx 3.7$ for all non-spiral, bulge dominated galaxies in the sample and 4 for $B/T > 0.7$.

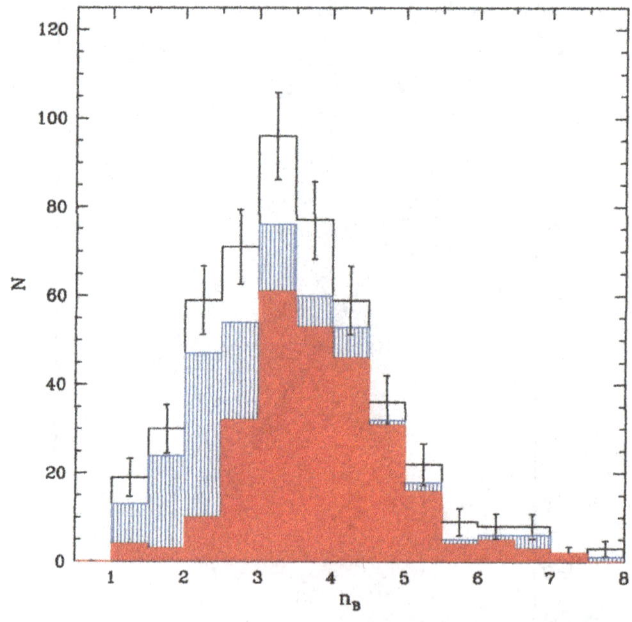

Fig. 11.14. The distribution of the n_B index in the Sersic profile, $I(R) = I_0 \exp(-kR^{\frac{1}{n}})$ from 558 early type galaxies in the EFAR survey. The solid red histogram shows all non-spirals with bulge to total ratios $B/T > 0.5$, the blue hatching shows all galaxies with no apparent spiral structure, and the top solid line denotes the whole sample, for which the median value is $n_B \approx 3.7$. (*Source*: de Jong R. S., Simard L., Davies R. L., Saglia R. P., Burstein D., Colless M., McMahan R., Wegner G. (2004), *MNRAS*, 355, 1155, "Structural properties of discs and bulges of early-type galaxies," Fig. 11, Reproduced by permission of Oxford University Press on behalf of the Royal Astronomical Society.)

Other quantities correlate with B/T. Hudson *et al.* (2010) studied properties of bulges and disks in galaxies varying from $B/T \sim 0.7$ (E) to ~ 0.3 (Sc). The disks are typically bluer by 0.24 mag., in the (B–R) colour. Bulge colours appear to be insensitive to environment, but star formation in disks may be quenched by cluster infall.

Other light profiles with additional fitting parameters can be applied (e.g., Binney & Tremaine, 2008). Generally, most galaxies require the combination of a central bulge like the $R^{\frac{1}{4}}$ and an exponential disk.

11.2.3. *Boxy and disky ellipticals*

The complexity of the Es is revealed by the presence of some gas and dust lanes in their centres (Fig. 11.15) and their isophotes, which can depart in different degrees from a general elliptical outline. These are usually slight distortions, but ellipticals can be classified as either *boxy* or *disky* as shown in Fig. 11.16. As an approximation, the brightness of the isophotes, $\Sigma(\theta)$, can be written as a Fourier series where the first terms are,

$$\Sigma(\theta) = \Sigma_0[1 + a_2\cos(2\theta) + a_4\cos(4\theta) + \cdots. \tag{11.25}$$

If $a_4 < 0$, the object is boxy and if $a_4 > 0$ it is disky. Luminous Es are often boxy. Midsized Es tend to be disky with the amount of the disk extending as a sequence into the S0s. The boxy objects have cuspy and lower density cores, while disky objects have steep higher density power law cores. Boxy ellipticals are slow rotators with moderately triaxial shapes and disky galaxies are faster rotators of nearly oblate shape and these differences are considered to be the result of their merger history. Despite these differences, the shape of the scaling relations used for distance determinations seems unaffected.

Although some ellipticals appear to have oblate shapes and be axial rotators, other modes are possible. Prolate systems would have an elongated sausage-like shape. Triaxial systems have three unequal axes. It is difficult to determine from their images which shapes many of these systems may have, but an indication is given by two other distortions in the fine structure of the isophotes shown in Fig. 11.16. Twists are observed where the projected angle of the major axis of the elliptical isophote changes with distance from the centre and the ratio of the axes of the isophotes change with distance. This also means that the orbital properties can differ in a complex way, e.g., stars follow tube orbits in axially symmetric systems, but they mainly follow box orbits in triaxial systems.

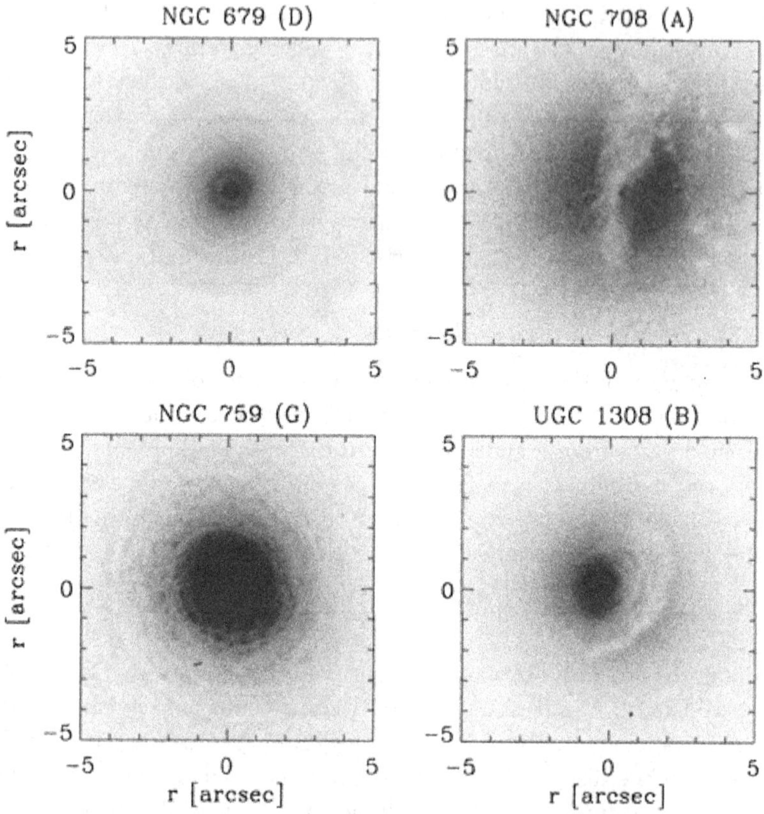

Fig. 11.15. The structures of four elliptical galaxies observed with the HST in the R band. These show examples of dust features observed in the centres of many ellipticals. (*Source*: Wegner G. A., Corsini E. M., Thomas J., Saglia R. P., Bender R., Pu S. B. (2012), *AJ*, 144, 78, "Further evidence for large central mass-to-light ratios in early-type galaxies: The case of ellipticals and lenticulars in the A262 cluster," Fig. 1, ©AAS Reproduced by permission.)

11.2.4. *The Faber–Jackson relation*

The earliest scaling relation for elliptical galaxies was the Faber–Jackson relation (Faber & Jackson, 1976). This relates the luminosity of ellipticals to the velocity dispersion

$$L_T \sim \sigma^x, \tag{11.26}$$

where $x \approx 4$. This is the same as the exponent in the TF relation (11.11) derived from the virial theorem. Consequently the projected velocity

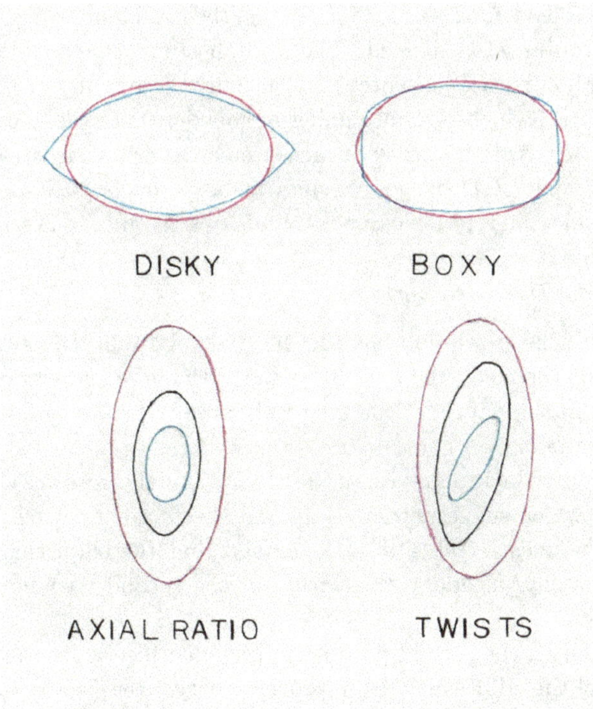

Fig. 11.16. Sketches of isophote distortions in elliptical galaxies. (Top) Disky and boxy profiles compared to an elliptical profile. (Bottom) Twists and ellipticity changes in the isophotes.

dispersion is

$$\sigma_p = 220(L_T/L_\star)^{\frac{1}{4}}\ \mathrm{kms}^{-1}. \tag{11.27}$$

L_T is determined from the growth curve, which gives the magnitude of the total light from the galaxy inside circular apertures of radius R,

$$L(R) = 2\pi \int_0^R I(r) r dr. \tag{11.28}$$

This can be evaluated analytically for the profiles such as the Hubble–Reynolds formula or the Sersic formula if n is an integer. For the $R^{\frac{1}{4}}$ relation,

$$L(R) = \frac{8!\pi I_e R_e^2 e^k}{k^8}\left[1 - \exp(-kR^{\frac{1}{4}})\left(\sum_{n=0}^{7}\frac{k^n}{n!} r^{\frac{n}{4}}\right)\right], \tag{11.29}$$

with $k = 7.67$. When $R \to \infty, L(R) \to L_T$, the total luminosity, but usually a limiting radius R is defined. The growth curve can be fitted using a number of photometric measurements (usually three or more) to obtain L_T as in Fig. 11.17, which plots a number of galaxies on the $R^{\frac{1}{4}}$ curve. Earlier determinations used photoelectric measurements through a series of circular apertures. Using CCD images, circular areas of precise size are employed. The Faber–Jackson relation can yield a galaxy's distance to $\Delta D/D \approx \pm 20\%$

11.2.4.1. *The D_n–σ relation*

The Faber–Jackson relation has the disadvantage that L_T is obtained by extrapolating the measured data fit to a growth curve and errors arise from imprecision in the data and the assumed curve, typically a $R^{\frac{1}{4}}$ relation or a Sersic relation.

In the Faber–Jackson relation, the observational errors are smaller than the scatter, which suggests that additional effects are involved. As already described Terlevich *et al.* (1981) concluded that the ellipticals are a two-parameter family, although the nature of the second parameter was not fully realized.

Subsequently, the second parameter was identified as the mean surface brightness of the ellipticals. An improvement over the Faber–Jackson relation defines a circular aperture, D_n, wherein the surface brightness has a specified value. For the 7S, this was 20.75 B mag arcsec^{-2}, which introduces the surface brightness as the second fitting parameter. Consequently, D_n gives a relative angular size distance rather than a luminosity distance. This relates to the velocity dispersion as

$$\log D_n = 1.2 \log \sigma - 2.462, \tag{11.30}$$

where the zero point is set relative to the well-observed Coma cluster at a redshift of 7203 km s^{-1}, D_n is in units of arcsec, and σ in km s^{-1} (Lynden-Bell *et al.*, 1988; Dressler *et al.*, 1987a,b). After the corrections to the observations already mentioned, K-term, cosmological effects including $(1+z)^4$ dimming, Malmquist effects, etc., the observed and calculated D_n ratio, $D_{\mathrm{Coma}}/D_{\mathrm{galaxy}}$ yields the galaxy's distance relative to the Coma cluster.

Compared to the Faber–Jackson relation, employing D_n–σ reduces the scatter as shown in Fig. 11.18 showing the two relations for the same galaxies. The relative distance error that can be obtained from the D_n–σ relation is about $\Delta D/D \approx 15\%$ compared to $\pm 20\%$ for the Faber–Jackson.

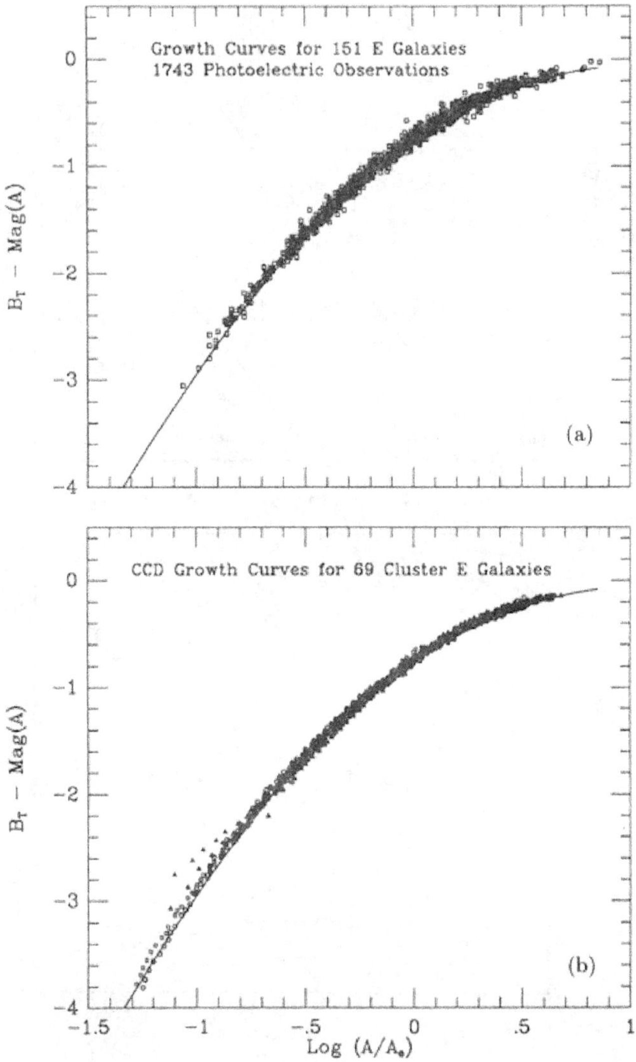

Fig. 11.17. Growth curves for elliptical galaxies fit to the $R^{\frac{1}{4}}$ law. (a) 151 field galaxies using individual photoelectric aperture measurements (b) Growth curves for 69 ellipticals from CCD measurements. The scatter from the CCD measurements can be seen to be smaller. (*Source*: Burstein D., Davies R. L., Dressler A., Faber S. M., Stone R. P. S., Lynden-Bell D., Terlevich R. J., Wegner G. (1987), *ApJS*, 64, 601, "Spectroscopy and photometry of elliptical galaxies. III. UBV aperture photometry, CCD photometry, and magnitude-related parameters," Fig. 2, ©AAS Reproduced with permission.)

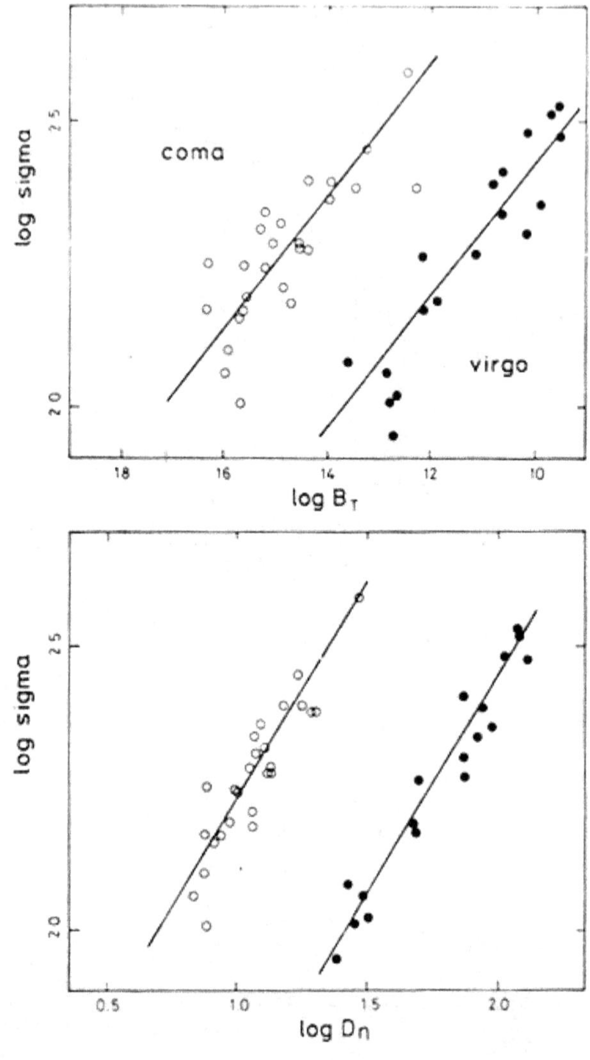

Fig. 11.18. Comparison of the Faber–Jackson and D_n–σ relations for ellipticals in the Coma and Virgo galaxy clusters. In the upper panel, the Faber–Jackson relation gives distance errors of 38% and 30% respectively for individual galaxies in Coma and Virgo. In the bottom panel (D_n in arcsec and σ in km s^{-1}), the D_n–σ relation gives 17% error in the distance of Coma and 15% in Virgo. Using the combined galaxies results in 3% distance errors for both clusters (*Source*: Lynden-Bell D. (1986), *QJRAS*, 27, 319, "How smooth is the hubble flow?" Fig. 5.)

11.2.5. *The fundamental plane*

A tighter relation than the FJ and D_n–σ relations that has a physical interpretation combines the central velocity dispersion with the effective radius, R_e, which contains half of the galaxy's total light and $\langle SB_e \rangle$, the average effective surface brightness within R_e. This defines the fundamental plane (FP), which is a narrow plane in $(\log R_e, \langle SB_e \rangle, \log \sigma)$ space (see Fig. 11.19). The FP was intensively investigated by numerous investigators starting with Djorgovski & Davis (1987), Dressler *et al.* (1987a), Jørgensen, Franz, & Kjaergaard (1996), Bender, Faber & Burstein (1992, 1997), and Burstein *et al.* (1997) and references therein for the earlier literature.

Early-type galaxies, except for the DM dominated faintest dwarfs are considered to be 'Dynamically Hot Galaxies'. They have a similar M/L ratio and closely homologous structures as already mentioned in Chapter 4.

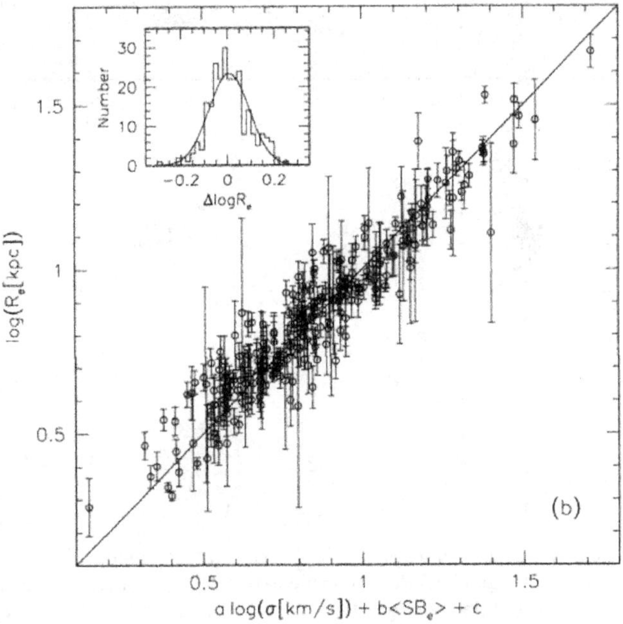

Fig. 11.19. The plot of $\log R_e$ vs. the best fit to EFAR data, $a \log \sigma + b\langle SB_e \rangle + c$ using 255 galaxies in 29 clusters with 6 or more members. The rms scatter about the fitted line, $\Delta \log R_e$, is 0.087 as seen in the inserted histogram and corresponds to an rms distance error of 20% for an individual galaxy. (*Source*: Colless M., Saglia R. P., Burstein D., Davies R. L., McMahan Jr. R. K., Wegner G. (2001), *MNRAS*, 321, 277, "The peculiar motions of early-type galaxies in two distant regions — VII. Peculiar velocities and bulk motions," Fig. 1(b), Reproduced by permission of Oxford University Press on behalf of the Royal Astronomical Society.)

In one formulation (Burstein *et al.*, 1997; Bender *et al.*, 1997), three orthogonal axes are defined in 'κ-space' in terms of observed quantities,

$$\kappa_1 \equiv (\log \sigma_0^2 + \log R_e)/\sqrt{2} \ \propto \ \log M, \tag{11.31}$$

$$\kappa_2 \equiv (\log \sigma_0^2 + 2\log\langle SB_e\rangle - \log R_e)/\sqrt{6} \ \propto \ \log(M/L)\langle SB_E\rangle^3, \tag{11.32}$$

$$\kappa_3 \equiv (\log \sigma_0^2 - \log\langle SB_e\rangle - \log R_e)/\sqrt{3} \ \propto \ \log M/L. \tag{11.33}$$

The thinness of the FP in κ-space is shown in Fig. 11.20 for the early-type galaxies. It is interesting to note that elliptical and spiral galaxies lie in similar parts of κ-space. This diagram indicates the locations of different

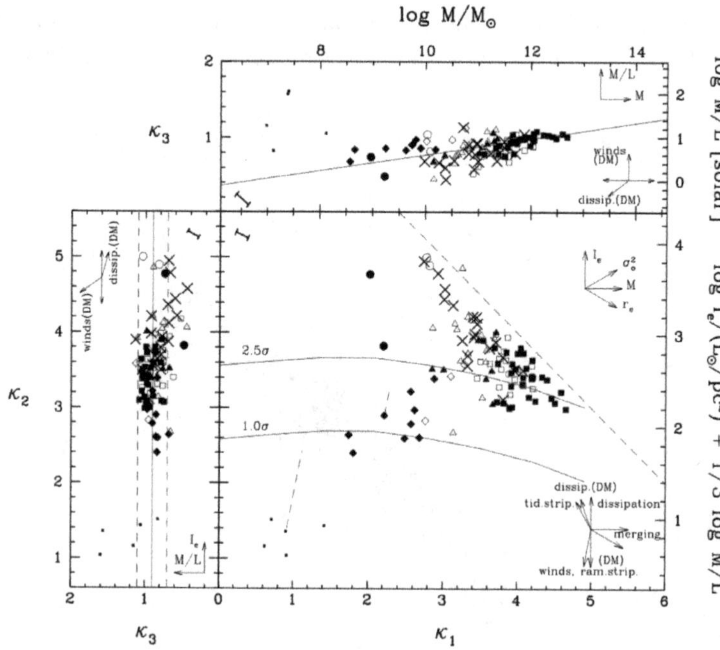

Fig. 11.20. Positions of the different types of early-type Dynamically Hot Galaxies in κ-space. Symbols: □ = giant E, △ = intermediate E, ○ = compact galaxy, ◇ = dwarf galaxy, × = bulge, tiny filled squares denote Fornax dwarf spheroidals. Open symbols are rotationally flattened galaxies, filled symbols denote anisotropic kinematics. Arrows show approximately how different processes (e.g., tidal stripping, ram pressure stripping, DM content, etc.) shift galaxies in κ-space. The zone of avoidance to the right of the dashed line in the κ_1–κ_2 plane contains no galaxies. The solid line in the κ_1–κ_3 plane is the FP fit to elliptical galaxies, $\kappa_3 = 0.15\kappa_1 + 0.36$. (*Source*: Bender, R., Burstein, D., Faber, S. M. (1997), "The fundamental-plane and Tully-Fisher relations viewed in kappa-Space", in *Galaxy Scaling Relations: Origins, Evolution and Applications*, proceedings from the ESO Workshop, November, 1996, eds. by Luiz da Costa and Alvio Renzini, Springer-Verlag, p. 95, Fig. 1, Reprinted with permission.)

types of dynamically hot galaxies and how they are affected by different physical processes. Consequently, the FP provides clues for the formation of the ellipticals as discussed below in further detail, but as the TF relation for spirals, it and the FJ relation can be derived approximately from the virial theorem. In both cases, the numerical coefficients are derived empirically.

Used as a distance indicator, the FP takes the form

$$\log R_e = a \log \sigma + b \log\langle \mathrm{SB}_e\rangle + c, \tag{11.34}$$

as shown in Fig. 11.19, where $a = 1.22 \pm 0.09$ and $b = -0.84 \pm 0.03$, which would produce an estimated distance error of $\Delta D/D = \pm 0.20$. These are representative values derived from a detailed fitting using 255 galaxies in 29 clusters in the EFAR project. Different investigators find similar numbers for a and b, depending on the colour band employed for the dataset and fitting method. The zero point, c, is determined using objects of known distance. A maximum likelihood method for determining the coefficients is described by Saglia *et al.* (2001) and Colless *et al.* (2001a). Other authors use different regression fitting methods, but results are similar. Colless *et al.* (2001a) and Magoulas *et al.* (2012) give lists of these coefficients for several published determinations, where $a = 1.14$ to $1.39, \mathrm{b} = -0.90$ to -0.79, and $\Delta D = 13\%$ to 21%. Some representative values are shown in Table 11.2 are sampled for different wavelength bands and numbers of galaxies.

The FP and D_n–σ method are closely related. Comparing the two, the σ terms nearly cancel

$$\begin{aligned} \log \frac{D_n}{R_e} &= (1.2 \log \sigma) - (1.22 \log \sigma - 0.84 \log\langle \mathrm{SB}_e\rangle) \\ &\approx 0.8 \log\langle \mathrm{SB}_e\rangle + \text{const.} \end{aligned} \tag{11.35}$$

Table 11.2. A sample of fundamental plane coefficients.

Band	N_{gal}	a	b	$\frac{\Delta D}{D}$	Fit[a]	Reference
B	97	1.33 ± 0.05	-0.83 ± 0.03	0.20	I	Dressler *et al.* (1987)
V	66	1.26	-0.82	0.17	F	Lucey *et al.*(1991)
g	5825	1.45 ± 0.06	-0.74 ± 0.01	0.20	M	Bernardi *et al.* (2003)
r	226	1.24 ± 0.07	-0.82 ± 0.02	0.17	O	Jørgensen *et al.* (1996)
R	255	1.22 ± 0.09	-0.84 ± 0.03	0.20	M	Colless *et al.* (2001a)
I	109	1.25 ± 0.02	-0.79 ± 0.03	0.20	m	Scodeggio *et al.* (1997)
J	8901	1.52 ± 0.03	-0.89 ± 0.008	0.30	M	Magoulas *et al.* (2012)
H	4589	1.56 ± 0.02	-0.80 ± 0.005	0.27	O	LaBarbera *et al.* (2010)
K	25	1.53 ± 0.08	-0.79 ± 0.03	0.21	O	Pahre *et al.* (1998)

[a]I = Inverse regression, F = Forward regression, M = Maximum likelihood Gaussian, O = Orthogonal regression, m = mean regression.

leaving the surface brightness. This shows that the D_n–σ relation is nearly an edge-on approximation to the FP. The derived distances from this simpler indicator are nearly as good as those from the FP ($\Delta D \sim 20\%$ vs. $\sim 15\%$), so some peculiar motion studies have preferred this simpler method.

Chapters 4 and 20 outline several newer and larger surveys to study the fundamental plane such as by the ATLAS^{3D} group (Cappellari *et al.*, 2013b) as well as the Tully–Fisher relation. The scaling relations is that they have been studied continuously by many investigators. These are empirical relations and the coefficients depend on several factors in the treatment of the observations and fitting methods. New observations and calibrations of the scaling relations continue to be appear in the literature.

It is interesting to note that although many new investigations have added larger numbers of galaxies to determining the scaling relations, the scatter in the new determinations has not been greatly reduced. The progress in improving accuracies has been in the determination of the zero points, e.g., using Cepheids, SNe Ia, and the TRGB.

Chapter 12

Nearby Peculiar Velocities

12.1. Definition of Peculiar Velocities

Given the total space three-dimensional velocity of a galaxy, **V**, except for the nearest objects only the radial component along the line of sight is observable. With current technology, galaxies are mostly too distant for the tangential component, $V_T = 4.74\,\mu/\pi\,\mathrm{km\,s^{-1}}$($\mu$ is the proper motion in arcsec yr^{-1} and π is the parallax in arcsec) to be detected. For the observed redshift, there are three major contributions to the radial component of the velocity. The first is the cosmological or Hubble expansion, $H_0 D_L$ expressed as a velocity in $\mathrm{km\,s^{-1}}$(where D_L is the luminosity distance defined in Chapter 8). The second is the Doppler shift due to the motion of the observer, V_o, relative to our comoving reference frame defined by the CMB dipole. The third is the peculiar velocity of the galaxy, V_{pec}. Other sources of apparent velocity include gravitational redshifts, the transverse Doppler effect, and pressure shifts, but these are usually small. Then, the observed redshift of a galaxy along the line of sight is

$$cz = H_0 D_L + V_{\mathrm{pec}} - V_o. \tag{12.1}$$

In solving this for the peculiar velocity, the total redshift, cz is measured to high accuracy, leaving $H_0 D_L$. If H_0 is known, the remaining step is to secure the distance to the galaxy as already discussed.

12.2. The Size of Peculiar Velocities

With the accumulation of large numbers of redshift measurements and N-body simulations, it is clear that the anisotropic filament-void structures contain higher density regions. Their gravitational attractions over the age of the universe generates velocity fields and these can be measured in the nearest galaxies. This process involves the details of gravitation, cosmological information, galaxy formation and evolution.

For gravitational instability produced by a density field, in the weak gravity linearized irrotational case, the velocity produced by a density contrast, $\delta(\mathbf{r}) = [\rho(\mathbf{r}) - \rho_0)]/\rho_0$, where $\rho(\mathbf{r})$ is the density and ρ_0 is the mean density of the universe, will be

$$\mathbf{V} = \frac{H_0 f(\Omega)}{4\pi} \int_0^\infty \frac{\delta(\mathbf{r}')(\mathbf{r} - \mathbf{r}')d\mathbf{r}'}{|\mathbf{r} - \mathbf{r}'|^3}, \tag{12.2}$$

H_0 is the Hubble constant and $f(\Omega)$ is the cosmic growth index. In terms of S the scale factor, the cosmic growth index is

$$f(\Omega) = \frac{S}{\delta}\frac{d\delta}{dS}. \tag{12.3}$$

Peebles (1980) showed that for a wide range of density, curvature, and Λ parameters, an accurate approximations is

$$f(\Omega) \approx \Omega^{0.6}, \tag{12.4}$$

where Ω_0 is the matter density relative to ρ_0. For ΛCDM, this exponent is 0.55 (Wang & Steinhardt, 1998; Linder, 2005).

The order of magnitude of the generated peculiar velocities can be estimated from a simple formula. At low redshifts, the expansion is $V = H_0 D$; gravitational attraction generates the additional peculiar velocity component. A galaxy's velocity near a large clump, or attractor, is observed as the radial component $V = H_0 D + \Delta V_{\mathrm{pec}}(D, s)$. The V_{pec} can be expressed in terms of the size and density enhancement of the attractor. In the case of a small mass at the edge of a spherical clump of radius R and density enhancement, $\delta\rho$ (Fig. 12.1), the infall velocity is

$$\Delta V_{\mathrm{pec}} \approx \frac{1}{3} H_0 R \Omega_0^{0.6} \frac{\delta\rho}{\rho}. \tag{12.5}$$

For a galaxy cluster with $R \sim 2\,\mathrm{Mpc}$ and mass $M \sim 10^{14}$–$10^{15}\ \mathrm{M}_\odot$, $\delta\rho/\rho \sim 20$, which produces $\Delta V_{\mathrm{pec}} \sim 100\ \mathrm{km\,s^{-1}}$for $H_0 \approx 70\,\mathrm{km\,s^{-1}\,Mpc^{-1}}$. Thus typical peculiar velocities produced by galaxy clusters is of order hundreds of $\mathrm{km\,s^{-1}}$. Much larger R or $\delta\rho$ are required to bring ΔV_{pec} up to thousand of $\mathrm{km\,s^{-1}}$.

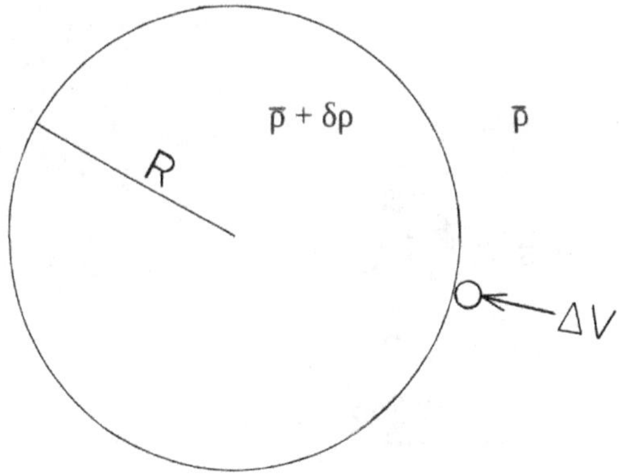

Fig. 12.1. The generation of the peculiar velocity ΔV of a small mass at the periphery of a spherical density enhancement of radius R.

12.3. Motions of the Local Group and the Local Supercluster

Detailed analyses of peculiar motions have been built up from the accumulation of many observations assembled from numerous investigations. Before considering more distant large-scale motions, it is important to look at motions of nearby objects, the Local Group, and the Local Supercluster.

Early investigations concentrated on mapping the distribution of galaxies near the Milky Way. Starting with the early all-sky visual and photographic coverage of the sky and redshift surveys, once an extragalactic distance scale became available, a map of the nearby galaxies began to come into focus. Our Milky Way has a system of satellites, surrounded by the Local Group (Fig. 12.2). In addition to our galaxy and the Andromeda galaxy M31, there are many smaller galaxies, notably the Large Magellanic Cloud (LMC) and the Small Magellanic Cloud (SMC), the smaller and more distant spiral M33 (NGC 598) and many others spread within a radial distance of order ~2 Mpc. The Local Group has been known since the early 20th century. Hubble (1936) listed eight members and recent investigations have found large numbers of ultrafaint dwarfs (e.g., Dooley *et al.*, 2017), increasing the number greatly. Compared to many more distant galaxy clusters, the Local Group is a rather poor example.

Beyond the Local Group, more galaxy groups are found. Many are dominated by one or two large galaxies (Fig. 12.3).

At a distance of about 15 Mpc, the Virgo cluster, which is the most massive in the region is reached. This is the nearest large galaxy cluster

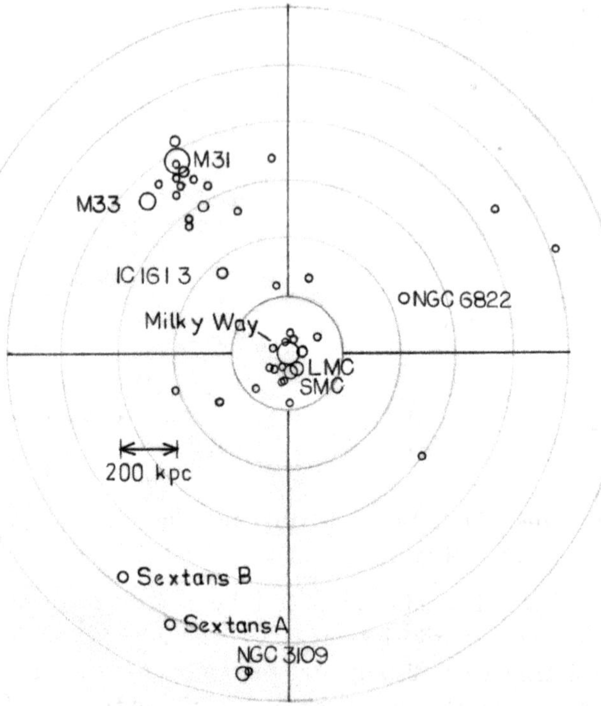

Fig. 12.2. Sketch of the Local Group projected into the galactic plane, showing some of the larger members. Galactic longitude $l = 0°$ is to the right and increases counterclockwise. This is based primarily on the compilation of Van den Bergh (2000).

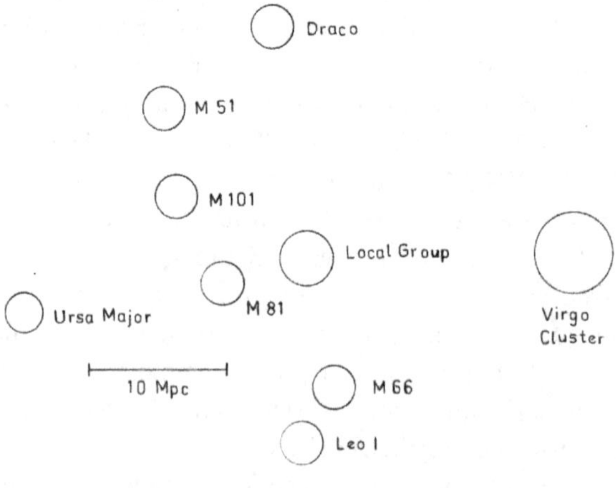

Fig. 12.3. Sketch of some of the groups in the local supercluster.

(centre $\alpha = 12^h 27^m$, $\delta = +12°43'$, $cz \approx 1500\,\mathrm{km\,s^{-1}}$). Tully & Fisher (1987) and others, e.g., Courtois *et al.* (2012), have compiled more detailed maps of these galaxy distributions, which are well worth looking at.

12.4. Motions of Satellites of the Milky Way

Measurements of the Milky Way's rotation and of a few bright nearby galaxies were among the earliest extragalactic dynamical studies. New progress is being made by discovering faint Milky Way satellite galaxies and searching for satellites of Milky Way analogs (e.g. Geha *et al.*, 2017; Nadler *et al.*, 2020; Mao *et al.*, 2021).

It was therefore natural that the motions of the satellite galaxies at a distance of about 50 kpc, the Large (LMC) and Small (SMC) Magellanic Clouds, would be studied to determine their orbits. Two results are important. The first is the Magellanic Stream and the second is the warp of the galactic disk, both found from 21 cm neutral hydrogen observations. The stream was revealed from high velocity 21 cm observations which show that the LMC and SMC lie near the head of a long filament, that runs for about 100° from the clouds through the galactic pole (Dieter, 1965; Mathewson, Cleary, & Murray, 1974; Nidever, Majewski, & Burton, 2008). The warp (Burke, 1957; Chen *et al.*, 2019) is often found in other galactic disks (Briggs, 1990; Nelson & Tremaine, 1996). Numerous models (see Lin, 1996 for a review) were proposed to fit these two effects. This was further complicated by the discovery of the Sagittarius dwarf in 1994, comparable in mass to the LMC and SMC but unknown due to being hidden by high dust extinction in the plane of the galaxy (Ibata, Gilmore, & Irwin, 1994). Figure 12.4 shows a picture of the Magellanic stream. Current interpretation (Bolokurov *et al.*, 2017; Diaz & Bekki, 2012; Besla *et al.*, 2012) is that the Magellanic stream is an ongoing merger with the Milky Way. Tidal interactions, mostly with the LMC have produced the tidal tail, but ram pressure stripping has contributed to the tail's formation. In addition, it might provide some clues about dark matter in the galactic halo.

Numerous other streams are now known, as pointed out by Lynden-Bell (1976) and Lynden-Bell & Lynden-Bell (1995), but currently widely studied (Belokurov, 2013; Carlberg & Grillman, 2013). The Sgr stream (Deason *et al.*, 2013) and the Orphan stream (Newberg *et al.*, 2010) are well studied. Tully (2013) and Hammer *et al.* (2013) find that small satellites of the Milky Way define thin planes, which may be related and Ibata *et al.* (2013) observed that these also appear in deep imaging of M31. Carlberg (2019) has described 'halo vibrations' in the stream velocities and modelling seems

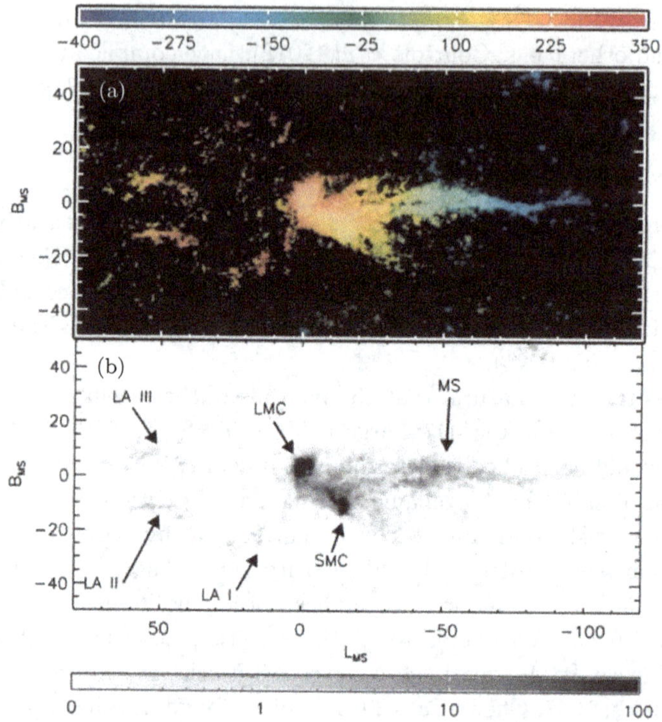

Fig. 12.4. The Magellanic stream and the LMC and SMC plotted on the sky. The top coloured panel gives the logarithmic hydrogen density, $\log N_H$. The lower panel labels the LMC and SMC and other features of the stream and gives the H I column density in units 10^{19} atoms cm^{-2}. (*Source*: Nidever D. L., Majewski S. R., Burton W. B. (2008), *ApJ*, 679, 432, "The origin of the magellanic stream and its leading arm," Fig. 8, ©AAS Reproduced with permission.)

to indicate that the streams may be useful for studying self-interacting DM in the galaxy.

12.5. Motions in the Local Group

Different investigators have studied the motions of the galaxies in the Local Group and the Local Supercluster. Neglecting the less massive and more distant members of the Local Group, the Milky Way and M31 are the most dominant players. The current distance to M31 is $R \approx 770$ kpc and its radial velocity is $cz \approx -123\,\text{km}\,\text{s}^{-1}$, indicating that it is approaching the Milky Way. Kahn & Woltjer (1959) and several subsequent investigators (e.g., Peebles, 1996) employed the radial infall model (discussed in

Chapter 18) that models the motion of two masses due to gravitational attraction in an expanding Einstein–de Sitter universe. The boundary conditions are taken as the present cz, and R, plus $t_0 \approx 15\,\text{Gyr}$, the age of the universe. This earlier result indicates that M31 and the Milky Way have nearly finished one period of a radial orbit and are approaching with a combined mass $M = 4$–$5 \times 10^{12} M_\odot$ for the pair. Since their total luminosity is $L = 5 \times 10^{10} L_\odot$, $M/L \approx 100 M_\odot/L_\odot$ and indicates the presence of DM.

The proper motion of M31 is now available from the HST and more detailed models using the structures of both galaxies and including M33 can be constructed using N-body and semi-analytic calculations (Sohn *et al.*, 2012; van der Marel *et al.*, 2012a,b and references therein). These authors derive a mass of $(3.17 \pm 0.57) \times 10^{12} M_\odot$ for the Local Group. M31 and the Milky Way are gravitationally bound and will undergo a nearly head-on collision. The first pericentre approach will be in about 4 Gyr and they will merge in about 6 Gyr to form one object resembling an elliptical galaxy. The sun's orbit will be strongly altered and could be placed in a more distant path.

Dynamical models that include more than two galaxies have been constructed by several methods including N-body simulations. Ideas for models have included whether the origin of the motions in the LG could be due to either gravitational or non-gravitational effects (Lynden-Bell & Lin, 1977). A further complication for modelling is that at least in the earliest stages, if the galaxies grew from a very smooth distribution of mass starting with primeval seeds, their masses would vary with time. However, if this process proceed rapidly, a point-mass approximation for computing the gravitation might be sufficient. This approach has been applied in recent models of the Local Group.

The model developed by Peebles (1993, 1995, 1996, 2011) to study the Local Group uses numerical integration of the action for the particles. This method can be applied to large numbers of galaxies treated as particles and as seen later, has been employed to model the Local Supercluster and beyond by, e.g., Tully and collaborators.

Starting with the action in terms of the kinetic energy and the potential energy, T, V, neglecting for the moment expansion,

$$S = \int_0^{t_0} dt(T - V) = \int_0^{t_0} dt \left[\sum_i \frac{m_i}{2} \left(\frac{dx_i}{dt} \right)^2 - V \right]. \tag{12.6}$$

Putting in $x_i \to x_1 + \delta x_i(t)$ for the infinitesimal changes in the orbit, gives δS, the change in the action. Including the expansion of the universe with a mean mass density $\rho(t)$, the action is (Peebles, 1993)

$$S = \int_0^{t_0} dt \left[\sum_i \frac{m_i a^2}{2} \left(\frac{dx_i}{dt} \right)^2 + \frac{G}{a} \sum_{i \neq j} \frac{m_i m_j}{|x_i - x_j|} + \frac{2}{3}\pi G \rho a^2 \sum_i m_i x_i^2 \right]. \tag{12.7}$$

The corresponding change in the action due to the infinitesimal change $\delta x_i(t)$ is

$$\delta S = \int_0^{t_0} dt \delta x_i \left[-\frac{d}{dt} m_i a^2 \frac{dx_i}{dt} + m_i a g_i \right] + \left[m_i a^2 \delta x_i \frac{dx_i}{dt} \right]_0^{t_0}, \tag{12.8}$$

where

$$g_i = \frac{G}{a^2} \sum_j m_j \frac{x_j - x_i}{|x_j - x_i|^3} + \frac{4}{3}\pi G \rho_b a x_i. \tag{12.9}$$

The boundary conditions assume the masses begin close together and move radially with the universe's expansion. Individual velocities can be given and orbits are adjusted to match the observed space motions now. Details of the derivations and numerical calculations are given in Peebles (1993, 1995, 1996) and Peebles, Tully & Shaya (2011).

As noted by Peebles, the solutions derived are not unique, but can be made consistent with observations and the boundary conditions by trial and error. Figure 12.5 shows the results of one calculation for galaxies in the Local Group by Peebles *et al.* (2011) with the Milky Way and M31 indicated.

12.6. The Virgo Cluster and Supergalactic Plane

Deviations from the smooth cosmological expansion, or 'Hubble Flow,' centred on the Milky Way produced by the attraction of the Virgo cluster can be detected by the anisotropy that it produces in the otherwise spherically symmetric Hubble flow, (Sandage & Tammann, 1984 and others), now known as the *Virgocentric infall* where deviations from the spherically symmetric expansion can be seen. This effect was sought by several of the early researchers in peculiar motions.

The Virgo cluster is the nearest massive cluster. It was noted by de Vaucouleurs (1953, 1956) that the Virgo cluster and the nearby groups form a somewhat flattened system as shown in Fig. 12.6 and is referred to as

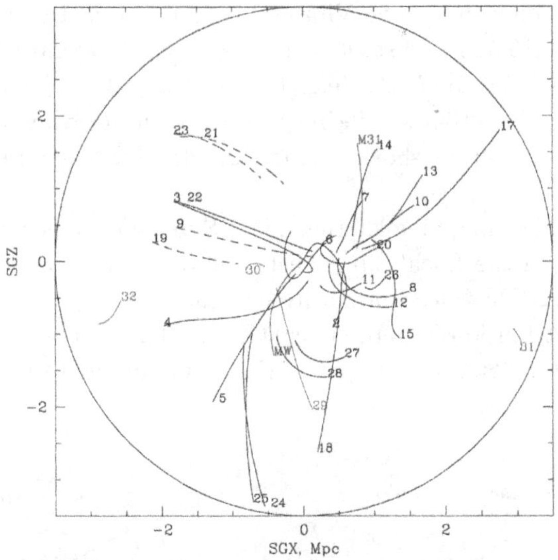

Fig. 12.5. Motions of 28 galaxies within 1.5 Mpc of the Local Group from the numerical action method in comoving galactic coordinates. Current positions are at the numbered locations. Objects labelled blue are thought to be outside the LG. Numbers correspond to objects, e.g., 6 = NGC 6822, 12 = IC 1613, 18 = M33, 24 = NGC 3109, 27 = SexA, 28 = SexB. The Milky Way and M31 are indicated as MW and M31. (*Source*: Peebles P. J. E., Tully R. B., Shaya E. J. (2011), "A dynamical model of the local group," 2011, arXiv:1105.5596, Fig. 4. Reproduced with author's permission.)

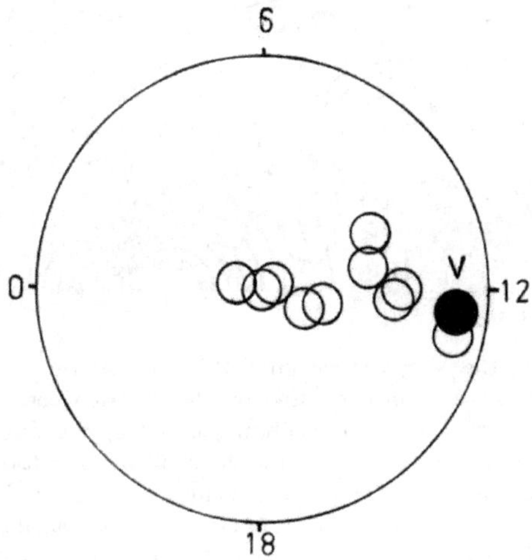

Fig. 12.6. Sketch of some of the objects in the local supercluster in equatorial coordinates showing the flattening of the system based on the data given in van den Bergh (1968). The Virgo cluster is the filled circle labelled **V**.

our 'Local Supercluster'. The Virgo cluster, shown in the Tully & Fisher (1987) atlas is 16.5 Mpc distant, contains about 1500 galaxies and has a mass of about $1.5 \times 10^{15}\ M_\odot$ (e.g., Foquè *et al.*, 2001; Mei *et al.*, 2007). This mass concentration significantly distorts the isotropy of the expansion in this direction, as shown in Fig. 12.7 and has been known for some time.

Some detailed mapping of the Local Supercluster is given in Lahav *et al.* (2000). In the Local Supercluster, flows can be found on relatively large distances. These lead primarily into the Virgo cluster with a smaller secondary contribution from the Local Group. This can be seen in Fig. 12.8 from Shaya *et al.* (2017) who applied the action method to galaxies in the region.

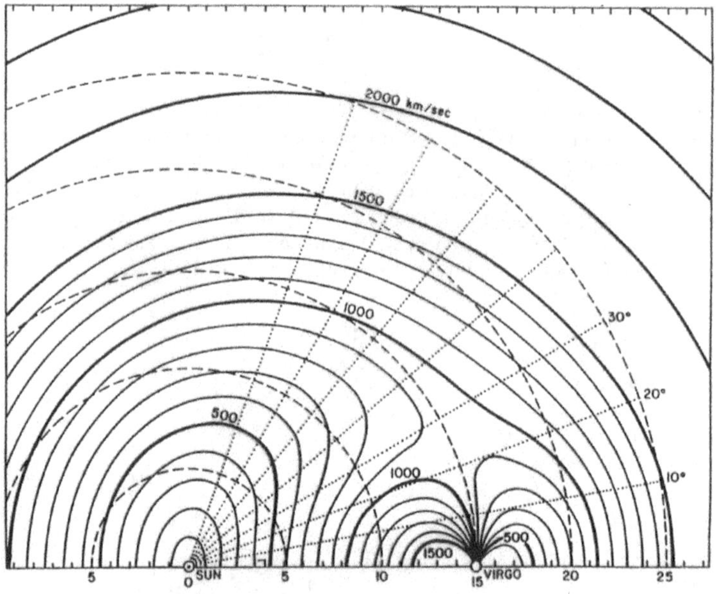

Fig. 12.7. Illustrating the Virgocentric infall. This shows the anisotropy of redshifts along lines of sight from the sun generated by the mass concentration of the cluster. Dashed circles are loci of constant redshift in an isotropic expansion assuming $V = H_0 D$ centered on the sun. Solid curves show the resulting deviations from the uniform expansion produced by the gravity of the Virgo cluster. (*Source*: Tonry J. L., Davis M. (1991), *ApJ*, 246, 666, "Velocity dispersions of elliptical and S0 galaxies. II. Infall of the local group to Virgo," Fig. 1, ©AAS Reproduced with permission.)

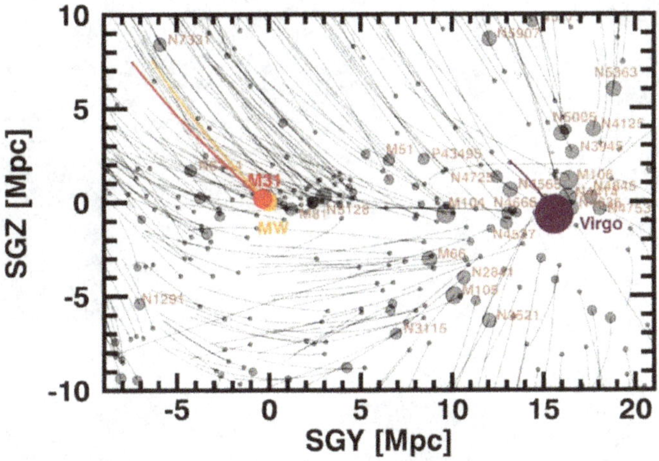

Fig. 12.8. Trajectories of galaxies in the Local Supercluster region, from the numerical action method, between the Local Group and the Virgo cluster, showing the flow of galaxies towards Virgo. The coordinates are supergalactic coordinates with a thickness of $-10 < \text{SGX} < +10\,\text{Mpc}$. (*Source*: Shaya E. J., Tully R. B., Hoffman Y., Pomarde D. (2017) *ApJ*, 850, 207. "Action dynamics of the local supercluster," Fig. 13, ©AAS Reproduced with permission.)

Extending peculiar motion studies to larger distances in the 1980s and 1990s, one realized motions directed towards larger more distant masses or attractors that were not known at the time.

Chapter 13

Measurements and Analysis Methods of Distant Peculiar Velocities

The peculiar motions measured for galaxies in the nearby universe are influenced by the net effect of additional structures and attractors. The measured directions of the dipole or bulk flow component of their motions varies with distance and is affected by the matter distribution. The CMB dipole, illustrated in Fig. 1.2, arises near a redshift $cz \approx 1100$ and results from a larger spacial volume than for the nearer galaxies and should result from all attracting matter within the last scattering surface. The observable galaxies define different dipoles. Inspection of Fig. 13.1 reveals this asymmetry of the motions in the $V_{\rm pec}$ of nearby Sc spiral galaxies. Although the direction and velocity of the flow nearby changes with distance, it should be expected to converge to the CMB value with growing distance. This contains cosmological information and is related to the power spectrum and other cosmological parameters. ΛCDM and other theories make definite predictions about the dipoles' behavior. Deviations from this would infer modifications to the cosmology, as some authors have claimed to have found, or even need for new physics.

It is assumed that the distance indicators, after proper corrections and calibration, measure the true distances to the galaxies. As already discussed, this appears to be a reasonable assumption that can be tested. The general models rely on the assumptions of the N-body simulations. Usually the FRW ΛCDM type of models, already described in Chapters 3 and 8, provide the background cosmology and sets of the parameters, e.g., H_0, Λ, and Ω are specified and extensively employed to interpret observations.

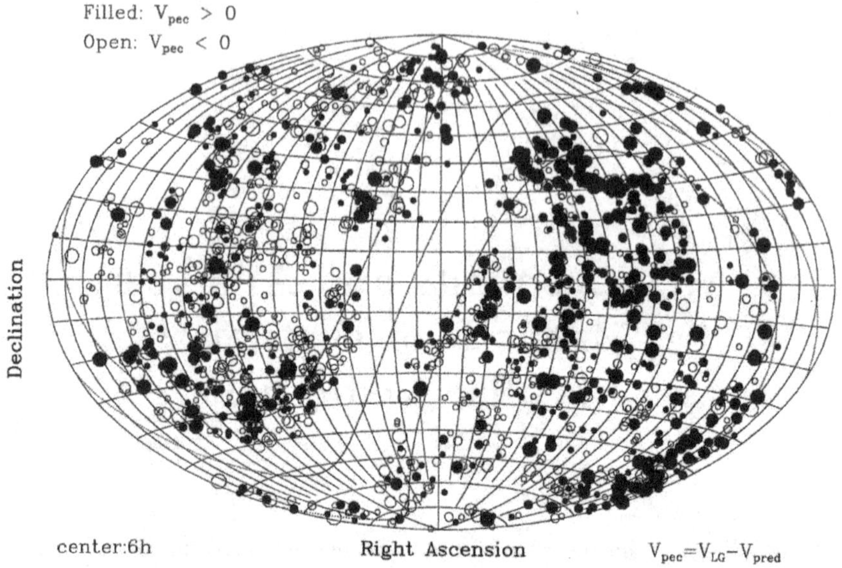

Fig. 13.1. Illustrating the asymmetry of peculiar motions on the celestial sphere relative to the Local Group from the SFI data due to the dipole motion. Filled circles denote $V_{\text{pec}} > 0$ and open circles, $V_{\text{pec}} < 0$. The sizes of the dots are proportional to the magnitude of V_{pec}. Negative peculiar velocities predominate on the left side and positive values dominate the right. The curves represent the position of the zone of avoidance in the galactic plane. (*Source*: Giovanelli R., Haynes M. P., Chamaraux P., da Costa L. N., Freudling W., Salzer J. J., Wegner G. (1996), *IAUS*, 168, 183, in *Examining the Big Bang and Diffuse Background Radiations, Proceedings from IAU Symposium 168.* Minas C. Kafatos, and Yoji Kondo, Kluwer Academic Publishers, Dordrecht. Meeting held in the Hague, The Netherlands, August 23–26, 1994, p. 183, "Spiral galaxies and the peculiar velocity field,", Fig. 1. Reprinted with permission by Springer Nature.)

Based on this model, the large-scale structure grows hierarchically by gravitational instability from a Gaussian random field of small fluctuations, possibly the result of an inflation phase that produced a scale invariant power spectrum, $P \propto k^n$ with $n \approx +1.0$, but other mechanisms have been suggested. The CMB defines the cosmological rest frame and the spectrum of density fluctuations relative to $\bar{\rho}$ the mean cosmological density, $\delta = (\rho - \rho_0)/\rho_0$, where ρ_0 is the mean cosmological density, grows with time during the radiation dominated era and depends on the type of dark matter (DM), which constitutes most of the mass density. This could be baryonic or non-baryonic, 'hot', 'cold', or other DM. As already explained, hot DM is generally ruled out as a major contributor, although warm and fuzzy DM forms have been proposed. The CMB and BAO constrain the amounts of these constituents with $\Omega_{\text{bar}} \approx 0.05$ and $\Omega_{\text{DM}} \approx 0.30$. In general

Ω depends on all the mass–energy components, as well if curvature, generally assumed to be not present, and $\Omega = \Omega_m + \Omega_R + \Omega_\Lambda$, where Ω_R is the radiation density, Ω_Λ, depends on the dark energy, and the matter density is $\Omega_m = \Omega_{\rm DM} + \Omega_{\rm bar} + \cdots$.

In this picture density fluctuations are initially weak, ($\delta^2 \ll 1$), and a linear velocity field is generated, which is expected to be irrotational. The observed galaxies' motions trace this velocity field with a small velocity bias factor related to the density bias factor. This results because the concentration of the galaxies does not exactly follow the clustering of the dominant dark-matter density field, or the bias $b = \delta_{\rm gal}/\delta$ is not precisely one (e.g., Desjacques, Jeong, & Schmidt, 2018).

13.1. Gravitational Instability

In regions where the density fluctuations grow with time, motions generated by gravitational instability can be described by linear and quasi-linear approximations. The velocities v and density fluctuations, δ can be related by the three equations (the continuity equation, Euler's motion equation for a fluid, and Poisson's equation) describing their evolution in a cosmologically expanding pressure less gravitating fluid defining the present scale factor $S \equiv 1$.

$$\begin{aligned} \dot{\delta} + \nabla \cdot v + \nabla \cdot (v\delta) &= 0, \\ \dot{v} + 2Hv + (v \cdot \nabla)v &= -\nabla\Phi, \\ \nabla^2\Phi &= \frac{3}{2}H^2\Omega\delta, \end{aligned} \tag{13.1}$$

where x and Φ are the position and gravitational potential; $H(t)$ and $\Omega(t)$ are the Hubble and cosmological density parameters.

In the linear case, these equations can be combined (Peebles, 1980, 1993) and yields

$$\ddot{\delta} + 2H\dot{\delta} = 4\pi G\bar{\rho}\delta. \tag{13.2}$$

More generally, the density can be replaced by $\bar{\epsilon}/c^2$, which depends on $w = p/\rho$ where, e.g., $w = \frac{1}{3}$ (radiation), w $= -1$ (Λ only, etc.) from which limiting cases are

$$\begin{aligned} \ddot{\delta} + 2H\dot{\delta} - \frac{3}{2}\Omega_m H^2\delta &= 0 \,(\text{matter only}), \\ \ddot{\delta} + \frac{\dot{\delta}}{t} &\approx 0 \,(\text{radiation only}), \\ \ddot{\delta} + 2H\dot{\delta} &\approx 0 \,(\Lambda \text{ only}). \end{aligned} \tag{13.3}$$

This second-order ordinary differential equation (13.2) has two solutions, the growing and decaying mode,

$$\delta(x,t) = \delta_1 D_1(t) + \delta_2 D_2(t). \tag{13.4}$$

The solution for the matter

$$\delta_m(t) \approx D_1 t^{2/3} + D_2/t, \tag{13.5}$$

with the growing mode is of interest, where

$$f(\Omega) = H^{-1}\dot{D}/D = \frac{d\ln D}{d\ln S} \approx \Omega^{0.6}, \tag{13.6}$$

which is Peebles' (1980) approximation for the dimensionless growth rate, D linear growth function and S is the scale factor. This exponent is insensitive over a range of cosmological parameters. Some investigators prefer the approximation, $\Omega^{0.55}$ (Linder, 2005). This will be discussed further on. In general, if one writes

$$f = \Omega^{\gamma}, \tag{13.7}$$

different theories of gravity predict different γ, but all are near 0.6. For the linear case, unlike high density regions such as galaxy clusters that collapse and form virialized systems that become strongly nonlinear and have erased any information of their initial conditions, the relation between density and velocity is

$$\delta = \delta_0 \equiv -(Hf)^{-1}\nabla \cdot v. \tag{13.8}$$

Dekel (1994) and others explain that this relation is reliable within the range of most v_{pec} samples (10 Mpc to 100 Mpc), but for large $|\delta|$ it breaks down. In the quasi-linear regime, interesting information results from the mild nonlinear effects. As in the N-body calculations (Chapter 3), the Zel'dovich (1970) approximation can be utilized, where the initial Lagrangian positions q are related to their Eulerian positions x and the growth factor at time t,

$$x(q,t) - q = D(t)\xi(q) = f^{-1}v(q,t), \tag{13.9}$$

which gives

$$q(x) = x - f^{-1}v(x), \tag{13.10}$$

as an approximation, if the flow motion is laminar. For a nearby galaxy, the peculiar velocity, $\mathbf{V}$, will depend on the density field, as discussed in the previous chapter.

13.2. Bulk Flows and Dipoles

As the large-scale structure grows according to the gravitational instability picture, gravity generates peculiar velocities. These are governed by the resulting power spectrum, $P(k)$, of the density fluctuations, $\delta\rho$. This and the direction of the dipole of motion compared to that of the CMB provide cosmological tests. For the linear case where again $\delta(r) = [\rho(r)-\rho_0]/\rho_0 \ll 1$ where ρ_0 is the mean mass density. Utilizing Chapter 12, the observed redshift of a galaxy along the line of sight will be

$$cz = H_o r + V_{\text{pec}} - V_O, \tag{13.11}$$

where $H_0 r$ is the expansion, V_{pec} is its peculiar velocity and V_O is the observer's peculiar velocity. The peculiar velocity is driven by gravitational attraction so

$$V_{\text{pec}} = \frac{H_0 f(\Omega_m)}{4\pi} \int d^3 r' \delta(r') \frac{r' - r}{|r' - r|^3}. \tag{13.12}$$

In Eq. (13.12) it can be seen that given H_0, the growth rate $f(\Omega_m) = \Omega_m(z)^\gamma$ defines the velocity scale. Instead of Ω_m being measured directly, it can be determined from the galaxy distribution but needs correction for other unseen mass components, e.g., DM. The density is then

$$\rho_m/\rho_0 = \Omega_m/b, \tag{13.13}$$

where b is the *linear biasing parameter*, the ratio of the galaxy fluctuations to the total mass fluctuations. This b cannot be measured directly from peculiar velocities, but as discussed by many investigators is of order one. Gaztañaga *et al.* (2005) used the three-point correlation function from the 2dF redshift survey and found $b = 0.93^{+0.10}_{-0.08}$. Westover (2007) determined that this depends on galaxy luminosity and obtained

$$b/b_* = (0.73 \pm 0.07) + (0.24 \pm 0.04)L/L_*,$$

where b_* is the bias of a L_* galaxy. The leading term in Eq. (13.12) is then written $H_0\beta/4\pi$ where the parameter $\beta = \Omega_m^\gamma/b$. The bulk motion of galaxies in spherical volumes of radius R surrounding the sun can also be determined from samples of galaxies.

In linear theory, the peculiar velocity of the Local Group, V_{pec} produced by the net attraction of all mass inhomogeneities, $\delta(r)$ within this sphere depends on the power spectrum and the one-dimensional rms produced will be (e.g., Borgani *et al.*, 1997; Watkins *et al.*, 2009)

$$V_{\rm rms} = \frac{H_0\Omega^\gamma}{\sqrt{3}}\left[\frac{1}{2\pi}\int_0^\infty dkP(k)W^2(kR)\right]^{\frac{1}{2}}, \tag{13.14}$$

where H_0 is the Hubble constant, Ω_0 is the cosmological density parameter, and $W(r,R)$ is a 'window function'. Different window functions have been employed, which have a width of R; two popular examples of the simplest ones are as follows:

$$W(r,R) = \begin{cases} 1, & |r| \le R\,\text{top hat}, \\ 0, & |r| > R, \end{cases} \tag{13.15}$$

$$W(r,R) = \exp\left(-\frac{r^2}{2R^2}\right)\ \text{Gaussian}. \tag{13.16}$$

Macaulay *et al.* (2011, 2012) have discussed the window function in detail.

In a homogeneous universe filled with density fluctuations, the apex direction and velocity of the bulk motion will bounce around with distance, but they are expected to converge to the CMB dipole values at sufficiently large distance, i.e., $\mathbf{V}_{\rm pec} \to \mathbf{V}_{\rm CMB}$ as $R \to \infty$. The distance where $|\mathbf{V}_{\rm pec} - \mathbf{V}_{\rm CMB}|$ grows small is referred to as the 'convergence depth', $d_{\rm conv}$, which can be computed using N-body simulation data and compared with observation (Fig. 13.2).

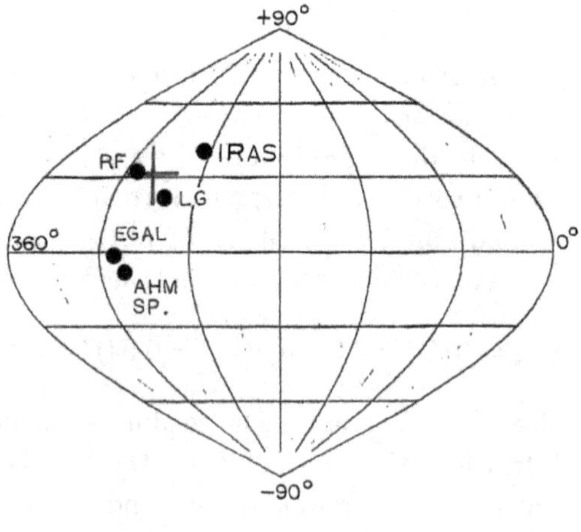

Fig. 13.2. Directions of the dipole from some early determinations of peculiar motions on the celestial sphere in galactic coordinates with respect to the CMB dipole. RF: Rubin *et al.* (1976b), LG = Local Group: (Fixen *et al.*, 1983; Lubin *et al.*, 1985), EGAL: Lynden-Bell *et al.* (1988), AHM: Aaronson *et al.* (1982), IRAS dipole (Yahil *et al.*, 1986), Red cross: CMB (Kogut *et al.*, 1993).

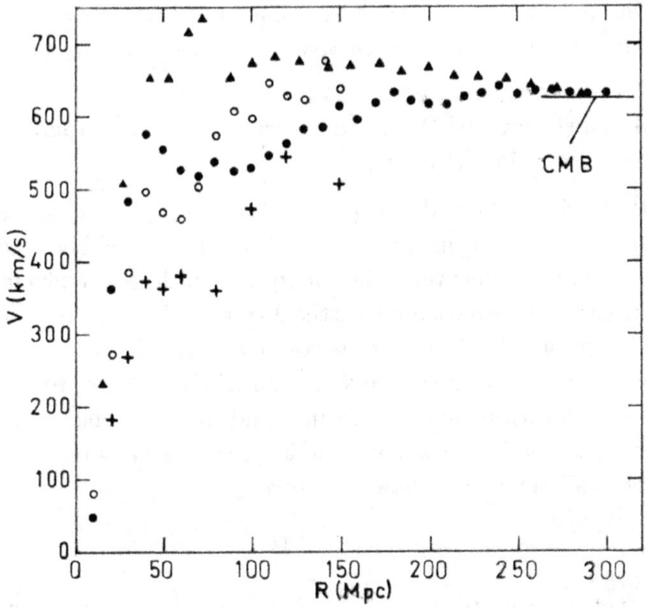

Fig. 13.3. Dipole amplitude velocities at different depths compared to the CMB value for different determinations: ○ Lawrence *et al.* (1999), ● Rowan-Robinson *et al.* (2000), ▲ Ergoğu *et al.* (2006), + Levaux *et al.* (2010). See the original references for the cosmological parameters employed.

Some determinations of the dipole velocity amplitude are shown in Fig. 13.3, where it settles down to near the CMB value beyond $R \sim 200\,\mathrm{Mpc}$. Giovanelli *et al.* (1998) concluded from the SFI data that the apex of the LG motion settles to the CMB dipole by $\sim 3500\,\mathrm{km\,s^{-1}}$. Dale *et al.* (1999a) use peculiar motions and find $d_{\mathrm{conv}} \approx 200h^{-1}\,\mathrm{Mpc}$. Sandage, Reindl, & Tammann (2010) use a combination of several different sources and conclude that most of the dipole motion is produced by matter within $5000\,\mathrm{km\,s^{-1}}$.

Summarizing this section, comparing the CMB and bulk motion dipoles provides a test for ΛCDM, etc. As seen further on, this continues to be the subject of study. The CMB background dipole moment is usually interpreted to be due to the Doppler shift of the sun's motion in the comoving reference frame. Its direction is influenced by the sun's net components of motion from galactic rotation ($220\,\mathrm{km\,s^{-1}}$), the galaxy's relatively small motion in the Local Group and the redshift of the Virgo cluster ($\sim 1100\,\mathrm{km\,s^{-1}}$). The desired parameters to be measured for the bulk motion and its dipole are the parameters β, σ_8, and b plus whether these can be explained by the observed matter distribution.

13.3. Measurement of the Apex Direction and Velocity from Peculiar Motions

Given the density field and the peculiar velocities, the question is how to measure their properties and compare them with N-body simulation modelling. Using peculiar velocities, or equivalent data, the two most basic parameters that can be measured are the direction, or apex, and the velocity of the bulk motion, **B**. These can be compared with the CMB dipole apex and velocity. One expects the bulk motion of the galaxies to converge to the CMB values with increasing distance and this is a fundamental test of the cosmological models producing the large-scale structure. The simplest model is that all velocities follow a vector,

$$\mathbf{V}(\mathbf{x}) = \mathbf{B}. \tag{13.17}$$

Analyzing velocities in spherical shells of different radii provides information on the flows of the galaxies.

So far, it appears that there is no significant quadrupole component to the peculiar velocity field. A strong detection of it might indicate the existence of large unseen attractors or different physics. Willick & Strauss (1998) found a weak (~10%) quadrupole, while more recently with larger measurements Pike & Hudson (2005) found no evidence for it.

From the peculiar velocities, V_{pec}, finding the location of the apex precedes much the same as the determination of the solar apex from stellar radial velocities (e.g., Smart, 1962), but including the expansion. If there is an overall anisotropy of the galaxies' motions, they will be directed away from a spot on the sky, the 'apex' at (α_0, δ_0), at a velocity $-\Delta V_B$ and converge toward an opposite place on the celestial sphere, the 'antapex' with velocity $+\Delta V_B$. The observed velocity of a galaxy along the line of sight will be

$$V_c = V_H - \Delta V_B + V_{\text{pec}}, \tag{13.18}$$

where V_H is the Hubble expansion, and

$$-\Delta V_B = V_B \cos\vartheta, \tag{13.19}$$

and ϑ is the angular distance of the galaxy from the apex.

For a galaxy at distance d in a rectangular coordinate system centred on the sun, in spherical coordinates, this is

$$\begin{aligned} x &= d\cos\delta\cos\alpha, \\ y &= d\cos\delta\sin\alpha, \\ z &= d\sin\delta. \end{aligned} \tag{13.20}$$

Differentiate with time to get the velocities,

$$\begin{aligned} \dot{x} &= \dot{d}\cos\delta\cos\alpha - \dot{\delta}d\sin\delta\cos\alpha - \dot{\alpha}d\cos\delta\sin\alpha, \\ \dot{y} &= \dot{d}\cos\delta\sin\alpha - \dot{\delta}d\sin\delta\sin\alpha - \dot{\alpha}d\cos\delta\cos\alpha, \\ \dot{z} &= \dot{d}\sin\delta + \dot{\delta}d\cos\delta. \end{aligned} \tag{13.21}$$

Here $\dot{d} = -\Delta V_B$. The $\dot{\alpha}$ and $\dot{\delta}$ correspond to the transverse or proper motions, which for these distances, are unavailable, so this gives for the velocities,

$$\begin{aligned} -X &= \Delta V_B \cos\alpha_0 \cos\delta_0, \\ -Y &= \Delta V_B \sin\alpha_0 \sin\delta_0, \\ -Z &= \Delta V_B \sin\delta_0. \end{aligned} \tag{13.22}$$

Solving for ΔV_B, gives

$$-\Delta V_B = aX + bY + cZ, \tag{13.23}$$

for each observed galaxy, where $a = \cos\delta\cos\alpha$, $b = \cos\delta\sin\alpha$, and $c = \sin\delta$. This system of equations is overdetermined, but it can be solved using the method of least squares or more sophisticated statistical methods.

For least squares, it is easiest to see how the solution is found. For each galaxy, i, this gives the three normal equations,

$$\begin{aligned} X\sum_i a_i^2 + Y\sum_i a_i b_i + Z\sum_i a_i c_i &= -\Delta V_B \sum_i a_i, \\ X\sum_i a_i b_i + Y\sum_i b_i^2 + Z\sum_i b_i c_i &= -\Delta V_B \sum_i b_i, \\ X\sum_i a_i c_i + Y\sum_i b_i c_i + Z\sum_i c_i^2 &= -\Delta V_B \sum_i c_1. \end{aligned} \tag{13.24}$$

Having obtained X, Y, and Z, the speed of the dipole motion is

$$V_B = \sqrt{X^2 + Y^2 + Z^2}, \tag{13.25}$$

and the apex direction is

$$\tan \alpha_A = \frac{Y}{X}, \tag{13.26}$$

$$\tan \delta_A = \frac{Z}{\sqrt{X^2 + Y^2}}. \tag{13.27}$$

Rubin *et al.* (1973; 1976a,b) determined the direction of the galaxy's motion. Redshifts of 96 galaxies in an all-sky sample of Sc I and Sc II galaxies were adopted. The apparent magnitude corrected for the inclination and linear diameter of the galaxies was used for the distance indicator. Reddening was corrected using the cosecant law and the Malmquist bias taken into account. Some early investigators considered observed bulk flows to be produced by observational errors, but some investigations at about this time seemed to confirm their reality, e.g., Visvanathan (1979) made new measurements and Clutton-Brock & Peebles (1981) discussed how density fluctuations could produce velocities of this magnitude. Subsequently, several measurements of the dipole were carried out using data as explained below. Some of these are shown in Fig. 13.2 and can be compared with more modern determinations discussed in Chapter 20. Although early determinations of bulk flow portrayed a rough agreement, they were not of high precision.

In order to determine the dipole motion realistically, the main problems include proper handling of the sources of error and noise in the observations and the biases. These vary according to the combination of Method I or Method II and the forward or inverse approach to the distances discussed in Chapter 9. Several combinations of these have been devised and further details discussed by Davis & Scrimgeour (2014), Watkins & Feldman (2015b), Nusser (2016). Some of the main methods follow.

13.3.1. χ^2 *minimization*

Used by Aaronson *et al.* (1982), for the bulk flow, the predicted redshift will be $cz_{\text{pred}} = \mathbf{B} \cdot \hat{\mathbf{r}}_i$ where $\hat{\mathbf{r}}_i$ is the unit vector in the ith galaxy's direction. From this, the difference in the observed and predicted distances will be $d_{\text{pred}} = (cz_i - cz_{\text{pred},i})h^{-1}/100\,\text{Mpc}$. For N observed galaxies, one minimizes,

$$\chi^2 = \sum_i^N \frac{(cz_{\text{obs}} - cz_{\text{pred}})^2}{2(\sigma_d^2 - \sigma_v^2)}, \tag{13.28}$$

with σ_d and σ_v the errors in the distance and observed line-of-sight velocity, cz.

13.3.2. *Maximum likelihood estimator (MLE)*

To model the errors in more detail, Kaiser (1988) treated the data as a convolution of a tensor window and a peculiar velocity field. Sarkar, Feldman, & Watkins (2007) also describe this method to determine the apex of the dipole. The flow components are U_i with observed errors σ_i. To this a random error, σ_* is added. The observed errors are assumed to be Gaussian and the likelihood function for the flow components is

$$L(U_i) = \prod_n \frac{1}{\sqrt{\sigma_n^2 + \sigma_*^2}} \exp\left[-\frac{1}{2}\frac{(S_n - \hat{r}_{n,i}U_i)^2}{\sigma_n^2 + \sigma_*^2}\right], \tag{13.29}$$

where $S_n = \hat{r}_n v_i(r_n) + \epsilon_n$, v_i is the observed velocity and ϵ_n is down from Gaussian of variance $\sigma_n^2 + \sigma_*^2$ of zero mean. The unit vector towards the galaxy is $\hat{\mathbf{r}}_\mathbf{n}$. The solution then has the form

$$U_i = A_{i,j}^{-1} \sum_n \frac{\hat{r}_{n,j} S_n}{\sigma_n^2 + \sigma_*^2}, \tag{13.30}$$

where

$$A_{i,j} = \sum_n \frac{\hat{r}_{n,i}\hat{r}_{n,j}}{\sigma_n^2 + \sigma_*^2}. \tag{13.31}$$

13.3.3. *Minimum variance method (MV)*

Watkins, Feldman, & Hudson (2009) and Feldman, Watkins, & Hudson (2010) describe the MV method in detail. It is a generalization of the MLE method and has been employed by Hong *et al.* (2014), Qin *et al.* (2018), and others. The method includes noise sources and cosmic variance. The window function estimate includes a model power spectrum, P, e.g., ΛCDM. Watkins *et al.* (2009) show that the components of the observed velocities and their relations to the weights can be expressed,

$$\sum_m (\langle S_n S_m \rangle) + \lambda P_{n,m} w_{i,m} = \langle S_n U_i \rangle, \tag{13.32}$$

where S_n, U_i are the same as for MLE above. In matrix form this is written,

$$(\mathbf{G} + \lambda \mathbf{P})\mathbf{w}_i = \mathbf{Q}_i, \tag{13.33}$$

where λ is a Lagrange multiplier to constrain the window function to 1/3 as $k \to 0$. Inverting this solves for the weights,

$$\mathbf{w}_i = (\mathbf{G} + \lambda \mathbf{P})^{-1} \mathbf{Q}_i. \tag{13.34}$$

Peery, Watkins, & Feldman (2018) describe an improved version of the MV method that should be more easily interpretable. Following Nusser (2014, 2016) as usual assuming an irrotational flow, the bulk flow depends on the radial component of the peculiar velocity. A constraint is added that eliminates dependence on the value of H_0 and a r^{-2} weighting scheme is employed.

13.4. IRAS All-Sky Galaxy Catalogue Measurements

A second method for determining the cosmic density field and the resulting peculiar velocities is to assume that light traces matter, based on the fact that both light and gravity obey the inverse square law with distance. An important early measurement of the dipole of the galaxies' motion was based on redshifts and the IRAS (Infrared Astronomical Satellite) operated in 1983 (Neugebauer *et al.*, 1984), covered 96% of the sky four times in four (12–100 μm) infrared photometric bands and reached faint enough to study galaxies within a distance of about 140 Mpc. About 250 thousand sources were catalogued from which several thousand galaxies can be separated from stars by their colours. The type of the galaxies detected turn out to contain dust and are brighter than expected from their appearances in the visual region. The 60 μm band was particularly useful from its low reddening effects, except near the galactic plane, after systematic effects in the data were eliminated, as discussed in the referenced papers. Several redshift surveys were subsequently conducted from this list of sources. These extensive data consequently provided an additional method of studying large-scale structure and peculiar motions (Strauss *et al.*, 1990; Rowan-Robinson *et al.*, 1991; Rowan-Robinson, 1993).

For the analysis, two assumptions are as follows: Firstly, light traces mass so that the usual density enhancement is

$$\delta(r) = [\rho(r) - \bar{\rho}]/\bar{\rho}, \tag{13.35}$$

where $\bar{\rho}$ is the mean density of the universe, and related to the total density enhancement by

$$(\delta\rho)_{\text{gal}} = b(\delta\rho)_{\text{total}}, \tag{13.36}$$

where b is the bias, which can be estimated various ways and is approximately one.

Secondly, a universal luminosity function must be assumed. For the IRAS galaxies, a two-power luminosity function gives a better fit than Schecter's (1976) because of a strong curvature and change in slope,

$$\eta(L) = CL^{-1}\left[1 + \frac{L}{\beta L_*}\right]^{-\beta} dL, \tag{13.37}$$

where $\log C \approx 7.12$) and $\beta \approx 2.4$, and $\log L_* \approx 11.3$ (Lawrence *et al.*, 1986).

Early analyses of the IRAS galaxies utilized only the fluxes and positions. Yahil, Walker, & Rowan-Robinson (1986) and Meiksin & Davis (1986) used these to measure the dipole anisotropy for the IRAS galaxies and compare it to the CMB. The dipole is obtained by fitting the light on the sky using spherical harmonics, similarly to the CMB. Since grav $\sim r^{-2}$ and light $\sim r^{-2}$, the light along the line of sight also measures gravity. The number of galaxies in the luminosity interval dL is then,

$$dN = \delta(r)d^3r\eta(L)dL, \tag{13.38}$$

and the smoothed surface brightness, σ, is related,

$$4\pi\sigma(S,\omega) = 4\pi S\frac{dN}{dSd\omega} = \int \delta(r)dr \int L\eta(L)\delta[S - (L/4\pi r^2)]dL. \tag{13.39}$$

From this, the peculiar velocity can be estimated using

$$V(r) \approx \frac{H_0\Omega_0^{0.6}}{4\pi}\int \frac{\delta(r')(r-r')dr'}{|r-r'|^3}. \tag{13.40}$$

Although in principle it is possible to invert these equations to get distance and velocity data, redshifts are needed. Combining the IRAS galaxy catalogues with redshift surveys provide a method of determining large-scale structure and peculiar velocities.

Figure 13.4 shows results for IRAS galaxies in the QDOT and PSCZ surveys with redshifts. The two largest concentrations in the direction of the Great Attractor and Perseus–Pisces are seen.

13.5. The POTENT Method for Density Reconstruction

Some of the first reconstructions of the peculiar velocity field used redshift surveys and assuming linear theory, iteratively solved for the gravity field (e.g., Yahil, 1988, Strauss & Davis, 1988; Hudson, 1994b). As in all peculiar motion analyses of this type, the analysis is made along the radial line

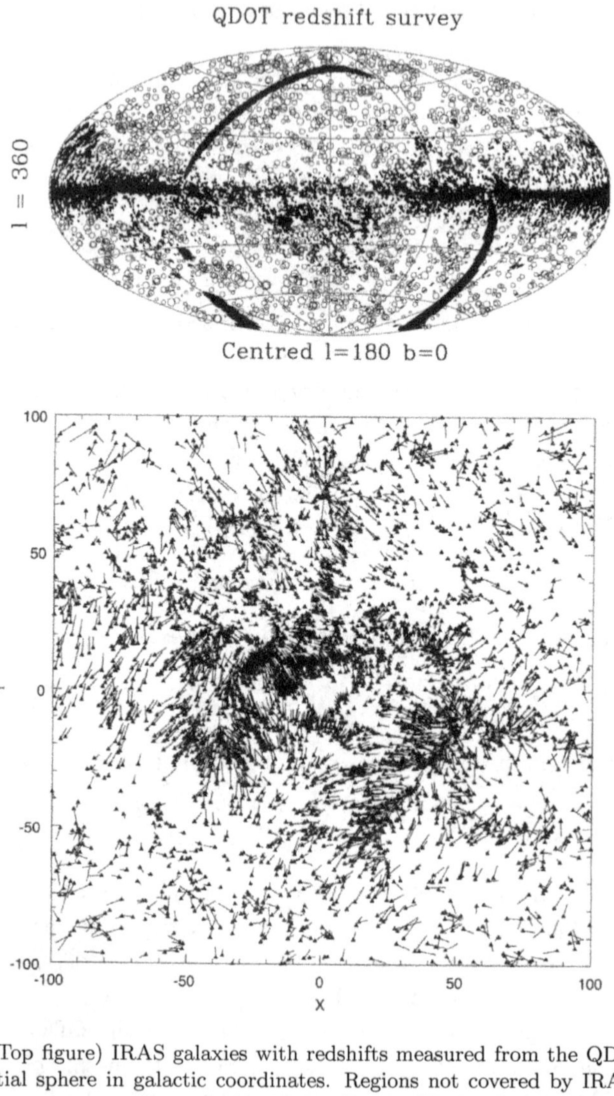

Fig. 13.4. (Top figure) IRAS galaxies with redshifts measured from the QDOT survey on the celestial sphere in galactic coordinates. Regions not covered by IRAS are seen as the long narrow arcs. The galactic plane where reddening is heavy leaving many galaxies out of the sample can be seen as the dark equatorial band. (*Source*: Lawrence A. *et al.* (1999), *MNRAS*, 308, 897, "The QDOT all-sky IRAS galaxy redshift survey," Fig. 1, Reproduced by permission of the Oxford University Press on behalf of the Royal Astronomical Society.) (Lower figure) The galaxy motions predicted from the PSCz survey of the IRAS data plotted for galaxies within ±22.4°of the supergalactic plane in the CMB frame. The concentrations in the directions of Perseus–Pisces are seen to the lower right and the GA upwards left of centre. (*Source*: Rowan-Robinson M. *et al.* (2000), *MNRAS*, 314, 375, "The IRAS PSCz dipole," Fig. 14, Reproduced by permission of the Oxford University Press on behalf of the Royal Astronomical Society.)

of sight. A more sophisticated treatment of the peculiar motion data was carried out using the POTENT method developed by Bertschinger & Dekel (1989); Dekel (1994), and Dekel, Bertschinger, & Faber (1990). This is a distance–distance technique, which uses the direct redshift independent distances and peculiar velocities. It allows recovery of the density field from the peculiar velocities. Starting with the radial component of the peculiar velocity and assuming the flows are irrotational ($\nabla \times \mathbf{v} = 0$),

$$V_r = cz - H_0 r, \tag{13.41}$$

and since the velocity is related to the potential,

$$V = -\nabla\Phi, \tag{13.42}$$

this can be solved for Φ

$$\Phi = -\int V_r dr. \tag{13.43}$$

If linear theory is valid, Poisson's equation gives the density:

$$\nabla^2\Phi = -4\pi G\rho. \tag{13.44}$$

Several versions of POTENT have been employed and although this procedure is straightforward, in practice one encounters complications that also appear in other peculiar velocity analysis methods. The most severe are smoothing the data and handling sources of noise. Ideally, one wishes to have a uniform smooth catalogue of galaxy motions, but this is not true for the observed sample, which also has regions of sparse or missing data such as in the highly obscuring galactic plane. For this reason, a tensor window function is employed in POTENT. This takes into account that the velocities have different directions. Assuming a model for the velocity field, $\mathbf{v}(\alpha_k, \mathbf{x})$, the free parameters can be obtained from a maximum likelihood fit,

$$-\log L \propto \sum_i W_i[\mathbf{u_i} - \hat{\mathbf{x}}_\mathbf{i} \cdot \mathbf{v}(\alpha_\mathbf{k}, \mathbf{x_i})]^2, \tag{13.45}$$

where W_i is a window function, the simplest model being the bulk flow model $\mathbf{v}(\mathbf{x}) = \mathbf{B}$.

The direct distance indicators are used, which requires correction for the Malmquist bias, but in addition, a window-bias and a sampling-bias must be accounted for.

Figure 13.5 shows the resulting density and flow direction reconstructions from Dekel (1994) in the supergalactic plane centered on the sun's position from POTENT. Most notable are the PP and GA regions, which also predominate in other subsequent analyses.

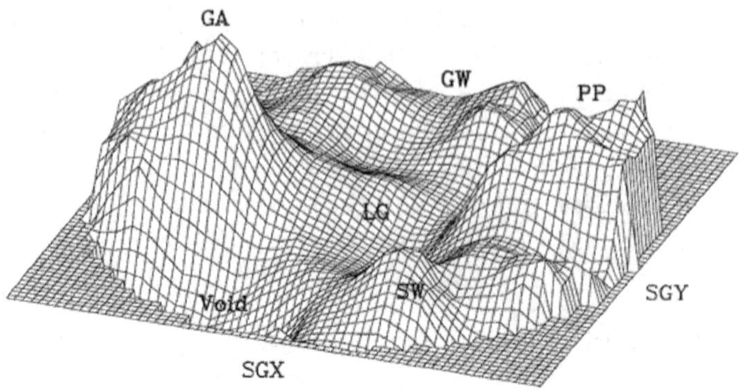

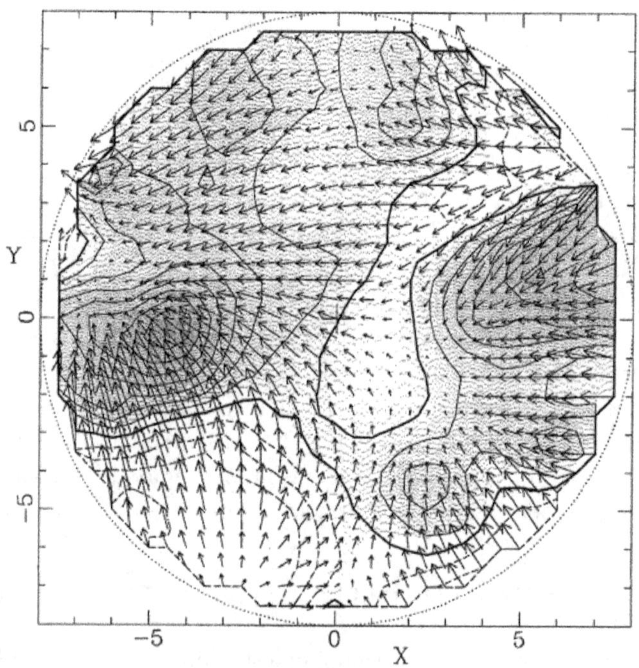

Fig. 13.5. POTENT reconstruction of densities and velocities in the supergalactic plane. Top: densities are denoted by height above the plane. GA and PP mark the Great Attractor and Perseus–Pisces regions, respectively. Other regions are: LG, the local group, GW, the great wall, etc. Bottom: arrows denote velocity vectors and corresponding densities are shaded. (*Source*: Dekel A. (1994) *ARA&A*, 32, 371, "Dynamics of cosmic flows," Fig. 3, ©1994 Annual Reviews www.annualreviews.org. Reproduced with permission.)

13.6. Wiener Filter Method

The Wiener filter technique of reconstruction works well for noisy all-sky surveys and has been utilized heavily in many recent studies. Several descriptions of this method can be found. An early result was by Kaiser & Stebbins (1991) and a more detailed account is Fisher *et al.* (1995). The density field is expanded using spherical harmonics. A Bayesian method is utilized that assumes a 'prior' model of the galaxy distribution. Using this, the statistical noise is evaluated, which can be processed using the Wiener filter and minimizes the variance between the derived field and the underlying true field. For astronomical applications, Rybicki & Press (1992) have reviewed the Wiener filter.

Dekel (2000) compared the reconstructed density distributions from POTENT and the Wiener filter methods using different databases. It can be seen that qualitative agreement is achieved, while several variances are also revealed in the final details. Later investigators have improved these maps using larger new peculiar velocity measurements, but some differences remain.

13.7. VELMOD Method

This is also referred to as the velocity–velocity or v–v method. The mass–density field is determined using redshifts. This is used to predict the velocity field and compared with observed peculiar velocities. VELMOD employs a parametric model for the selection function, luminosity function, and the distance indicator, which are evaluated by maximum likelihood. Both forward VELMOD and inverse VELMOD can be used (e.g., Davis, Nusser, & Willick, 1996; Willick & Strauss, 1998). Branchini *et al.* (2001) used this to compare SFI peculiar velocities with the PSCz gravity field within $cz = 6000\,\mathrm{km\,s^{-1}}$and found $\beta = 0.42 \pm 0.04$ within $cz = 4000\,\mathrm{km\,s^{-1}}$.

13.8. Additional Methods to Reconstruct Initial Conditions

13.8.1. *MAK method*

More recently, several other methods to reconstruct initial conditions that can also follow galaxy trajectories have been devised. Mohayee *et al.* (2006) apply the MAK (Monge–Ampère–Kantorovich) method. This starts with a constraint,

$$\rho(x)dx = \rho_0(q)dq, \tag{13.46}$$

where ρ and x are the current density and coordinate and ρ_0 and q are the initial values. For this, a quadratic cost function is found to minimize,

$$I = \int_q \rho_0(q)|x - q|^2 d^3q = \int_x \rho(x)|x - q|^2 d^3x. \tag{13.47}$$

The peculiar velocities and positions are then calculated using the Zel'dovich approximation described in Chapter 3.

Additional developments have been described, e.g., Jasche & Wandelt (2013) use Bayesian reconstruction with second-order Lagrangian perturbation theory. Schmittfull, Baldauf, & Zaldarriaga (2017) utilize an iterative Lagrangian method. The larger databases of the future methods of these types will be used to study large-scale structure.

13.8.2. *BBGKY analysis*

Juszkiewicz *et al.* (2000) have demonstrated how to use the relative velocities of pairs of galaxies, based on Bogoliubov–Born–Green–Kirkwood–Yvon (BBGKY) hierarchy in kinetic theory (Davis & Peebles, 1977; Peebles, 1980; Binney & Tremaine, 2008). Defining a two-particle distribution function, the two-particle correlation function is found, which relates the velocity of a pair, v_{12} and the two-point correlation function, $\xi(r)$:

$$v_{12} = -\frac{2}{3} H r \Omega^{0.6} \bar{\bar{\xi}}[1 + \alpha \bar{\bar{\xi}}(r)], \tag{13.48}$$

with

$$\bar{\bar{\xi}}(r) = \frac{3\int_0^r \xi(x) x^2 dx}{r^3[1 + \xi(r)]}. \tag{13.49}$$

Using the peculiar velocities in the Mark III database (Dekel *et al.*, 1999), a matter density of $\Omega_m = 0.35 \pm 0.15$ was found, which at the time was evidence for a low-density universe, and is consistent with current CMB results. Further analysis of peculiar motion surveys by Feldman *et al.* (2003) pointed out that this method is independent of any prior assumptions. This approach continues to have applications. Wang *et al.* (2018) have used it to investigate cosmic variance and Dupuy, Cortois, & Kubik (2019) have incorporated it to estimate the growth rate, $f(\Omega) = d\ln D/d\ln a \approx \Omega^\gamma$.

13.8.3. *Fisher matrix*

Utilization of the Fisher matrix method in astrophysics and cosmology stems from the field of Fisher information, a major concept in statistical inference, originally developed by R. A. Fisher early in the 20th century. It makes use of Bayesian inference and is being used extensively to analyze large datasets (e.g., Verde 2010; Coe 2009).

13.9. Redshift Space Distortions (RSD)

For sources where distances are unavailable, it is still possible to extract information from peculiar velocities and redshifts about β, the linear redshift distortion parameter. Peculiar velocity measurements between pairs of galaxies, $v_{12}(r) = c\Delta z$, separated by a distance r are related by the two-point correlation function of the mass–density fluctuations, $\xi(r)$, providing estimates of the parameters Ω_m and σ_8 (Juszkiewicz *et al.*, 2000). Feldman *et al.* (2003) found $\Omega_m = 0.30^{+0.17}_{-0.07}$ using this method (which translates to $\beta = 0.43$).

Utilizing redshift surveys, the peculiar velocities provide a similar measurable effect (Kaiser, 1987; Hamilton, 1998; Peacock *et al.*, 2001). In redshift space, the redshifts, along the line of sight, are distorted by the peculiar motions, but the perpendicular transverse motions on the plane of the sky are unaffected. This is related to the well-known 'finger of god' effect which is produced in the redshift distributions of galaxy clusters. The 'redshift distance' is

$$s = \frac{cz}{H_0} = D + \frac{V_{\text{pec}}}{H_0}, \tag{13.50}$$

and differs from the actual distance, D. Defining $\pi = c\Delta z/H_0$ as the radial pair separation and $\sigma = cz\Delta\theta$ as the transverse pair separation, one can determine a two-point correlation function. One such estimator that is employed is

$$\xi(\sigma, \pi) = \frac{\text{DD} - 2\text{DR} + \text{RR}}{\text{RR}}, \tag{13.51}$$

where DD is a normalized sum of weights of galaxy–galaxy pairs, RR is the same for random–random pairs in a random galaxy catalogue, and DR is the normalized sum of galaxy–random pairs. The shape of $\xi(\sigma, \pi)$ depends on the arrangement fo the peculiar velocities as shown in Fig. 13.6 for example models where the contours are circular if there is not distortion of the expansion, a collapsing region, which shows flattened contours, etc.

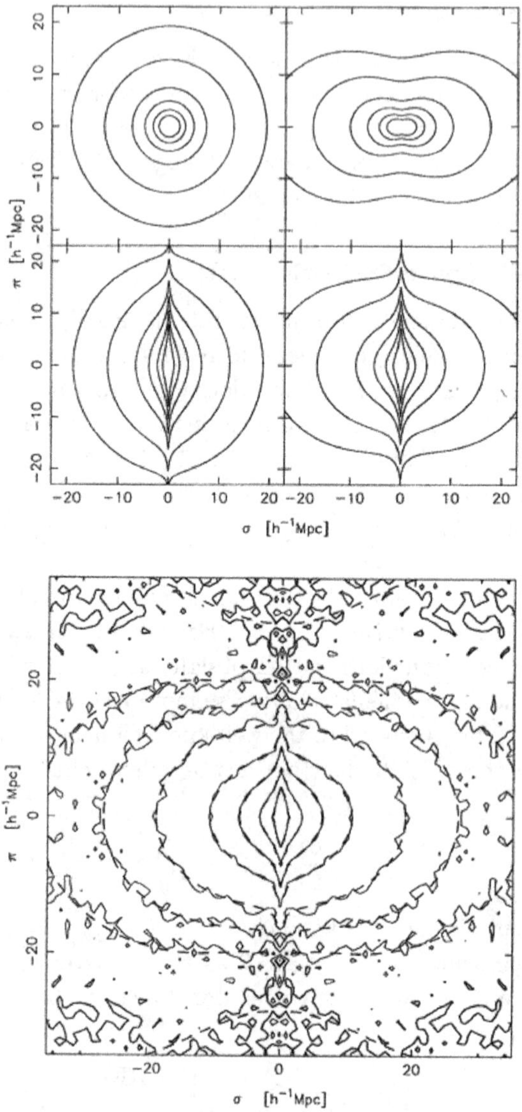

Fig. 13.6. Model and observed contours for the RSD. Top section: Model contours of constant $\xi(\sigma, \pi)$. Top left, the circular contours are for undisturbed expansion; top right is for a coherent infall region; bottom left is the pattern of purely random pairwise velocities; bottom right is for both random motions and infall. Bottom section: $\xi(\sigma, \pi)$ observed contours by the 2dF galaxy redshift survey and (dashed curves) the best fitted model. (*Source*: Hawkins E. and The 2dFGRS team (2003), *MNRAS*, 346, 78, "The 2dF Galaxy Redshift Survey: correlation functions, peculiar velocities and the matter density of the Universe," Figs. 12 and 22. Reproduced by permission of the Oxford University Press on behalf of the Royal Astronomical Society.)

The observed $\xi(\sigma, \pi)$ from the 2dF galaxy redshift survey (Hawkins *et al.*, 2003) show a mixture of effects from small scale 'fingers of god' to larger scale flattening. These contours can be fitted using different methods including models and the quadrupole of the distribution, yielding $\beta = 0.47^{+0.19}_{-0.16}$ for these data.

13.10. The Integrated Sachs–Wolfe (ISW) Effect

Although not directly using peculiar velocities, the integrated Sachs–Wolfe (ISW) effect is a direct probe of the linear rate of structure formation. The simple Sachs–Wolfe effect (Sachs & Wolfe, 1967) is seen as the temperature fluctuations produced by the gravitational redshift at the surface of last scattering (White & Hu, 1997). This is related to the gravity of density perturbations in the potential, Φ, that generated temperature fluctuations. In the case of a flat Friedmann model, this is

$$\frac{\Delta T}{T} = -\frac{1}{3}\Phi. \tag{13.52}$$

The ISW is produced at later times and can be generated by scalar, vector, and tensor-mode fluctuations, but the scalar mode is dominated by the evolution of the matter density fluctuations in the expanding large-scale structure. It is applied to CMB and large-scale structure measurements such as redshift observations and provides determinations of density structures and tests for cosmological models (Giannantonio *et al.*, 2008; Granett, Neyrinck, & Szapudi, 2008; Nishizawa, 2014). In ΛCDM theory, this is related to the time variation of Φ along the line of sight and is a probe of dark energy. A number of studies including the Planck Collaboration (Ade *et al.*, 2016) and Dong *et al.* (2020) have combined CMB data with galaxy catalogues.

Chapter 14

The Great Attractor and Peculiar Velocity Studies

14.1. The Great Attractor and the Seven Samurai Study

Investigations of the motions of the galaxies and their flows were conducted by several investigators in the last decades of the 20th century. Following Rubin *et al.* (1976a,b), Lynden-Bell *et al.* (1988),[1] in one of the largest early collaborations, concluded that the observed flow of about 570 $\mathrm{km\,s^{-1}}$ for the 400 elliptical galaxies shown in Fig. 14.1 is produced by a spherical mass concentration of $M \approx 3 \times 10^{16} M_\odot$, given the name 'Great Attractor (GA)', at a distance corresponding to $cz = 4350\,\mathrm{km\,s^{-1}}$ in the direction of ($l = 307°$, $b = 9°$) towards the Hydra-Centaurus constellations. This was roughly in agreement with the IRTF Sc I measurements (Aaronson *et al.*, 1981, 1982, 1986) and a re-analysis of Rubin *et al.* (1976b). Different mass models could be considered for the GA, e.g., Faber & Burstein (1988) adopted a somewhat different model. This revealed that the observed elliptical galaxies obey a large-scale coherent flow towards the GA, but raised several questions at the time as no mass concentration of this size could be identified and whether such large bodies composed entirely of DM can exist as produced by cold dark matter. Burstein (1990) reviews details of the history and developments up to 1990.

The problem for identifying the source of the Great Attractor (GA) is that it falls into the galactic plane and consequently it is heavily obscured

[1] This and related papers are sometimes referred to as the 'Seven Samurai' or 7S study due to its seven authors.

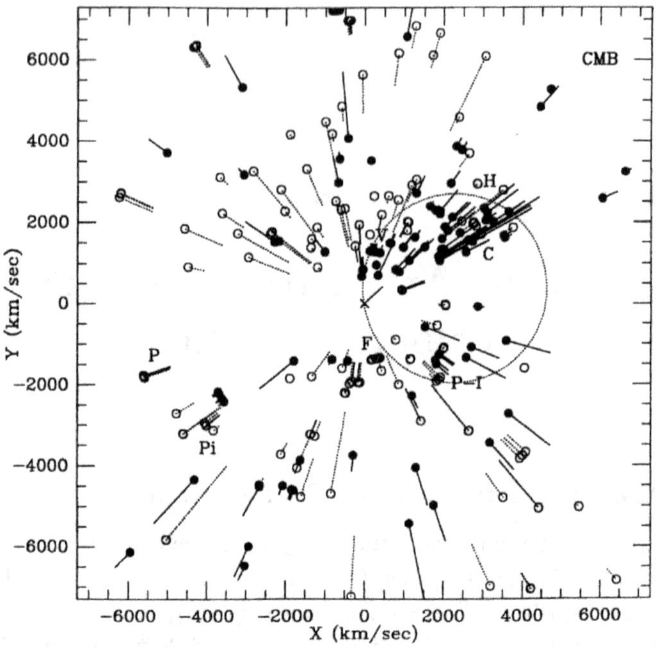

Fig. 14.1. Radial peculiar velocity vectors for all elliptical galaxies in the 7S sample within 45°of the supergalactic plane. Open dots denote motion towards the sun at the origin and solid dots are objects moving away. The location of the GA region is indicated by the dotted circle (*Source*: Lynden-Bell D., Faber S. M., Burstein D., Davies R. L., Dressler A., Terlevich R. J., Wegner G. (1988), *ApJ*, 326, 19, "Photometry and spectroscopy of elliptical galaxies. V. Galaxy streaming toward the new supergalactic center," Fig. 6, ©AAS Reproduced with permission.)

by dust in that direction. Figure 14.2 by Woudt & Kraan-Korteweg (2001) shows the location of the assumed GA squarely in the highly obscured galactic plane. Nevertheless, there were a number of attempts to identify it.

14.2. Peculiar Motions Beyond the GA

One way to locate the GA and to test if it is large spherical mass is to find a back flow of the peculiar motions of galaxies beyond it. Were the GA a single massive object, the galaxy flow observed in the peculiar velocities directed towards it would reverse direction at distances beyond its centre corresponding to $cz \approx 4500\,\mathrm{km\,s^{-1}}$, or distances greater than about $50h^{-1}$ Mpc. Dressler & Faber (1990a,b) and Dressler, Faber, & Burstein (1991) looked for the velocity reversal effect by obtaining additional

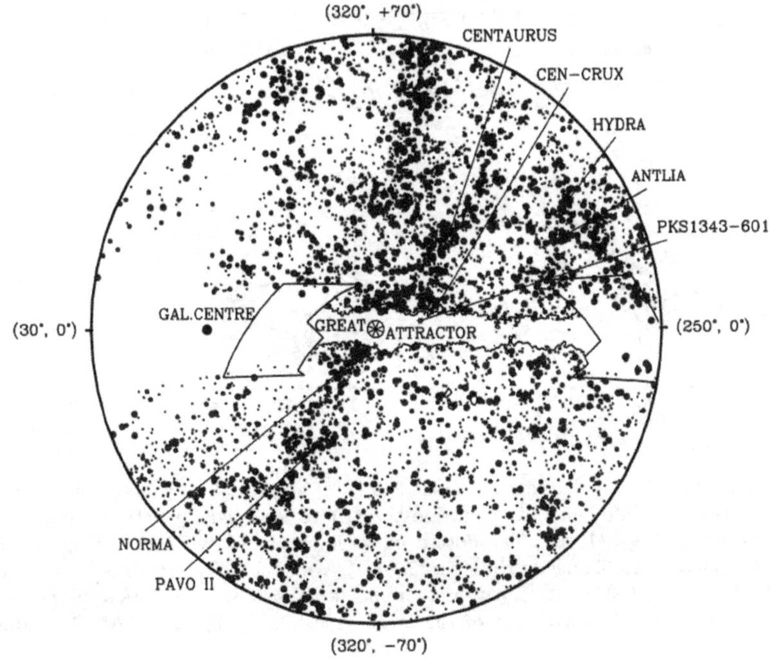

Fig. 14.2. Map of a portion of the celestial sphere in visible light centred on the Great Attractor's direction. The lack of galaxies in the horizontal zone of avoidance due to high dust absorption in the galactic plane is visible. The projection of the local supercluster can be seen by the strip of an enhanced number of galaxies running nearly vertically. Some notable cluster regions are labelled (*Credit*: Woudt P. A., Kraan-Korteweg R. C. (2001), *A&A*, 380, 441, "A catalogue of galaxies behind the southern Milky Way. II. The Crux and Great Attractor regions (l289° to 338°)," Fig. 16, ©ESO reproduced with permission.)

observations of elliptical galaxies using D_n–σ distances and spirals utilizing Tully–Fisher distances. The results of their survey are shown in Fig. 14.3 which, they concluded, confirms the backfall on the far side of the GA.

D. Mathewson and collaborators (Mathewson, Ford & Buchhorn, 1992 a,b; Mathewson, 1992) secured I- band Tully–Fisher distances of 1355 spirals measuring 21 cm and optical Hα rotational widths. Inclination corrections, the Tully–Fisher relation, etc., were independently determined from these observations. This dataset was subsequently enlarged to include 2400 galaxies (Mathewson & Ford, 1994; Mathewson, 1996). Their measurements are shown in Fig. 14.4, which are displayed analogously to Fig. 14.3 above. This plot does not appear to show the expected S-curve in the

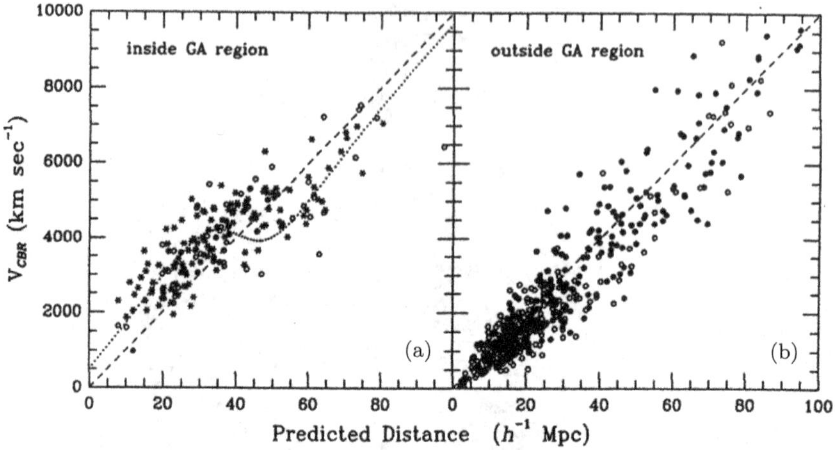

Fig. 14.3. The Hubble diagram for about 253 E (∗) and spiral (o) galaxies in the region towards the GA (left panel) and outside the GA region (right panel) from the study of Dressler & Faber (1990). The dotted curve is the infall and reversed outfall relation predicted by the Faber & Burstein (1988) GA model. (*Source*: Dressler A., Faber S. M. (1990), *ApJ*, 354, 45, "New measurements of distances to spirals in the Great Attractor: Further Confirmation of the large-scale flow," Fig. 2, ©AAS Reproduced with permission.)

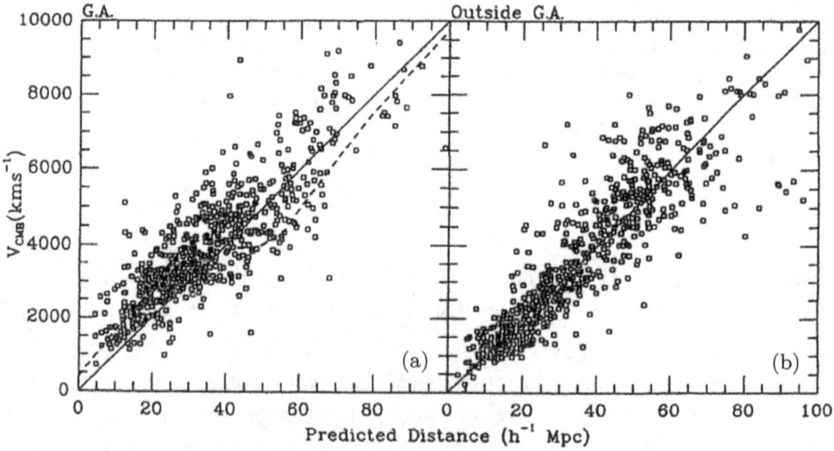

Fig. 14.4. The Hubble diagram for 657 spiral galaxies in the direction of the GA (left panel) and outside the GA region (right panel) from Mathewson *et al.* (1992). Similar to Fig. 14.3, the dashed curve is the velocity relation predicted by the single Faber & Burstein (1988) GA model. (*Source*: Matthewson D. L., Ford V. L., Buchhorn M. (1992), *ApJS*, 81, 413, "A southern sky survey of the peculiar velocities of 1355 spiral galaxies, Fig. 12, ©AAS Reproduced with permission.)

velocities and it was concluded that while they detected the flow of galaxies, it cannot be modelled by Faber & Burstein's (1988) GA Model.

At this time, other investigators looking for the backflow into the GA reached differing conclusions. Landy & Szalay (1992) considered a general analysis of the Malmquist bias and applied it to the D_n–σ relation. They argued that the backside infall could be a product of the Malmquist bias. Bothun *et al.* (1992) used the TF relation and 48 spirals. Although they found some indication of a reversed flow beyond the GA, they discussed how sensitive the result is to the Malmquist correction and the need for more data over a wider spread of sky. They suggested that the GA is not one object, but more likely a collection of objects. Ekholm & Teerikorpi (1994) considered the analysis of the Mathewson *et al.* (1992b) TF velocity — distance diagrams, comparing results for the direct and inverse relations using this dataset. They used different models and also showed that the results are sensitive to the slope and scatter of the TF relation, concluding that the Mathewson *et al.* (1992b) data do not present evidence against the presence of the GA. Schmoldt *et al.* (1999) used the PSCz galaxy survey to reconstruct the velocity field and found evidence for back infall towards Hydra-Centaurus. The resolution of the problem began to come with more data and the realization of the complexity of the structure in this region.

14.3. The Shapley Cloud and the Norma Cluster

Shapley (1930), using photographic plates from the 24-inch Bruce telescope in Arequipa, Peru, discovered a cloud of galaxies approximately in the direction of what is now the GA in the southern sky. He described this as having about ten times the concentration of the northern Coma–Virgo Cloud. Scaramella *et al.* (1989) proposed that this might be the source of the GA. Observations place this object well beyond the GA with a redshift $cz \approx 14,000\,\mathrm{km\,s^{-1}}$. Raychaudhury (1989) compiled a catalogue of 16,963 galaxies and concluded that the Shapley Cloud is unlikely to be the location of the GA as it would cause a rise in the peculiar velocities in this direction beyond the distance of the GA. This was not subsequently found starting with the study by Aaronson *et al.* (1989) using TF data and later motion studies.

This prompted several galactic plane searches for galaxies in the zone of avoidance using 21 cm observations (e.g., Kerr & Henning, 1987; Stavely-Smith *et al.*, 2000; Kraan-Korteweg *et al.*, 2005) and optical searches (e.g., Weinberger, Saurer, & Seeberger, 1995; Weinberger *et al.*, 2000).

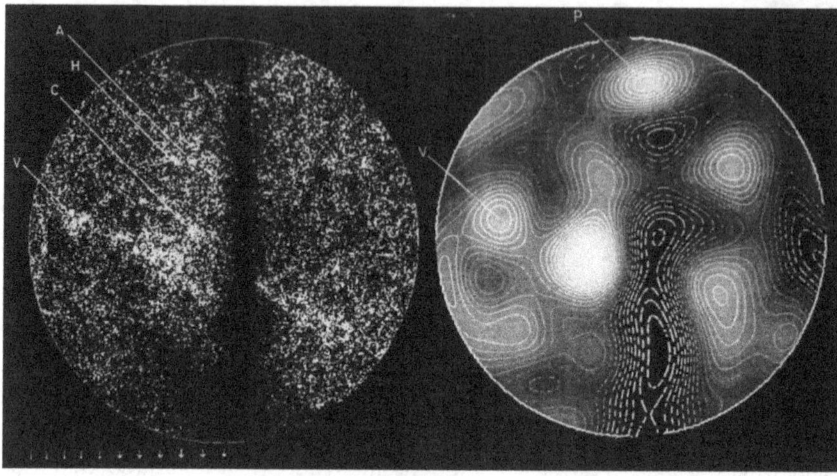

Fig. 14.5. Optical and IR galaxies towards the Great Attractor centred on the 7S direction to the GA ($l = 307°$, $b = 9°$). Top: optical galaxies from the UGC, ESO, and MCG catalogues (Lynden-Bell, 1987). Lower: spherical harmonics reconstruction from the IRAS $60\,\mu$ sources. Dark strip through both plots is from heavy dust absorption in the galactic plane. A, H, C, and V denote the Antila, Hydra, Centaurus, and Virgo clusters. The most prominent IR feature is towards the GA and Cen region, also the farther Shapley cloud. The IR feature, P, missing in the optical picture towards Puppis lies behind the galactic plane (*Source*: Scharf C., Hoffman Y., Lahav O., Lynden-Bell D. (1992), *MNRAS*, 256, 229, "Spherical harmonic analysis of IRAS galaxies: implications for the Great Attractor and cold dark matter," Plate 1. Reproduced by permission of the Oxford University Press on behalf of the Royal Astronomical Society.)

Using IRAS $60\,\mu$ data, which better penetrates dust, Scharf *et al.* (1992; 1993) produced spherical harmonic maps of the smoothed galaxy distribution shown in Fig. 14.5, in galactic coordinates. The upper plot of galaxies shows the projection of the supergalactic plane running diagonally across it and the dark zone of avoidance from the dust in the galactic plane. In the lower infrared map, the smoothed contours show that an object lies in the direction of the direction of Centaurus and the Shapley Cloud. This had been proposed as the identification of the GA by several investigators.

The great Norma galaxy cluster (Abell 3627) is the most massive object in this direction ($M \sim 5 \times 10^{15} M_\odot$) at $cz = 4871\,\mathrm{km\,s^{-1}}$ in the GA direction, and has been proposed as the core of the GA (Kraan-Korteweg *et al.*, 1996; Woudt *et al.*, 2007; Mutabazi *et al.*, 2014). Falling in the Hydra–Cen–Norma supercluster, this region appears to be the source of the GA, although it has a structure more complicated than initially envisioned. Several other investigations concluded that the original single GA model does not fit the peculiar motions and propose that it is due to the complicated structure. Kocevski & Ebeling (2006) employed REFLEX X-ray (Böhringer *et al.*, 2004) and other data and concluded that the attraction is comprised by 44% from the GA and 56% from more distant attractors. Bolejko & Hellaby (2008) constructed a Lemaître–Tolman model for the GA and obtained a mass, $M = 4\text{–}6 \times 10^{15} M_{\odot}$ from this idealized case.

14.4. The Mark III Catalogue

This landmark project's goal, like several that have followed, was to combine all available published peculiar velocity data and produce a larger homogeneous set of non-redshift derived distances. Six spiral TF galaxy samples along with D_n–σ data were combined yielding about 3300 individual measurements. The sources include data from Mould *et al.* (1991, 1993), Han & Mould (1992), Willick (1991), and Courteau *et al.* (1993). Details and references are given in Willick *et al.* (1995–1997). The main difficulties included combining the diverse TF width measurements that were from 21 cm and optical rotation curves and datasets that only covered limited sky regions.

These data were used by Dekel 1994 and Dekel *et al.* (1999) employing POTENT, already described, to study the flows and the density field. As shown at that time in Fig. 13.5, the region in the vicinity of the Local Group is dominated by lower density and farther away by the GA and PP mass enhancements.

14.5. Lauer & Postman Abell Cluster Inertial Frame (ACIF) Measurement

Lauer & Postman (1994) (LP) studied the velocity of the Local Group relative to an all-sky sample of 119 galaxy clusters within a redshift $cz = 15{,}000\,\mathrm{km\,s^{-1}}$, which gave an effective depth of 8000–11,000 $\mathrm{km\,s^{-1}}$ at the time at greater distances than most surveys. LP found a dipole motion for the inertial frame defined by these clusters to be 689 $\mathrm{km\,s^{-1}}$ towards ($l = 343°$, $b = +52°$), which is about 75° away from the Local Group CMB apex and concluded that the CMB dipole is generated by the gravitational attraction of mass concentrations beyond the $cz = 15{,}000\,\mathrm{km\,s^{-1}}$ limit of their survey. It is surprising that such a large volume could move relative to the CMB rest frame and this would place severe restrictions on standard ΛCDM cosmology, which already was strongly indicated from available data including correlation functions (see Chapter 5). Feldman & Watkins (1994) and Strauss *et al.* (1994) had shown that the ACIF dipole is inconsistent with many versions of CDM and HDM cosmological models, but Colless (1995) completely reanalyzed LP data and verified their result not finding any serious error.

Although LP used R-band photometry and made the usual corrections (discussed in Chapter 8), e.g., absorption, homogeneous Malmquist bias, etc., the one feature unique to the project was the choice of the older first-ranked or brightest cluster galaxy (BCG) distance indicator (Hoessel, 1980 and references therein). This assumes that a BCG's absolute magnitude, L_m, measured through an aperture of fixed, i.e., a metric, size is given by a relation

$$\alpha \equiv \mathrm{d}\log L_m / \mathrm{d}\log r_m, \tag{14.1}$$

where the aperture size was chosen to be $r_m = 10h^{-1}\,\mathrm{kpc}$ and α is determined from the photometry. Hoessel (1980) showed that this relation provides galaxy absolute magnitude errors of $\pm 0.21\,\mathrm{mag.}$, but mentions that a galaxy's absolute magnitude also correlates with the cluster's properties. PL determined the $L_m - \alpha$ relation for $\alpha \approx 0.2$ to 1.

This result has received criticism. Hudson *et al.* (1997) studied 249 galaxies in 16 clusters using the FP. They compare 11 clusters in common with ACIF and find significant disagreement. Müller *et al.* (1998) studied about 200 galaxies using the FP in three regions with $V < 20{,}000\,\mathrm{km\,s^{-1}}$ and exclude ACIF to the 99.8% level. The SFI survey (Giovanelli *et al.*, 1996) using TF distances found no support for the ACIF. Finding the correct BCG might be an additional problem; Lynam *et al.* (2000) conducted an enlarged X-ray BCG survey using ~250 clusters and the ROSAT all-sky survey to improve the α calibration, but they concluded that BCGs are relatively poor distance indicators. Of the 49 clusters in common with PL, ~30% had different BCGs that were not coincident with the X-ray centroids.

14.6. The Field Spirals in *I*-band (SFI) Project

This was an all-sky project to measure peculiar velocities for spiral galaxies using the TF relation. I band data were obtained from 1988 to 1992. About 75% of the images were obtained at the MDM Observatory on Kitt Peak and the rest from KPNO and CTIO. In all 1727 Sbc and Sc galaxies were observed. Haynes *et al.* (1999a) describes the observing and analysis techniques and presents the data. In addition, 21cm HI observations were used for 1201, mostly type Sc, were secured (Haynes *et al.*, 1999b) from the Arecibo 305 m, Greenbank 91 m and 43 m, Nançay, Kraus-type, and Effelsberg 100 m radio telescopes. Giovanelli *et al.* (1997c) describes the effect of the scatter in the TF relation from measurement of the velocity widths. This project and the related SCI and SCII in the next section, generated a number of papers, not only on peculiar velocities, but also different aspects of spiral galaxies and the TF relation, some already mentioned (discussed in Chapters 7, 11, and 13). Notable results included a determination of $H_0 = 69 \pm 5\,\mathrm{km\,s^{-1}\,Mpc^{-1}}$ (Giovanelli *et al.*, 1997a), a comparison of the SFI peculiar velocities with the IRAS 1.2 Jy gravity field (da Costa *et al.*, 1998), and a comparison of the observed peculiar velocities with CDM (Borgani *et al.*, 1997). Da Costa *et al.* (1996) estimated the SFI mass distribution and found a less massive GA and that the Perseus–Pisces region has a significant flow. In all of the determinations of the dipole, lower velocities were found (Giovanelli *et al.*, 1998c). Branchini *et al.* (2001) compared the SFI peculiar velocities with the PSCz gravitational field using the VELMOD v–v technique and found $\beta_I = 0.42 \pm 0.04$

14.7. The Cluster *I*-Band (SCI, SCII, and SFI++) Projects

These cluster projects built on the SFI data (Giovanelli *et al.*, 1998) using the I-TF relation. New data were secured, but mostly optical line widths were measured instead of radio (Dale *et al.*, 1999a–c, 2000). The convergence depth of the cluster dipole within $200\,h^{-1}\,\mathrm{Mpc}$ was estimated from this sample, from which it was concluded that the Local Group has a slow motion of about $200\,\mathrm{km\,s^{-1}}$ relative to the CMB dipole. The SFI++ database added additional photometric I-band data (Springob *et al.*, 2007).

14.8. Streaming Motions of Abell Clusters (SMAC)

Hudson *et al.* (1997) used the inverse-FP to measure the peculiar velocities of clusters in the Perseus–Pisces region, which were found to be small. In Hudson *et al.* (1999) this was enlarged to the SMAC sample, which also used distances from the FP and determined the peculiar velocities of 699 early-type galaxies in 56 clusters within $\sim$12,000 $\mathrm{km\,s^{-1}}$. They found a dipole velocity, $V = 630 \pm 180\,\mathrm{km\,s^{-1}}$, but in a different direction from ACIF. Spectroscopic and photometric observations and reductions of the SMAC survey are detailed in Smith *et al.* (2000, 2001) from which a combined dataset including literature measurements was completed (Hudson *et al.*, 2001). Detailed analysis by Hudson *et al.* (2004) determined a slightly revised bulk flow for the SMAC sample, $V = 687 \pm 203\,\mathrm{km\,s^{-1}}$ towards $(l, b) = (260°, 0°)$. This is consistent with the PSCz result, but more mass seems needed to produce the flow. The flow found from SMAC does not drop with distance and thus cannot be explained by the earlier simple GA models, and it is concluded that more massive attractors at large distances, including the Shapley Cloud, should generate the observed flow.

14.9. The Far-Elliptical (EFAR) Project

Observations for the EFAR project began in 1986 and completed in 1993. It did not have all-sky coverage. It was intended to use the FP to determine the peculiar velocity fields for early-type galaxies in two dynamically independent volumes in the opposite directions of the flow as it was known at that time, namely towards Hercules–Corona Borealis and Perseus–Pisces–Cetus and at the greater distances with redshifts between 6000 and 15,000 $\mathrm{km\,s^{-1}}$. This was to test whether the large-scale flows found locally are part of bulk motions in the universe (Wegner *et al.*, 1996). The R-band photometric data were obtained for 776 galaxies (Saglia *et al.*, 1997a). The photometric fitting employed a seeing-convolved $R^{1/4}$ plus an exponential component fitting algorithm described in Saglia *et al.* (1997b). Spectroscopic observations were mostly long-slit spectra on 2 m telescopes as well as 4 m telescope multi-fiber measurements. A total of 1319 spectra were secured and velocity dispersions and line strengths of Mg_b and Mg_2 were obtained for 676 galaxies (Wegner *et al.*, 1999). The Mg line strengths were analyzed in Colless *et al.* (1999) and discussed in Sec. 7.2. Saglia *et al.* (2001) describe the maximum-likelihood method used to fit the FP and other relations analyzed by EFAR. Colless *et al.* (2001) present an analysis of the peculiar motions.

14.10. Redshift-Distance Survey of Nearby Early-Type Galaxies (ENEAR)

This was a spectroscopic and R-band survey of nearby early-type galaxies, brighter than $m_B = 14.5$ and $cz \leq 7000\,\mathrm{km\,s^{-1}}$. The objects are in two subgroups, ENEARm the magnitude limited field galaxy sample, and the ENEARc galaxies in 28 clusters. The selection details are described by da Costa *et al.* (2000a). The main thrust for the ENEARm was to extend the 7S survey using the D_n–σ and FP distance indicators and roughly trebling the number of objects. The spectroscopic observations are presented in Wegner *et al.* (2003) where central σ and Mg_2 line indices are given for about 1178 E and S0 galaxies; 984 had no previous measurements. The corresponding R-band circular aperture photometry for 1332 objects is given in Alonso *et al.* (2003). The D_n–σ relation for the ENEARm and ENEARc are derived by Bernardi *et al.* (2002a,b). Zaroubi *et al.* (2001) get the power spectrum and bulk flow. Nusser *et al.* (2001) compares ENEAR with PSCz gravity field. Da Costa *et al.* (2000b) find low bulk flow, $V = 220\,\mathrm{km\,s^{-1}}$, ($l = 304°$, $b = 25°$). Borgani *et al.* (2000) find that ENEAR and SFI flows consistent with each other and flows being generated by known clusters.

14.11. The Sunyaev–Zel'dovich-Effect

A new promising method to determine peculiar velocities may become perfected using the Sunyaev–Zel'dovich (1980) (SZ-effect) proposed about 1971 and first observed by Birkinshaw & Gull (1983). Details can be found in, e.g., Longair (2011) and Birkenshaw (1999). A typical galaxy cluster ($R \sim 1.2\,\mathrm{Mpc}$) will have a total mass of $10^{14}\,\mathrm{M_\odot}$ and contain a large amount of hot ($T \sim 10^8$–$10^7\,\mathrm{K}$ or $\sim$6 keV) ionized gas and due to the inverse Compton effect, the radiation from the CMB is altered in an observable way.

14.11.1. *The thermal SZ-effect*

When a photon from the CMB passes through the cluster, it can collide with one of the hot electrons, be scattered, and gain energy by the inverse Compton effect. Since the CMB is blackbody radiation, as described by the Kompaneets equation, this shifts the energy distribution in a direction looking through the cluster to higher frequency by the Compton y-factor

$$\Delta T/T = -2y. \tag{14.2}$$

Along the line of sight, x, this depends on the cluster's gas temperature and electron density, T_g and n_e:

$$y = \int dx \frac{kT_g}{m_e c^2} \sigma_T n_e, \tag{14.3}$$

where σ_T is the Thomson cross-section for electron scattering. T_g and n_e need to be modelled in the cluster and usually a β-model of the form

$$n_{\mathrm{gas}} = n_{\mathrm{gas}}(0) \left[1 + \left(\frac{r}{r_c}\right)^2\right]^{-3\beta/2}, \tag{14.4}$$

has been employed for the gas density where r_c is a core radius (e.g., Arnaud, 2009). This can be used to construct maps of galaxy clusters in the radio 1–2 mm region (Birkenshaw, 1999; also the Planck Collaboration). Integrating y over the surface assuming a spherical cluster yields the integrated SZ-effect

$$\int \mathrm{d}^2\theta y = \frac{1}{D_A} \int \mathrm{d}^2 R y \propto \frac{1}{D_A^2} \int \mathrm{d}V n_e T_g \propto \frac{M_g T_g}{D_A^2}, \tag{14.5}$$

where D_A is the angular diameter distance and M_g is the mass of the gas. Both M_g and T_g can be independently measured from X-ray data and combining this with the radio measurements provides a distance determination to the cluster and has been used for an independent estimate of H_0, e.g., Birkenshaw, Hughes, & Arnaud (1991) for an early discussion and Bonamente *et al.* (2006) who obtained $H_0 = 77 \pm 10\,\mathrm{km\,s^{-1}\,Mpc^{-1}}$.

14.11.2. *The kinetic SZ-effect*

If a galaxy cluster has a peculiar velocity, v_p, then the photons scattered by the electrons will produce a spectral signature differing from the thermal SZ-effect that can measure the peculiar velocity (e.g., Holzapfel *et al.*, 1991).

$$\frac{\Delta T_{KSZ}}{T} = -\int \mathrm{d}\ell \frac{v_p}{c} \sigma_T. \tag{14.6}$$

If there is a bulk flow, V_{bulk}, this will depend on the angle from the apex, θ, and

$$\frac{\Delta T_{\mathrm{KSZ}}}{T} = \frac{V_{\mathrm{bulk}}}{c} \tau \cos\theta, \tag{14.7}$$

where τ is the optical depth in the cluster. Typical values for the detection of the thermal SZ-effect give $\Delta T_{\mathrm{TSZ}}/T \sim 10^{-4}$ and for the kinetic SZ-effect $\Delta T_{\mathrm{KSZ}}/T \sim 10^{-5}$ for a peculiar velocity of order 300 $\mathrm{km\,s^{-1}}$. This means that

$$\frac{\Delta_{\mathrm{KSZ}}}{\Delta_{\mathrm{TSZ}}} \sim 0.1, \tag{14.8}$$

so the observational problem is that the kinetic SZ-effect is at least an order of magnitude smaller than the thermal SZ-effect and consequently greatly more difficult to detect. The kinematic SZ-signal can be separated from the background thermal SZ because of their different spatial properties. The CMB power tends to peak on degree scales, but the clusters will typically peak on arc minute scales.

14.12. Summary of the Flows and their Relation with ΛCDM from Early Measurements

By the beginning of the 21st century, peculiar motions found from the several investigations began to present a consistent picture. The main details of the nearby flows are illustrated in Figs. 14.6 and 13.3, which show the two largest mass concentrations, but with relatively low resolution. The position of the CMB dipole apex on the sky discussed in Chapters 13 and 20 shows scatter, but seems to converge with distance, although its position and magnitude are not fully explained. A summary of the comparison between the observed bulk motions and ΛCDM discussed in this chapter is given in Fig. 14.7. On the whole, it appears that most of the results are consistent with ΛCDM within the errors of the data, or about one σ if the three

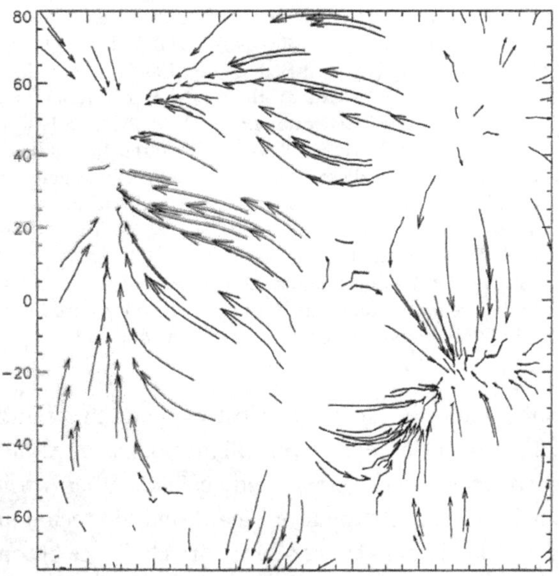

Fig. 14.6. The velocity field reconstructed from the ENEAR data in the supergalactic (X, Z)-plane. Distances on the axes are in Mpch^{-1}. The flow lines are proportional to the velocities at the starting points and run tangent to the local velocity field. The GA region in the upper left near (−60, +40) and the Perseus–Pisces region on the right near (+50, −20) are visible. (*Source*: Zaroubi S., Bernardi M., da Costa L. N., Hoffman Y., Alonso M. V., Wegner G., Willmer C. N. A., Pellegrini P. S. (2001), *MNRAS*, 326, 375, "Large-scale power spectrum and structures from the ENEAR galaxy peculiar velocity catalogue," Fig. 8, Reproduced by permission of the University of Oxford Press on behalf of the Royal Astronomical Society.)

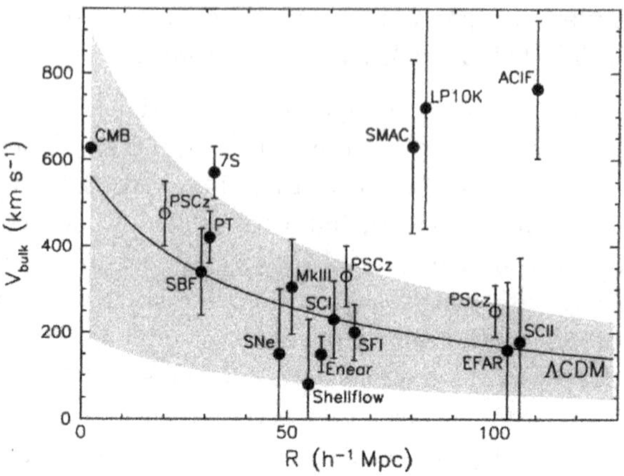

Fig. 14.7. Bulk motions of whole samples given their estimated approximate effective depth compared to the cosmic microwave background. Symbols denote: CMB, (Kogut *et al.*, 1993), PSCz (Saunders *et al.*, 2000; Dekel, 2000), SBF (Tonry *et al.*, 2000), PT (Pierce & Tully, 2000), 7S (Lynden-Bell *et al.*, 1988), SNe (Riess, 2000) MKIII (Dekel *et al.*, 1999), Shellflow (Corteau *et al.*, 2000), Enear (da Costa *et al.*, 2000), SCI (Giovanelli *et al.*, 1998b), SFI (Giovanelli *et al.*, 1998a), SMAC (Hudson *et al.*, 1999), LP10K (Willick, 1999), EFAR (Colless *et al.*, 2001), SCII (Dale *et al.*, 1999a), ACIF (Lauer & Postman 1994; Colless 1995). The curve and shaded region, denoting a 90% cosmic scatter range, are the bulk motion expected within a sphere of radius R using a ΛCDM model with $\Gamma = 0.25$, $\sigma_8 = 1.0$, $H_0 = 70$ km s^{-1} Mpc^{-1}). (*Source*: Colless M., Saglia R. P., Burstein D., Davies R. L., McMahan Jr. R. K., Wegner G. (2001), *MNRAS*, 321, 277, "The peculiar motions of early-type galaxies in two distant regions — VII. Peculiar velocities and bulk motions," Fig. 14, Reproduced by permission of the University of Oxford Press on behalf of the Royal Astronomical Society.)

most discrepant points are excluded. Most of these investigations indicate that $\Omega_m \approx 0.3$, close to the more recent higher precision values in Chapter 20. Consequently, to lower the errors and achieve higher precision, surveys using more precise distance estimators, new analysis techniques, and considerably larger numbers of galaxies are required. This is becoming possible with new generations of instruments in the present century.

In the meantime, remarkable new discoveries in astronomy moved some of the spotlight away from peculiar motion studies and opened new avenues of investigation. Before discussing later 21st century results for peculiar motions, these new developments will be outlined in the following five chapters in order to place them in context.

Chapter 15

Revolution by the CMB Studies

15.1. Discovery of the Cosmic Microwave Background

Since around the turn of the 21st century, studies of the Cosmic Microwave Background (CMB) have changed the landscape in all branches of astrophysics and cosmology. Kolb & Turner (1990) give a still useful summary of late 20th century cosmology. Although the basic ideas were long known, predicted, e.g., by Alpher & Herman (1948) and others, the discovery of the CMB by Penzias & Wilson (1965), generated considerable interest, but it was not until after its launch in 1989 when the COBE satellite produced an accurate spectral energy distribution and the acoustic power spectrum. This instrument was followed by the increasingly powerful WMAP and PLANCK missions. Results published in 2003 and 2013, accompanied by several important ground-based and balloon experiments, began the era of 'precision cosmology' in earnest.

The analyses of these data have eliminated many earlier cosmological models that are unable to explain the main observational data (e.g., steady state, Alfvén's plasma-universe, Milne's kinematical relativity, Segal's chronometric cosmology, etc.) and led to the astronomical community to focus on ΛCDM.

In mid-20th century there were only two generally accepted solid facts about the large-scale universe: the dark sky or Olbers' paradox and the redshift-distance law. Now many generally accepted observational facts about the universe have emerged (Longaire 1992) and continue to mount that are used to test cosmological models and include:

(1) Dark sky or Olbers' paradox.
(2) The universe expands (Hubble's law).

(3) Isotropy. This was assumed in the standard models, but was studied (e.g., Condon, 1991; Gregory & Condon, 1991) from optical and radio galaxies, is now an accurate approximation from the CMB, but not as well determined from galaxy and quasar counts.
(4) The spectrum of the CMB is black body ($T = 2.735\,\mathrm{K}$) and implies matter and radiation were in thermodynamic equilibrium at recombination at $z \sim 1100$.
(5) General relativity is a good theory of gravity. All solar system tests are consistent. Binary pulsars, black holes, and gravity wave detections provide strong gravity tests.
(6) In the present (standard) ΛCDM models the ages of the oldest objects, such as globular clusters, are consistent.
(7) The universe has evolved in the past. At $z \gg 1$ there are more blue galaxies peculiar galaxies, QSO's, etc.
(8) The dynamics of all objects indicate dark matter. This can be observed with gravitational lensing. The expansion indicates the existence of Λ.
(9) The abundances of the light elements, D, He, Li, are convincing evidence for the big bang.
(10) The large-scale structure of filaments, voids, baryonic acoustic oscillations (BAOs) and peculiar velocity flows appear consistent with an inflation era.
(11) To these, new data including CMB power spectrum, polarization and additional redshift surveys are being added.

The theoretically anticipated all-sky picture of the temperature fluctuations of the CMB was first realized with COBE and subsequently mapped with increasing signal-to-noise and resolution by WMAP and PLANCK, and supplemented by a number of earth-based experiments. The resulting famous all-sky image of the CMB temperature fluctuations (see, e.g., https://www.cosmos.esa.int/web/planck/picture-gallery) is ubiquitous and need not be repeated here. Several steps are required in the data reductions, elimination of foreground sources of radiation from the Milky Way and the dipole anisotropy are important. This is accomplished by observations in the wavelength bands of these instruments and the definite wavelength dependence of these sources, subtracted from the observed CMB data. The accurate blackbody distribution of the CMB is a remarkable feature providing an important constraint for theories of its origin. The power

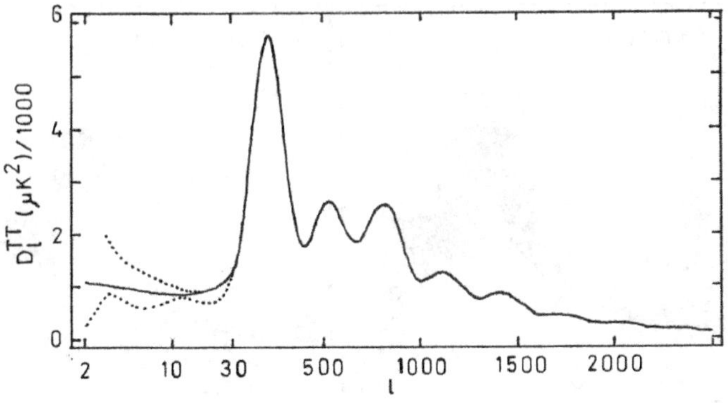

Fig. 15.1. The CMB temperature power spectrum based on PLANCK VI (2020), *A&A*, 641, A6, Fig. 1 data for the different multipoles, ℓ. The dotted curves for $\ell < 30$ depict roughly the range of error bars. For $\ell \geq 30$, the errors are almost all smaller than the width of the curve.

spectrum of the temperature fluctuations contains a considerable amount of information and are of order $\Delta T \sim 10^{-5}$ K.

The power spectrum, e.g., in Fig. 15.1 measured by PLANCK is obtained from spherical harmonic fits and its peaks for higher multipoles contain information on cosmology. Figure 15.2 illustrates the sensitivity of the CMB temperature power spectrum with variations in some of the cosmological parameters of the model including dark energy and H_0. The position of the first peak near $\ell = 200$ is determined by curvature and baryons, Ω_b, and tells that the universe is quite close to being flat. The second peak is affected by the matter density and the third peak is sensitive to DM density as well. Multipoles less than the first peak are sensitive to large scales and errors increase as ℓ decreases.

Several CMB codes exist for analyzing the observations and modelling the CMB. Two that have been popular and publicly available are CMBFAST (Seljak & Zaldarriaga, 1996) and CMBEASY (Doran, 2005), and many representing the satellite groups have developed ever more sophisticated models.

Although the analyses involve complicated many parameter fits to large amounts of data, the WMAP and PLANCK CMB measurements give remarkably consistent values for the cosmological parameters, considering the different instruments and the codes to analyze them. Table 15.1 is a comparison of partial data derived from independent programs and

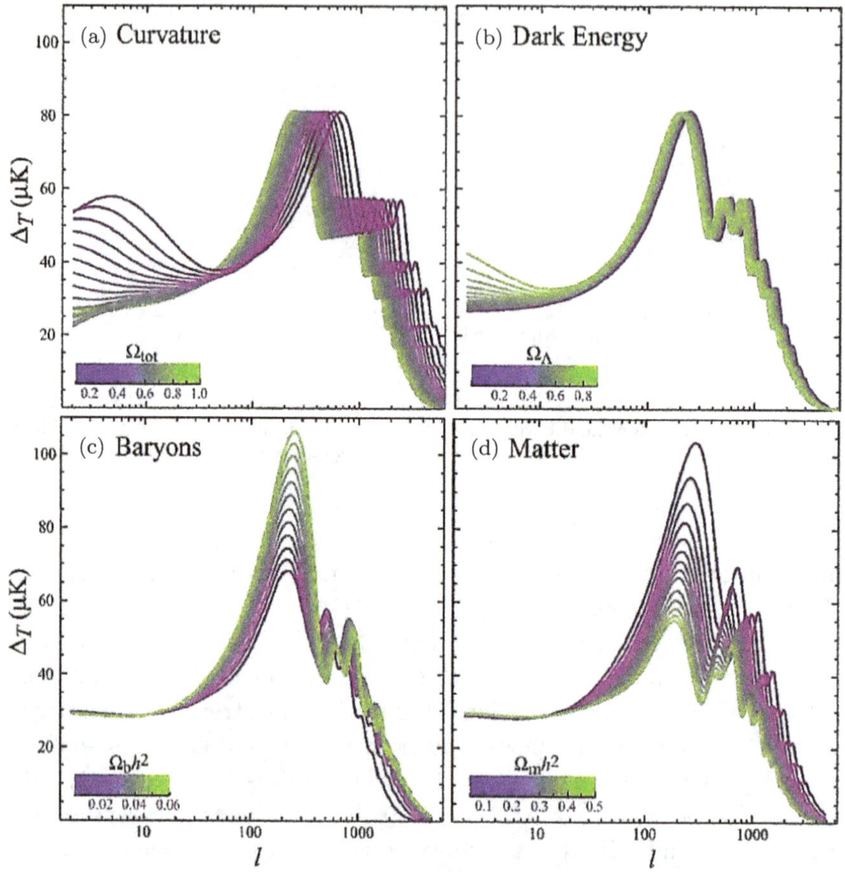

Fig. 15.2. How the shape of the CMB acoustic power spectrum depends on cosmological parameters. Curvature shifts the peaks in ℓ. Dark energy (Λ) constrains the shift. Baryons strongly affect the height of the first peak and total matter density changes the widths and positions of the peaks. (*Source*: Hu W., Dodelson S. (2002), *ARA&A*, 40, 171, "Cosmic microwave background anisotropies," Fig. 4, ©2002 Annual Reviews www.annualreviews.org with permission.)

datasets. It is intended to illustrate the agreement in cosmological parameters found independently from different studies and also a comparison with some early peculiar motion results already described. Komatsu *et al.* (2011) used WMAP data and the Plank Consortium (2018) obtained their results from the PLANCK satellite.

Different evaluations of the parameters are reached by utilizing additional non-CMB datasets and combining with different independent

Table 15.1. Sample of some cosmological parameters for the CMB from PLANCK and WMAP.

Parameter[a]	PLANCK Planck Consortium (2018)	WMAP7 + BAO + H_0 Komatsu *et al.* (2011)
$\Omega_b h^2$	0.0224 ± 0.0001	0.02255 ± 0.054
$\Omega_d h^2$	0.120 ± 0.001	0.1126 ± 0.036
τ	0.054 ± 0.007	0.088 ± 0.014
n_s	0.965 ± 0.004	0.968 ± 0.012
σ_8	0.811 ± 0.006	0.816 ± 0.024
$N_{\rm eff}$	2.99 ± 0.17	$4.34^{+0.76}_{-0.80}$
w_0	-1.03 ± 0.03	-0.93 ± 0.12
Ω_k	0.001 ± 0.002	-0.0109 ± 0.0088
Ω_Λ	0.685 ± 0.007	0.725 ± 0.016
Ω_m	0.315 ± 0.007	0.275 ± 012
H_0	67.4 ± 0.05	70.2 ± 1.4
t_0	13.80 ± 0.02	13.76 ± 0.11

[a]Ω_b is baryonic density, Ω_d is dark matter density, Ω_k is curvature, Ω_Λ is dark energy, Ω_m is total matter density, H_0 is the Hubble constant in $\mathrm{km\,s^{-1}\,Mpc^{-1}}$ and t_0 is current age of the universe in Gyr.

surveys. Figure 15.3 shows diagrams using supernovae, SNe, Baryonic oscillations, BAO, and the CMB. The individual surveys do not uniquely pin down the cosmological parameters, but the intersection of the overlapping bands they produce constrains them. These differences seem small compared to the sizes of the standard errors, although some argue that they are in part significant. Table 15.1 compares typical values and their errors from two studies. As a result, one popular set of 'concordance' parameters, widely seen in the literature is: $\Omega_m = 0.3$, $\Omega_\Lambda = 0.7$, $\Omega_k = 0$, $w = -1$, and $H_0 = 70\,\mathrm{km\,s^{-1}\,Mpc^{-1}}$.

As the ℓ decreases, the accuracy of the measured power spectrum decreases. This is where peculiar motions can contribute as a probe of large scales. In the CMB power spectrum, the first peak at $\ell \approx 200$ has an angular size on the sky of $\alpha \approx 0.5°$. For an open universe with $H_0 = 70\,\mathrm{Mpc^{-1}\,km\,s^{-1}}$, $\Omega_m = 0.2$, and $\Omega_\Lambda = 0.7$, the angular size distance is $D_A = 12.4\,\mathrm{Mpc}$ and the scale is 0.2 Mpc/deg (Wright, 2006), so the scale of the first peak corresponds to $\ell_o = \alpha D_A$ or $\ell_o \sim 0.1\,\mathrm{Mpc}$. At the current epoch, this is $(1+z)$ larger, or about 110 Mpc. Since peculiar motion studies reach depths of 100–200 Mpc, they can sample scales specified by $\ell \sim 1$–2. As discussed earlier, modified gravity and higher dimensional gravity might show effects on these scales.

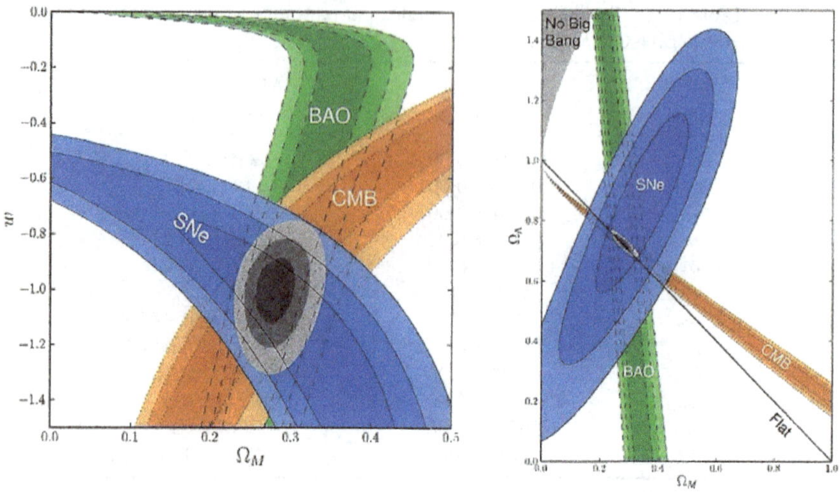

Fig. 15.3. Examples of determining cosmological parameters using the combination of three different sets of data. The grey regions at their intersections indicate the resulting constraints. (*Source*: Amanullah R. *et al.* (2010) *ApJ*, 716, 712, "Spectra and hubble space telescope light curves of six type Ia supernovae at $0.511 < z < 1.12$ and the Union2 compilation," right halves of Figs. 10 and 11, ©AAS Reproduced with permission.)

15.2. CMB Polarization

The polarization of the CMB provides additional information. The total polarization of the CMB is about 10%. The polarization can be divided into two types, linear or E-mode and circular or B-mode, which is about an order of magnitude smaller signal. The E-mode was first detected by the Degree Angular Scale Interferometer (DASI) at the South Pole (Kovac *et al.*, 2002). It is caused by Thomson scattering of photons by free electrons at the last scattering surface of the CMB in the ionized plasma mostly composed of hydrogen, which is recombining to form neutral gas and detailed maps have been constructed.

Polarization by gravitational lensing of the CMB has been observed. Raghunathan *et al.* (2019) have demonstrated detection of the lensing of about 18,000 galaxy clusters with data from the South Polar Telescope and Dark Energy Survey. Using a CMB Stokes Q/U polarization estimator, the polarization measurements suffer less from background than other methods and it is estimated that a significant numbers of accurate cluster masses will be obtained.

The circular or B-mode is of great interest as a test for inflation. Although other processes might produce it, B-mode polarization is expected to be formed by inflation forming a background of gravitational waves that would produce observable circular polarization. Although a positive detection was reported (Ade *et al.*, 2014, 2021) by the Background Imaging of Cosmic Extragalactic Polarization (BICEP2) experiment, subsequent analyses using PLANCK data (BICEP2/Keck Collaboration 2015) found that the signal was produced by effects of galactic dust along the line of sight and to date, B-mode circular polarization has not been found in the CMB. Gravitational lensing by sources along the line of sight can also induce circular polarization. Nevertheless, the search continues with improved and larger instruments. The Keck array came into operation in 2012. BICEP3 and Spole experiments began in 2016. The BICEP array will start operating in 2020 and the Japanese Space Agency plans to operate the LiteBIRD satellite in the 2020s and one may anticipate an eventual detection.

15.3. CMB Anomalies

The isotropy of the CMB is taken as confirmation of the fundamental assumption of the isotropy of the universe in the ΛCDM cosmological picture. However, a number of effects have been noted in the WMAP and PLANCK data that raise questions about this notion. The origins of these effects, which remain controversial include: (1) the 'axis of evil' — temperature fluctuations in the CMB appear to be in a plane on the sky; (2) the giant cold spot — a coherent blob in the southern hemisphere slightly cooler than the mean CMB; (3) lopsided quadrupole — the total size of fluctuations on half of the celestial sphere is slightly larger than on the other half.

Since these effects are at a low signal level after substantial correction for the galaxy, etc., analysis methods of the data have been considered. Mask processing, the treatment of bad pixels, is carefully examined (Rasst *et al.*, 2014; Santos *et al.*, 2015). The Planck Collaboration, VI (Aghanim *et al.*, 2018) and VII (Akrami *et al.*, 2019) find that the ΛCDM model explains the structure of the CMB in detail with good consistency using six parameters, also confirm some of the anomalies.

Possible natural causes include, confusion in the galactic background, astrophysical effects in the galactic halo and the motion through the CMB. The removal of secondary large-scale cosmological signals reduces some of the anomalies. The cold spot is also visible on a synchrotron map of Faraday rotation depth. It could be due to an ionized region in or near the galaxy rather than a cosmological effect. Whether or not this hints at the need for new cosmologies will doubtless be debated further.

15.4. Big Bang Nucleosynthesis (BBNS) of the Light Elements

Although BBNS is outside the direct pursuit of peculiar motions, cosmic abundances of the isotopes of the light elements, mainly deuterium, H^2, helium, He^3, He^4, and lithium, Li^7, etc. are generally agreed to be a major clue about the early universe. They provide observational information about a time stretching from a few seconds to minutes in the life of the Big Bang (BB) model compared to about ~400,000 years for the CMB. Elements with nuclei heavier than Li^7 are mostly formed later in stellar interiors and violent explosive events. The light elements formed as the universe cooled and their abundances can provide the baryon to photon ratio, $\eta \sim 6 \times 10^{-10}$ which translates to Ω_b. Calculations show that as a function of η He^4 reaches a steady value of about 0.24 of the total mass, but H^2, He^3, and Li^7 are varying and their measured cosmic abundances yield $\Omega_b h^2 \approx 0.0214$. This can be compared with $\Omega_b h^2 = 0.0224 \pm 0.0001$ deduced from the acoustic peaks of the CMB (Table 15.1). So although the values from the two independent methods appear consistent, at high precision the agreement is not perfect and some questions remain.

Detailed reviews of the different determinations of light element abundances are available (e.g., Steigman, 2007; Cyburt *et al.*, 2016; Mathews, Kusakabe, & Kajino, 2017). Measurement of each isotope requires different methods and presents difficulties, which are being improved. Deuterium (Cook, Pettini, & Steidel, 2018) and helium (Cook & Fumagalli, 2018) are easier to measure, but lithium is difficult to measure, because it is easily destroyed in stellar atmospheres, and seems discrepant. Although this could hint new physics, the situation may also change with future observations. Mossa *et al.* (2020) have measured a new cross-section for the $D(p,\gamma)^3He$ reaction which further improves agreement between CMB and BBNS at the LUNA (Laboratory for Underground Nuclearphysics at Grand Sasso) in Italy and places stronger limits on non-standard particle physics.

Chapter 16

Baryonic Acoustic Oscillations (BAOs)

In the early 21st century, an important discovery for cosmology is that of the Baryonic Acoustic Oscillations (BAOs). This feature in the cosmic power spectrum is being extensively pursued for its great potential to provide additional cosmological information from combining the massive simulations and observational surveys now planned or underway. This subject has advanced from the combination of CMB and redshift data. With the increase in the number and depth of redshifts and galaxy counts, the structures of the correlation function and thus the power spectrum could be investigated in greater detail.

The early measurements primarily, using correlation functions, are discussed in Chapter 4. The peculiar motions could be employed to compare the power spectrum with versions of ΛCDM as in Fig. 16.1 and show reasonable agreement with ΛCDM. However, the resolution of the $P(k)$ curve achieved is not great. Large redshift surveys have determined the power spectrum in greater detail, such as in Fig. 16.2, which compares results from the SDSS and the 2dFGRS and CDM (Percival, *et al.*, 2007).

New data using redshifts with increased resolution gave an indication that a 'wiggle' or 'bump' feature appears in the power spectrum (e.g., Broadhurst *et al.*, 1990; Peacock & Dodds, 1994; Einasto *et al.*, 1997a,b; Gaztanaga & Baugh, 1998). The PSCz galaxy catalogue (Tadros *et al.*, 1999) and subsequently the 2dF redshift survey (Cole *et al.*, 2005) and the SDSS (Percival *et al.*, 2007) considerably improved knowledge of the power spectrum. Although initially appearing indistinct, current data establish

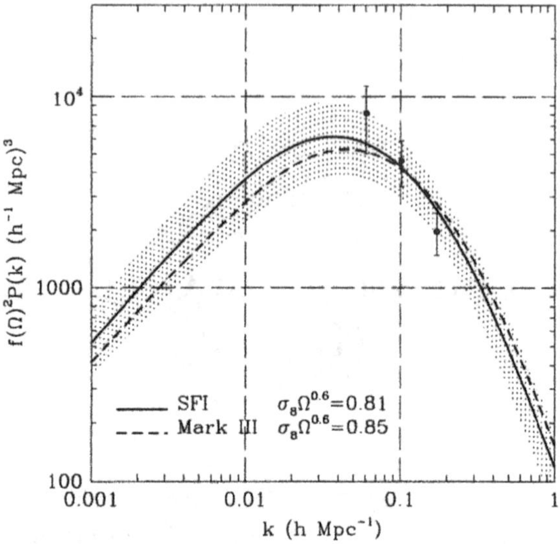

Fig. 16.1. Power spectra from peculiar motion data employing maximum likelihood fits to SFI (solid curve) and Mark III (dashed curve) data using tilted ΛCDM. The 90% confidence level for the SFI data is the shaded region. The three dots and error bars are from Kolatt & Dekel (1997) using Mark III data. (*Source*: Freudling W., Zehavi I., da Costa L. N., Dekel A., Eldar A., Giovanelli R., Haynes M. P., Salzer J. J., Wegner G., Zaroubi S. (1999), *ApJ*, 523, 1, "Large-scale power spectrum and cosmological parameters from SFI peculiar velocities," Fig. 11, ©AAS Reproduced with permission.)

this feature with high significance to be at a scale near 150 Mpc in the power spectrum (Fig. 16.3).

Eisenstein *et al.* (1998a,b), Blake & Glazebrook (2003), Eisenstein, Seo, & White (2007), Beutler *et al.* (2011) and other investigators realized that CDM cosmology produces this feature and it provides a major cosmological clue. If the density of the baryons is a non-negligible fraction of the total matter density, then the large-scale power spectrum will show finer features as shown in Fig. 16.3.

Baryon Acoustic Oscillations (BAOs) are artefacts frozen into the large-scale structure formed before the recombination time of the universe. Density perturbations of dark matter and baryons (Chapter 13) formed in the initially ionized plasma of mostly electrons and baryons (mostly protons and light nuclei). Electron scattering was a strong opacity source and in that early era photons could not flow freely, producing a high pressure baryon–photon fluid. Two effects competed. The first was that gravity pulled gas into the perturbations composed mostly of DM. The second effect was the

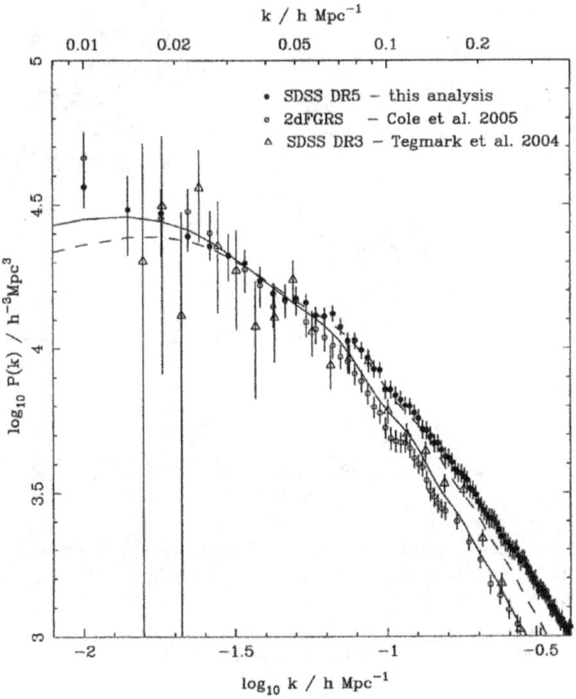

Fig. 16.2. Comparison of the power spectra derived from the SDSS DR3 and DR5 and the 2dFGRS redshift data. The dashed curve is the fit to a CDM model $0.01 < k/h\,\text{Mpc}^{-1} < 0.06$ and the continuous curve is the best fit for $0.01 < k/h\,\text{Mpc}^{-1} < 0.15$ (*Source*: Percival W. J. *et al.* (2007), *ApJ*, 657, 645, "The shape of the sloan digital sky survey data release 5 galaxy power spectrum, "Fig. 13, ©AAS Reproduced with permission.)

growing perturbations compressed the gas and increased the pressure causing an expansion. With the radiation pressure this created acoustic waves. Each overdensity was then surrounded by an expanding spherical shell of a size governed by the 'sound horizon', $r_s = c_s t_{ls}$, where c_s is the sound speed and t_{ls} is the time of last scattering at recombination ($z \approx 1100$). Then the baryons and electrons decoupled and the sound waves ceased expanding. The electron scattering stopped and photons expanded freely to produce the CMB. The BAO remain detectable from the higher density where galaxy formation was enhanced leaving the universe filled with a network of overlapping spherical shells of size $r_s \approx 150$ Mpc in the nearby universe. This size scale repeats across the sky and can be found from the spatial correlation function, $\xi(r)$, using redshift surveys.

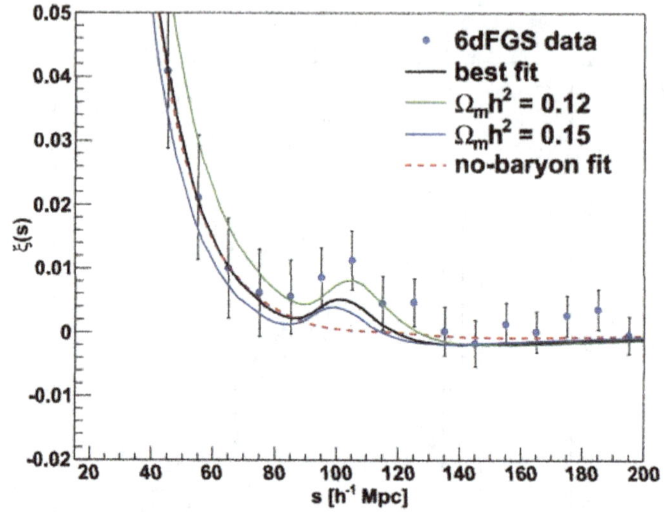

Fig. 16.3. The large-scale correlation function determined by the 6dFGS survey showing sensitivity to the matter parameter, for which the best fit was $\Omega_m h^2 = 0.138 \pm 0.020$. (*Source*: Beutler F. *et al.* (2011), MN, 416, 3017, "The 6dF Galaxy Survey: baryon acoustic oscillations and the local Hubble constant," Fig. 2, Reproduced by permission of the Oxford University Press on behalf of the Royal Astronomical Society.)

Linear theory predicts a series of bumps or 'wiggles' in the power spectrum. This can be seen from a simple model. The correlation function, $\xi(r)$ and the power spectrum $P(k)$ form a Fourier transform pair (Chapter 5)

$$P(k) \propto \int_0^\infty \xi(r) \sin krrdr. \tag{16.1}$$

Figure 16.4 shows how correlation functions of different widths produce the wiggles in the power spectrum. At early times, the radii of the acoustic waves will be very small. Approximating $\xi \sim \delta(r)$, the corresponding power spectrum, $P(k) \sim \sin k$. At late times, the acoustic waves have greatly expanded producing a widened correlation function. Assuming a Gaussian function, $\xi(r) \propto \exp\left[-(r - r_s)^2/2\sigma^2\right]$, as an illustrative approximation shows, that the higher frequency, k wiggles get progressively damped down.

The form of this curve has been predicted (e.g., Eisenstein *et al.*, 1997a,b, 2007) and discussed by many authors. This was observed by Beutler *et al.* (2011) in redshift data in the 6dFGRS survey. Larger data sets are being employed, e.g., Ross *et al.* (2015) used SDSS DR7 data; Alam *et al.* (2017) analyzed 1.2 million redshifts in the SDSS BOSS survey.

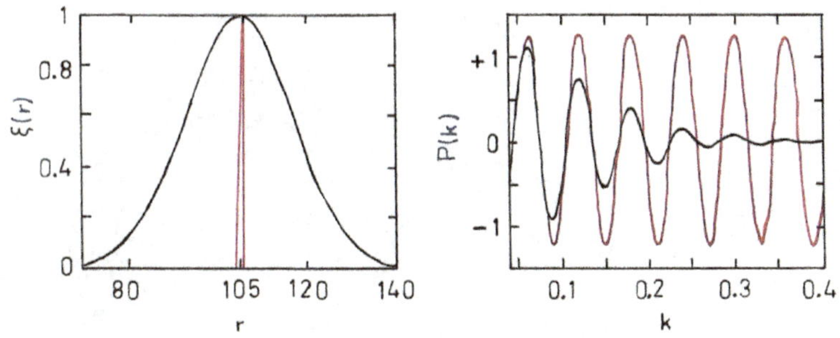

Fig. 16.4. Schematic diagram for the baryon acoustic signal in the Fourier pairs $\xi(r)$ and $P(k)$ for different widths of the correlation function starting from a δ-function and increasing the width. As the width is increased, the acoustic oscillations become damped for increasing k (Numerical values based on Bassett & Hlozek, 2009; Fig. 1.16).

Gil-Marín *et al.* (2018), Hou *et al.* (2018), Zarrouk *et al.* (2018), and Alam *et al.* (2021) analyzed the SDSS IV BOSS and DR14Q 148,659 quasar sample and have further studied cosmological parameters at higher redshifts. The year 3 of the Dark Energy Survey (DES) combined weak lensing and photo-Zs for about 226 million galaxies (e.g. Amon *et al.*, 2021). In general these yield no significant inconsistency with the PLANCK CMB studies.

The BAOs are an independent, although statistical, 'standard ruler' (Chapter 8) that can be measured at different redshifts or thus as a function of cosmic time. This provides the means to determine cosmological parameters. Knowing the standard size, d_0, the angle that it subtends at the angular diameter distance, D_A, is related to the Hubble parameter $H(z) = \dot{S}/S$, where S is the scale factor,

$$\begin{aligned} \Delta\vartheta &= d_0/D_A, \\ D_A(z) &\propto \int_0^z \frac{dz'}{H(z')}. \end{aligned} \tag{16.2}$$

When measurements of the subtended angle are measured at different redshifts, separated by Δz, the relation between the redshift and the Hubble parameter is

$$c\Delta z = d_0 H(z), \tag{16.3}$$

where $H(z)$ depends on the adopted cosmological model. The modelling and extension of this standard ruler to higher redshifts with the surveys now promised or underway will doubtless produce new advances in large-scale structure, peculiar motions and cosmology.

Chapter 17

Gravitational Lensing

Gravitational lensing provides a means to map mass distributions in several kinds of objects. When perfected these will be compared with with other independent methods such as peculiar motions. The bending of light by the sun's gravity is one of the three important early classical tests and achievements of General Relativity (GR). The deflection angle of a photon's path as it skims the limb of the sun is

$$\omega = \sqrt{\frac{4GM_\odot}{c^2 R_\odot}} = 1''\!.7. \tag{17.1}$$

The idea of light deflection was considered in the 19th century, but Newtonian physics gives the angle that the trajectory of a 'corpuscle' is bent as it passes the limb of the sun found by Seldner (1804) is $\omega = 0''\!.84$ (using the physical constants of his time). The factor of two difference, was confirmed, albeit with large error, in the famous 1919 May 29th eclipse measurement of Dyson, Eddington, & Davidson (1920), who found $\omega = 1''\!.98 \pm 0''\!.18$. Mitchell (1932) gives a detailed account of several early eclipse results. Many subsequent measurements using eclipses and other means have improved the determination of ω. Current radio measurements of quasars aligning with the sun using Very Long Baseline Interferometry (VLBI), (e.g., Fomalont *et al.*, 2009; Shapiro *et al.*, 2004) confirm the bending to higher precision of about $\pm 0.03\%$. The bending parameter compared to GR measured using the Cassini spacecraft is $\gamma = 1 + (2.1 \pm 2.3) \times 10^{-5}$ (Bertotti, Iess, & Tortora, 2003), all persuasive support of GR.

Compared to the relatively weak solar gravity, greater light bending occurs in strong gravity, such as near collapsed stars, massive galaxies, galaxy clusters, quasars, and black holes that produce gravitational lenses.

Einstein (1936) had considered the details of simple lensing, but with the technology available in his day, concluded that the effect would be undetectable. Refsdahl (1964) worked out details of gravitational lensing in greater detail and the first gravitational lens was discovered by radio observations of a twin quasar by Walsh, Carswell, & Weymann (1979). More complicated lenses produce additional images, such as a four lobed 'Einstein cross', of which several examples are now known (e.g., Fig. 17.2 and Jee *et al.*, 2019).

For a point mass acting as the lens, the geometry is shown in Fig. 17.1 where the different paths of the images are indicated. In the top left of Fig. 17.1, the geometry is shown for a weak or 'thin' lens. D_d is the distance of the lens and the observer, D_s is the distance of the source from the

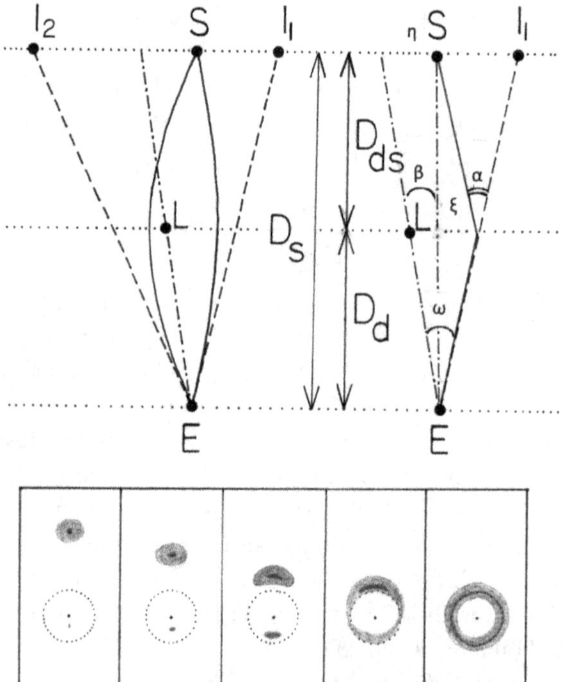

Fig. 17.1. Spatial arrangement of a simple gravitational lens at L and the formation of the Einstein ring. Top left: the paths of light from the imaged source beyond the lens at S and locations of the images I_1 and I_2 seen from earth, E. Top right: geometrical relations for the case of a weak lens. Bottom: appearance of the two images at different alignment angles as seen by the observer. The source is moved closer to the lens from left to right until the nearly perfect alignment forms the Einstein ring.

observer, and D_{ds} is the distance between the lens and the source. ξ is the distance the light ray passes from the lens centre, and β is the true direction to the lens. The angle ω is the direction of the lens as seen from the earth.

The image of a point source behind the lens splits into two images and is amplified as shown in the bottom of Fig. 17.1. As the angle between the lens and the imaged source decreases, the images grow and when they are precisely on the line of sight, they merge into the 'Einstein ring'. The ring's angular radius is given by

$$\theta_E = \sqrt{\frac{4GM}{c^2}\frac{D_{dS}}{D_d D_S}}, \tag{17.2}$$

and typical values for a galaxy and a cluster of galaxies are $\theta_E \sim$ 1–2 arcsec and $\theta_E \sim$ 30 arcsec, respectively. Figure 17.2 displays a collection of Einstein rings observed with the HST.

Four images are commonly produced by non-point-like gravitational potentials. Light amplification of the images is produced when β is near θ_E and this has been used to determine structures of distant lensed galaxies.

Since the light-paths of the images differ in length, a fluctuation in the brightness of the source, will arrive at different times, which depends on the Hubble constant. The time delay, τ, is a combination of H_0, the angular

Fig. 17.2. Examples of Einstein rings observed with the HST. These are produced by galaxies, which are the lenses, at the centre of each image, surrounded by the blue images of distant background galaxies. (*Credits*: NASA/ESA/SLACS Survey team: A. Bolton (Harvard/Smithsonian), S. Burles (MIT), L. Koopmans (Kapteyen), T. Treu (UCSB), L. Moustakas (JPL, Caltech.))

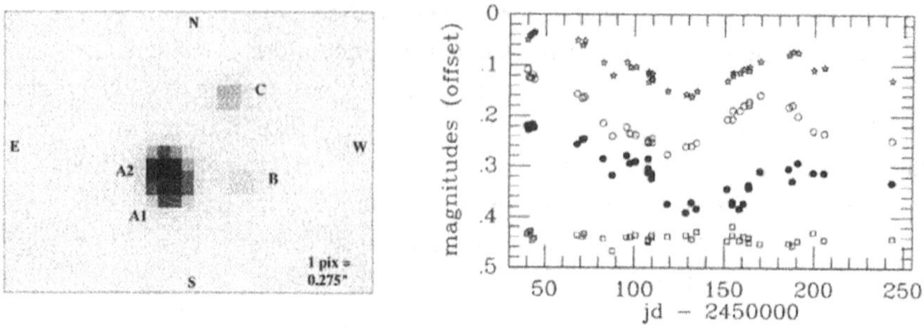

Fig. 17.3. Showing the lensed components of the quasar PG 1115+080. Component C (○) leads B (•) by about 23.7 days and components A1+A2 (⋆) by 9.4 days. (*Source*: Schechter P. L. *et al.* (1997), *ApJ*, 475, L85, "The quadruple gravitational lens PG 1115+080: time delays and models," Fig. 2, ©AAS Reproduced with permission.)

diameter distance, $D_A \approx \frac{cz}{H_0}\left(1 - \frac{3+q_0}{2}z\right)$, and $\varphi_{\rm lens}$, which depends on the lens' structure:

$$\tau = \frac{1}{c} D_A \varphi_{\rm lens}. \tag{17.3}$$

The differences in the arrival times are seen in the light curves in Fig. 17.3 for the four component quasar PG1115+080 (Schechter *et al.*, 1997).

Besides fitting the individual light curves of the components to secure τ, a major problem is knowing the shape of the gravitational potential of the lens; e.g., a point mass and the singular isothermal sphere give quite different values of H_0, 84 and 42 km s^{-1}Mpc^{-1}, respectively, for PG1115+080. Schechter (2005) discusses these problems and finds $H_0 = 61 \pm 7$ km s^{-1}Mpc^{-1} for ten sources with lens delays. With time, the precision of the determination of H_0 has improved. Jee *et al.* (2019) obtained accurate determinations of D_A from the angular diameter distances of two well observed quasars, RXJ1131-1231 and B1608+656, and used this to calibrate the relative distances of 740 SN Ia to derive $H_0 = 82^{+8.4}_{-8.3}$km s^{-1}Mpc^{-1}.

17.1. Further Applications of Gravitational Lensing

The degree of lensing can be subdivide into three classes:

(1) **Microlensing**: Microlensing is produced by individual masses in a galaxy. This has been used to probe the extent of dark stars in the halo of the galaxy by several surveys looking for massive compact halo

objects, or machos. These surveys include MACHO (Alcock *et al.*, 2000), OGLE (Udalski, 2003), EROS-2 (Tisserand *et al.*, 2007), and WeCAPP (Lee *et al.*, 2015), but there are many more underway. These investigations each consumed several years of investigation in order to look through the galactic halo for lensing events of stars in the Magellanic clouds and Andromeda galaxy. If a lensing object aligns with a star in one of these background objects, it can be observed by the light amplification and Einstein ring effect. The light curve can be distinguished from variable stars as the lensing produces a monochromatic curve, while variable stars change colour. Although these investigations have found thousands of new variables and a large number of extra-solar planet suspects, the results have produced a surprisingly low number of lensing events. In addition to cosmological effects, searches for planets by microlensing are being conducted (Wamsganss, 2016).

(2) **Strong lensing**: Luminous arcs have now been observed for many galaxies and clusters of galaxies (such as Abell S1063 in Fig. 18.1). Some of the arcs have been employed to reconstruct the structures of imaged background galaxies. Masses can be obtained from the strong lensing. Geller *et al.* (2014) compared masses determined for the Abell 383 galaxy cluster using the dynamical caustic technique with strong and weak lensing, which agree well, $M_{200}(\text{caus}) = (5.07 \pm 0.09) \times 10^{14}\text{h}^{-1}\text{M}\odot$ and $M_{200}(\text{lens}) = 4.6^{+1.1}_{-0.8} \times 10^{14}\text{h}^{-1}\text{M}\odot$, respectively.

(3) **Weak lensing**: These effects of the lensing appear as minute arclets that generally require large amounts of data. This has several important applications to cosmology in determining mass structures of objects on different scales from individual galaxy clusters to the construction of weak-lensing maps. Gravitational lensing depends neither on the type of matter nor its dynamical state as light passes through it along its path and hence weak lensing directly measures the total mass density along the line of sight and does not require using biased mass tracers such as galaxies.

The techniques for converting the arclets to convergence diagrams and hence mass distributions were first given by Kaiser & Squires (1993) and Bartelmann & Schneider (2001) and used in many studies of galaxy clusters as discussed in Chapter 18. Tyson, Valdes, & Wenk (1990) first observed the effect in the galactic clusters Abell 1689 and CL 1409+52. However, wide-field convergence maps have only recently become possible due to the construction of high quality wide field imagers on large telescopes.

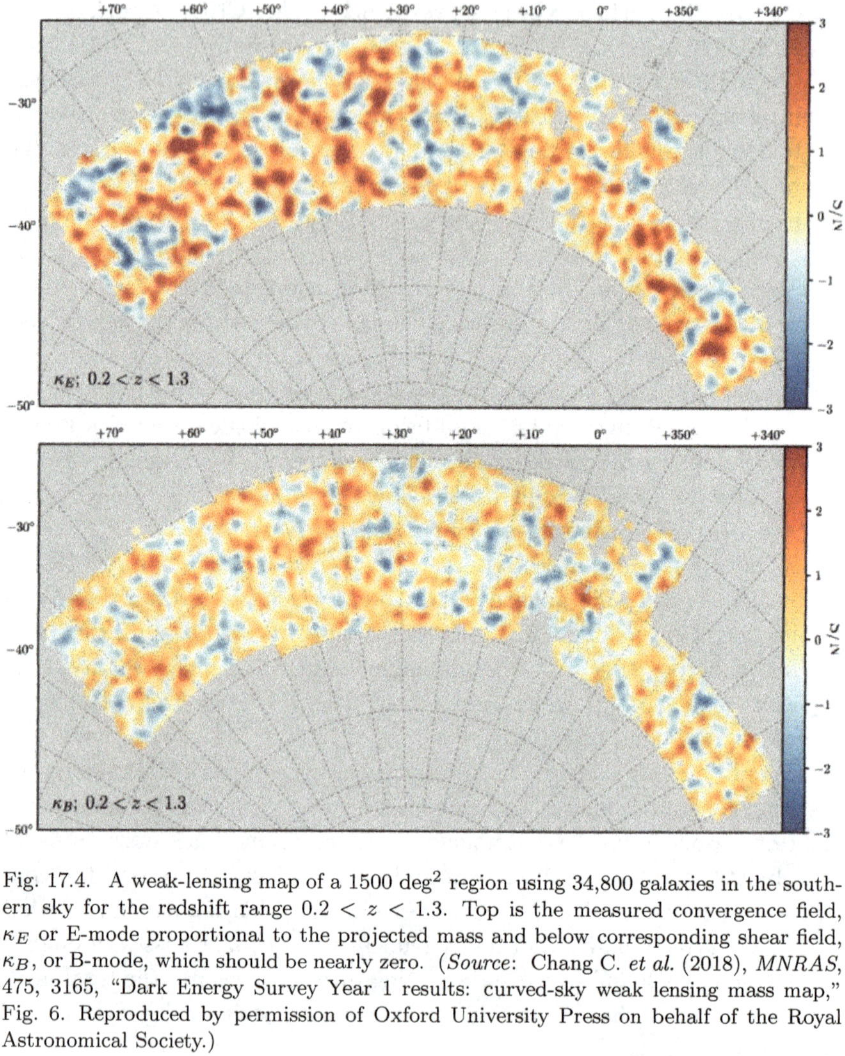

Fig. 17.4. A weak-lensing map of a 1500 deg^2 region using 34,800 galaxies in the southern sky for the redshift range $0.2 < z < 1.3$. Top is the measured convergence field, κ_E or E-mode proportional to the projected mass and below corresponding shear field, κ_B, or B-mode, which should be nearly zero. (*Source*: Chang C. *et al.* (2018), *MNRAS*, 475, 3165, "Dark Energy Survey Year 1 results: curved-sky weak lensing mass map," Fig. 6. Reproduced by permission of Oxford University Press on behalf of the Royal Astronomical Society.)

Van Waerbeke *et al.* (2013) produced one of the first large mass maps from the CFHTLenS survey. Geller *et al.* (2005; 2010) and Kurtz *et al.* (2012) compared mass maps from weak lensing with extensive redshift surveys in the same sky regions. Weak lensing is becoming a method to map matter distributions. Starikova *et al.* (2014) compared weak lensing selected galaxy clusters with X-ray detections and found the lensing method more efficient for high-mass systems. Extensive observational surveys are underway by several large collaborations, e.g., using the Hyper Suprime-Cam (HSC) on the 8.2 m Subaru telescope (Oguri *et al.*, 2018; Aihara *et al.*, 2018) and by the Dark Energy Survey (Chang *et al.*, 2018). Figure 17.4 shows a portion of the convergence map from the Dark Energy survey. Compared to the mass maps derived from peculiar motions, these maps reach higher redshifts. The resolution on the sky is lower (3.44 arcmin DESY1) and they still are not all-sky.

Chapter 18

Motions of Binary Galaxy Clusters

18.1. Galaxy Clusters in General

Motions of matter generated in different objects complement peculiar motions and provide important insight into gravity as they contain the same baryonic and dark matter, etc. constituents. Galaxy clusters have been studied extensively in the 20th and 21st centuries. The matter in motion of the components within galaxy clusters and the collisional motions of the clusters themselves are of considerable interest for galaxy evolution and cosmology. In the ΛCDM picture of structure formation, galaxy clusters grow hierarchically, typically along the intersections of the filaments from the mergers of smaller higher density knots. Although the centres of these regions become highly nonlinear and the early histories of the individual systems subsequently become erased by virialization, the mergers of the clusters themselves produce motions that can probe aspects of the collisional dynamics involving gravitation, dark matter, and galaxy evolution in dense environments.

Typical galaxy clusters have radii ~2–10 Mpc and contain $\sim 10^2$–10^3 galaxies visible in the optical, but have much larger total masses of $\sim 10^{14}$–$10^{15}\,M_\odot$. Early X-ray imaging revealed the hot gas and structures of galaxy clusters (Forman & Jones, 1982). The abundances of their components are similar to the universe as a whole; directly detectable baryonic matter consists of the galaxies and hot X-ray emitting gas ($T \sim 10^7$–10^8 K) that contribute roughly 1% and 9% respectively of the total mass. Dark matter makes up the remaining 90%. Investigations of galaxy clusters employ X-ray satellites such as Chandra for the gas properties, HST and ground-based surveys for imaging of the galaxy distributions and properties, strong and weak lensing, and MOS measurements of redshifts.

Fig. 18.1. HST image of the galaxy cluster Abell S1063 ($z = 0.351$) showing lensing features and intergalactic light. (*Image credit*: NASA, STScI, ESA, and M. Montes (University of New South Wales).)

Special surveys of galaxy clusters with HST using several wavelength passbands have been conducted in the CLASH (Cluster Lensing and Supernova survey with Hubble) for 25 clusters (Postman *et al.*, 2012) and the Frontier Fields survey (Lotz *et al.*, 2018) of six additional clusters using nearly an order of magnitude longer exposures to reach about 1.5 magnitudes fainter. Figure 18.1 is a deep optical image of the single galaxy cluster, Abell S1063 obtained with the HST Frontier Survey. The brightest cluster galaxy dominates the centre of the picture and is surrounded by smaller cluster members, luminous arcs from strong lensing of background sources as well as weak lensing and intergalactic light. Montes & Trujillo (2018) have studied intergalactic light from six massive clusters in the Frontier Fields survey and conclude that it is an excellent tracer of the DM.

Many galaxy clusters in general and including double clusters can be studied from combined optical, radio, and X-ray observations and utilizing the powerful technique of weak lensing to reconstruct their mass distribution, which can reveal substructures. Weak lensing was first applied by

Tyson, Valdes, & Wenk (1990) for galaxy clusters and numerous examples of the analysis are given, e.g., in Okabe & Umetsu (2008), Umetsu *et al.* (2012), and Medezinski *et al.* (2013).

If the cluster galaxies formed by redshifts $z \sim 2$, much of the stellar content in the large ellipticals will have evolved following an early burst of star formation, which consumes available gas for star formation. The massive blue main sequence stars evolve rapidly and will be gone. In later stages, the light from these galaxies is dominated by stars red giant branch stars in the HR diagram. Lower mass spirals that have not consumed their gas supply can still form young stars and have bluer colours. In the nearby universe, most galaxies lie on the famous Hubble (1926) "tuning fork" diagram, but by redshifts $z > 0.3$, this morphological pattern changes and it is nearly gone by $z > 1.0$ (e.g., Abraham & van den Bergh, 2001). Consequently, the effects of evolution can be seen in the colour-magnitude diagrams of galaxy clusters. Many galaxy clusters show a distinct red sequence such as in Fig. 18.2. Other galaxy clusters may show a second blue sequence below the red sequence in this diagram due to the star forming galaxies.

In addition to this colour separation, the corresponding morphological classes of galaxies differ in spatial distribution as in Abell 576 (Fig. 18.3) where spirals are spread more compared to the E and SO galaxies. These effects and others to be mentioned offer clues to the formation, dynamical evolution, and constituents in the galaxy clusters.

18.1.1. *Spherical infall model and galaxy cluster evolution*

The spherical infall or halo model is often used to provide a general order-of-magnitude analytic description of the formation and evolution of galaxy clusters and galaxies in spherical halos. This has been studied in great detail with N-body simulations already described, but this model provides a useful approximation. Since dark matter dominates the matter, this really refers to *dark matter halos* and the galaxies and baryonic gas follow the main process. This model is based on $\Lambda = 0$ CDM and Einstein–de Sitter cosmology (e.g., Peebles, 1993; Padmanabhan, 1993; Binney & Tremaine, 2008) where the scale factor is $S \sim t^{2/3}$ and it utilizes linear perturbation theory. The spherical geometry is also an idealization. Frenk (1991), Efstathiou (1990), and others have discussed the model's qualitative agreement with numerical CDM simulations. In the expanding universe, the mean matter density decreases as $\rho_0 = 1/(6\pi G t^2)$. A density enhancement, δ will expand more slowly due to its larger gravity. If its original density contrast is above

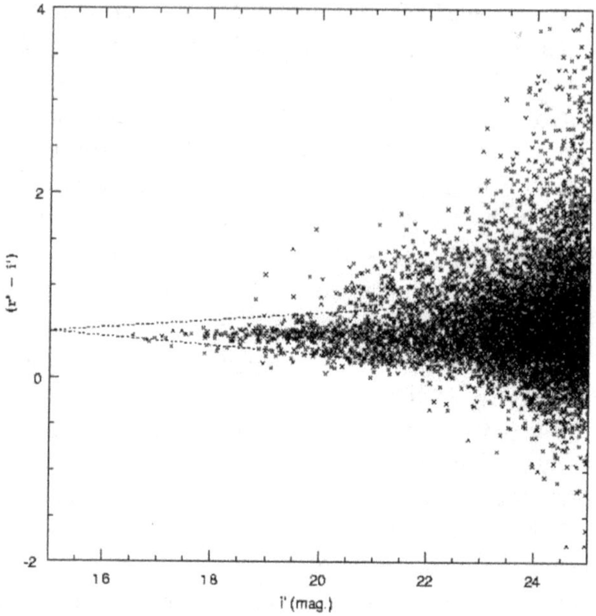

Fig. 18.2. The red sequence for non-stellar objects in the galaxy cluster Abell 2465 ($z = 0.245$) separated using the SEXTRACTOR code (Bertin & Arnouts, 1996). The red sequence members are outlined by the dotted triangle. Many of the faint red objects in the upper right part of the diagram will be background objects. (*Source*: Wegner G. (2011), *MNRAS*, 413, 1333, "The double galaxy cluster Abell 2465 — I. Basic properties: optical imaging and spectroscopy," Fig. 3, Reproduced by permission of Oxford University Press on Behalf of the Royal Astronomical Society.)

a critical value,

$$\Delta = \rho/\rho_0 \approx \frac{1.69}{(1+z)}, \tag{18.1}$$

it grows to a maximum radius until 'turn around', after which it collapses until it reaches virial equilibrium when the total energy is twice the potential energy, $W = 2E$. The density contrast at this stage is $\Delta = 18\pi^2 = 178 \approx 200$ and the radius of the resulting sphere of mean density, $\Delta\rho_0$, is often defined as the 'virial radius', $R_{\rm vir} \approx \sqrt[3]{3M_{200}/(4\pi \times 200\rho_0)}$. Although a homogeneous density profile is assumed for this model, it will not be constant and different approximations such as the NFW or Einasto profiles (Eqs. (18.5) and (18.7)) would hold.

The halos grow hierarchically from mergers and the Press–Schechter (Press & Schechter, 1974; Longair, 1998) approximation resulting from a

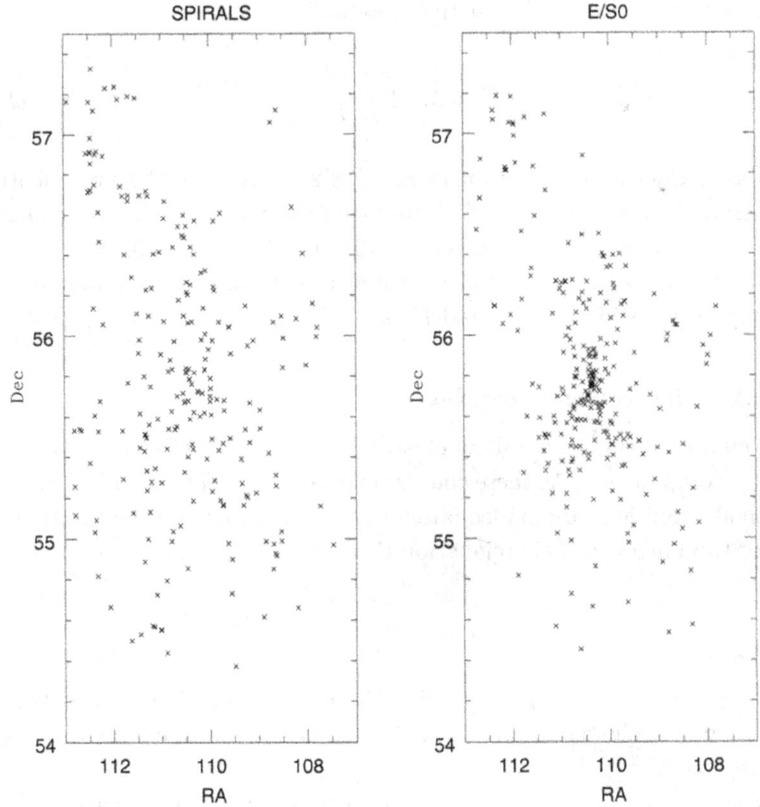

Fig. 18.3. The distributions of spiral and elliptical/S0 galaxies in the galaxy cluster Abell 576 ($z = 0.0389$) showing the concentration of the gas-poor early-type galaxies towards the centre compared to the more widely spread less massive gas-rich spirals. From the author using the POSS-2 R images and cluster member redshifts in Rines *et al.* (2000).

scale-free power spectrum of density fluctuations ($P(k) \sim k^n$), describes the number distribution of systems of different mass, $N(M)$, by a formula of the form,

$$N(M) = \frac{1}{\sqrt{\pi}}\left(1+\frac{n}{3}\right)\frac{\rho_0}{M^2}\left(\frac{M}{M^*}\right)^{(3+n)/6} \exp\left[-\left(\frac{M}{M^*}\right)^{(3+n)/3}\right], \tag{18.2}$$

where M is the mass and M^* is a lower cutoff for the formation of structures. The time dependence of $N(M)$ results from the time variation of M^*

relative to the value of M^* at the present time, t_p

$$M^* = M_p^* \left(\frac{t}{tp}\right)^{4/(3+n)}. \tag{18.3}$$

The analytical approximation in Eq. (18.2) illustrates the main features of hierarchical clustering. It describes that the most massive systems form later and massive galaxy clusters ($\sim 10^{14}$ to 10^{15} $M_\odot$) will become more numerous at low z. The low mass systems ($\sim 10^9$ $M_\odot$) have a large number density over a wide range of redshifts.

18.1.2. *Relaxation processes*

The final structure of a galaxy cluster is related to the baryonic and dark matter components. As these constituents relax to a final equilibrium state, two-body collisions are inadequate for them to relax in a time less than the age of the universe. This relaxation time is

$$\tau_{\rm relax} \approx \frac{0.1N}{\ln N}\tau_{\rm cross}, \tag{18.4}$$

for N particles and $\tau_{\rm cross} \sim R/V$ (Binney & Tremaine, 2008). For the typical galaxy cluster, $\tau_{\rm relax} \sim 2 \times 10^{10}$ years and for an elliptical galaxy it is $\sim 10^{11}$ years.[1]

Thus, a number of other processes could speed up relaxation. These include: (a) *dynamical friction* can affect the orbits of the galaxies as they pass through the cluster, but since the main DM component of the clusters and the galaxies are collisionless, dissipationless processes that produce relaxation should operate. These include: (b) *phase mixing*, orbits originally in phase get increasingly out of phase and separate, approaching a steady state; (c) *violent relaxation* (Lynden-Bell, 1967; Kandrup, 1998; Antoniazzi *et al.*, 2007), from evolution in the gravitational potential, e.g., during early collapse, changes a galaxy's orbit; and (d) chaotic mixing and Landau damping. The baryonic matter can have an effect on the DM orbits via *baryonic contraction* (e.g., Gnedin *et al.*, 2004), where the cooling of the baryons sinks them towards the centre of the DM halo leading to adiabatic contraction which steepens the radial profile in the central regions.

[1]This type of argument was used in the early 20th century to estimate a larger age of the universe than what is now accepted.

18.1.2.1. *NFW and other profiles*

Numerical simulations show that the growth of DM halos leads to an almost universal density law that can be described by a number of formulae. The approximation often used is the NFW (Navarro, Frenk, & White, 1997) profile,

$$\rho(r) = \frac{\rho_0}{\left(\frac{r}{R_S}\right)\left(1 + \frac{r}{R_S}\right)^2}, \tag{18.5}$$

where the 'scale radius', R_S, is related to the virial radius by $R_{\text{vir}} = cR_S$ and c is the 'concentration parameter'. Łokas & Mamon (2001) give a useful compilation of the properties of the NFW profile. Equation (18.5) does not have finite total mass as $r \to \infty$, so the object is generally taken to be truncated at R_{vir}. This mass at the virial limit is,

$$M = 4\pi \int_0^{R_{\text{vir}}} \rho(r) r^2 dr = 4\pi\rho_0 R_s^3 \left[\ln(1 + c) - \frac{c}{1 + c}\right]. \tag{18.6}$$

The named profiles are not derived from first principles and are fit empirically to N-body simulations. The Einasto profile (Merritt *et al.*, 2006; Retana-Montenegro *et al.*, 2012) is defined by a power-law logarithmic slope

$$\gamma(r) = -\frac{d \ln \rho(r)}{d \ln r} \propto r^{-1/n}. \tag{18.7}$$

It provides a good fit and has finite mass. The Hernquist, profile (Hernquist, 1990) has also been employed. The NFW profile describes galaxy cluster and many early-type galaxy profiles well. Typical predicted values for c are $c = 3$–20 for galaxies and $c = 4$–6 for galaxy clusters, with the larger c occurring at low redshift (Zhao *et al.*, 2003).

For low-mass dwarf galaxies, Eq. (18.5) conflicts with observations, known as the *cuspy halo problem* (e.g., Moore, 1994; de Blok, 2010). N-body simulations predict that CDM halos follow a steeply rising central density law, but the central regions of dwarf galaxies have a nearly constant central density, possibly indicative of a modification of their formation physics.

18.2. Colliding Galaxy Clusters

A convenient approximation to the motion of two colliding galaxy clusters, as well as other physical situations, that has been extensively employed by several investigators is the 'radial infall model' applied by Beers, Geller, & Huchra (1982) to the Abell 98N and Abell 98S pair. The two clusters are taken to be mass points of total mass $M = m_1 + m_2$ that move on a linear trajectory. Their relative velocity and separation are V and R at the time t. The cosmology is approximated by the Einstein–de Sitter model where the flat case expansion has the scale factor $S(t) \sim t^{2/3}$. V and R are related to the observed separation, R_p and velocity, V_r, projected on the sky, which depend on the projection, or inclination, angle φ as

$$R_p = R\cos\varphi \text{ and } V_r = V\sin\varphi. \tag{18.8}$$

It can be shown that the condition for the clusters to be bound is given by

$$V_r^2 \leq 2GM\sin^2\varphi\cos\varphi. \tag{18.9}$$

The motion is expressed in terms of the parameter η and if the two clusters are bound,

$$R = \frac{R_m}{2}(1 - \cos\eta), \tag{18.10}$$

$$t = \sqrt{\left(\frac{R_m^3}{8GM}\right)}(\eta - \sin\eta), \tag{18.11}$$

$$V = \sqrt{\left(\frac{2GM}{R_m}\right)}\frac{\sin\eta}{(1 - \cos\eta)}, \tag{18.12}$$

where R_m is the maximum separation of the clusters.

For the unbound case, the solution is

$$R = \frac{GM}{V_\infty}(\cosh\eta - 1), \tag{18.13}$$

$$t = \frac{GM}{V_\infty^3}(\sinh\eta - \eta), \tag{18.14}$$

$$V = V_\infty\frac{\sinh\eta}{(\cosh\eta - 1)}, \tag{18.15}$$

where V_∞ is the asymptotic expansion velocity.

The three equations can be solved numerically by adding one additional piece of information, $t = t_o$, the age of the universe at the cluster's observed redshift.

For the bound case, there are three solutions, which are the infalling and the outgoing cases. Figure 18.4 gives the motion of the double cluster

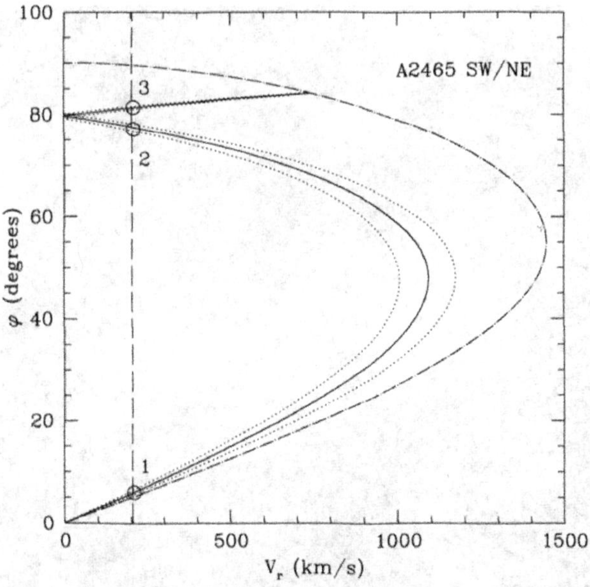

Fig. 18.4. The three solutions of the bound case of the radial infall model for the galaxy cluster Abell 2465. The solid and dotted curves are for masses of $M = 8 \times 10^{14} M_{\odot}$. The vertical broken line shows the measured $V_r = 205\,\text{km s}^{-1}$. The rightmost dot-dashed curve is the bound solutions limit. Solutions 1 and 2 denote the outgoing and 3 the infalling cases. (*Source*: Wegner G. (2011), *MNRAS*, 413, 1333, "The double galaxy cluster Abell 2465 — I. Basic properties: optical imaging and spectroscopy," Fig. 18, Reproduced by permission of Oxford University Press on Behalf of the Royal Astronomical Society.)

Abell 2465 as an example. Several investigators have used the radial infall model to study the dynamics of double galaxy clusters, including, Boschin *et al.* (2010) for Abell 2345, Girardi *et al.* (2011, 2014) for Abell 2254 and Abell 1351, Maurogordato *et al.* (2011) for three bimodal clusters, Barrena *et al.* (2013) for Abell 1914, and Owers *et al.* (2014) for Abell 1882.

Although the radial in fall model is a useful approximation, it doesn't describe the individual behaviors of the baryonic gas, the stars, and the dark matter in the galaxy clusters. The 'Bullet Cluster' (IE0657-56) has been extensively studies (e.g. Clowe *et al.*, 2006). Additional clusters of this type are known such as MACS J00254.4-1222 (Fig. 18.5, also Fig. 18.9). Including gas, dark matter, non-radial orbits, etc. can be studied by N-body models, e.g., by Roettiger, Loken, & Burns (1997), Takizawa (2000), Ricker & Sarazin (2001), Poole *et al.* (2006, 2008), and Molnar, Hearn, & Stadel

Fig. 18.5. Post collision double galaxy clusters showing the separation of the gas and collisionless components. Combined optical and X-ray picture of the merging double galaxy cluster MACS J00254.4-1222. The optical galaxies and dark matter determined from weak lensing are shown in blue. The distribution of the hot gas visible in X-rays lies between the two clusters. (*Credit*: NASA, ESA, CXC, M. Bradač & S. Allen.)

(2012), of the cluster collisions give insight into how the major constituents, the dark mater, galaxies and stars, and the baryonic gas behave during the interaction. The modelling predicts, and is borne out by observation. Dark matter and gas separate; the collisionless dark matter and galaxies pass through each other and form two lobes, while the hot shock heated baryonic gas is left in the middle. The effect begins to be felt even before the closest approach of the two clusters (Vijayaragharan & Ricker, 2013). Figure 18.6 shows some sequences of equal $10^{15}M_{\odot}$ cluster collisions by Poole *et al.* (2006) that illustrate the surface evolution of the gas.

Rather than assuming the total binding mass, Eqs. (18.10)–(18.12) can be combined to give M_G, and the age of the system, t_H in terms of the observed projected distance between the two galaxy clusters, $\Delta\ell_\perp$ and their

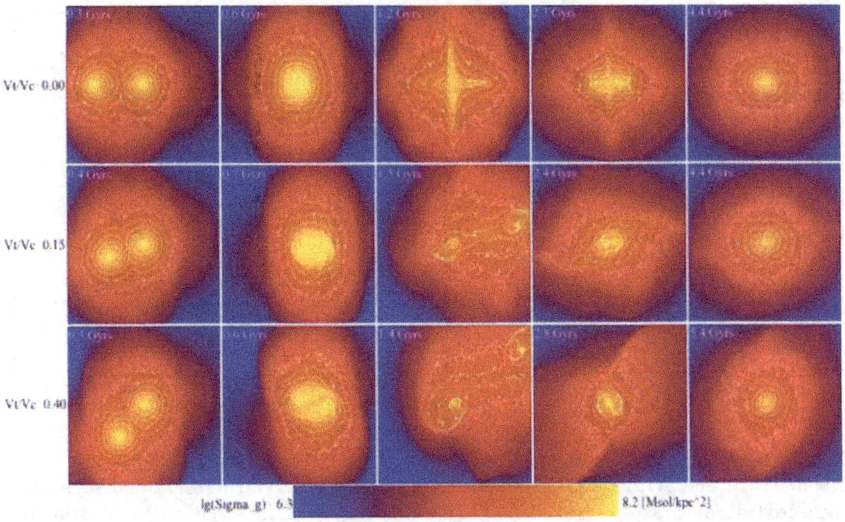

Fig. 18.6. The surface density of gas and simulated *Chandra* isophotes in the collisions of two equal 10^{15} $M_\odot$ galaxy clusters that follow a NFW profile looking perpendicularly to the orbit paths. The squares are 3 Mpc on edge, V_t/V_c is the ratio of the tangential to central velocity of the collision. (*Source*: Poole G. B., Fardal M. A., Babul A., McCarthy I. G., Quinn T., Wadsley J. (2006), *MNRAS*, 373, 881, "The impact of mergers on relaxed X-ray clusters — I. Dynamical evolution and emergent transient structures," Fig. 1, Reproduced by permission of Oxford University Press on behalf of the Royal Astronomical Society.)

velocity difference, $\Delta V_{\rm los}$,

$$t_H = 9.8 \times 10^8 \text{yrs} \left[\frac{(\eta - \sin\eta)\sin\eta}{(1-\cos\eta)^2}\cot\varphi\right]\left(\frac{\Delta\ell_\perp}{1\text{Mpc}}\right)\left(\frac{1000\text{km/s}}{\Delta V_{\rm los}}\right), \tag{18.16}$$

and

$$M_G = 2.3 \times 10^{14} \text{M}_\odot \left[(1+\cos\eta)\cos^2\varphi\sin\varphi\right]^{-1}\left(\frac{\Delta V_{\rm los}}{1000\text{km/s}}\right)\left(\frac{\Delta\ell_\perp}{1\text{Mpc}}\right), \tag{18.17}$$

with $\tan\phi = \Delta_{\rm los}/\Delta\ell_\perp$ and assuming t_H, allows Eq. (18.16) to be solved for η and ϕ. The result is insensitive to the assumed age; $t = 12$ and 17 Myr produce similar results. M_G in Eq. (18.17) is a function of φ. Even if $\Delta V_{\rm los}$ cannot be established, these curves have a minimum mass as seen in Fig. 18.7. This M_G, or at least its minimum can be compared to the virial masses, M_1, M_2, of the two sub-cluster components, measured from

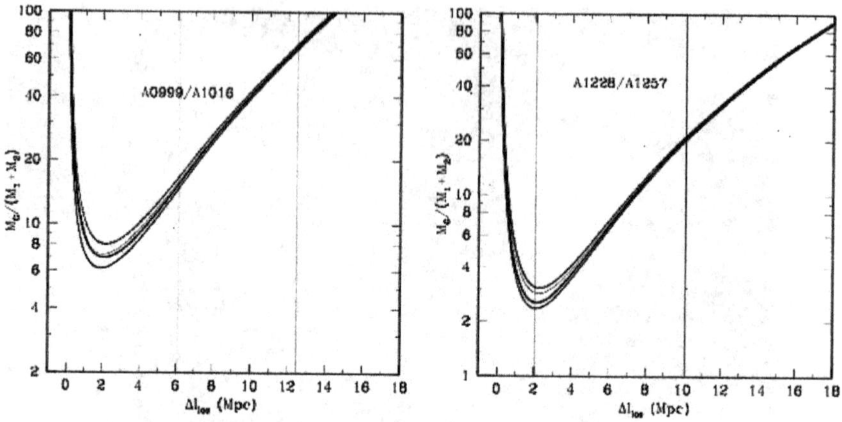

Fig. 18.7. Ratios of the merger mass, M_G, to the virial mass for the two cluster pairs, A999/A1016 and A1228/1257, as a function of $\Delta_{\rm los}$ using Eq. (18.17) and separately measured virial masses, M_1 and M_2. The nine curves include the range of estimated errors in $\Delta_\perp$ and $V_{\rm los}$. $t_H = 13.3$ Gyrs was used; differences in age have only small effects and were not included. Left-hand vertical lines represent derived $\delta_{\rm los}$ and right-hand lines is the 0.7 confidence level (Wegner, unpublished).

their redshifts (e.g., Heisler, Tremaine, & Bahcall, 1985), giving the ratio of the total mass and the virial masses, $M_G/(M_1 + M_2)$.

Mohr & Wegner (1997) studied the double cluster Abell 2626 ($z \sim 17500\,{\rm km\ s^{-1}}$) using this method. With $\Delta V_{\rm los} = 2350\,{\rm km\ s^{-1}}$and $\Delta \ell_\perp = 0.707 \pm 0.052)h^{-1}\,{\rm Mpc}$, they employed the fundamental plane to measure $\Delta_{\rm los} = (14 \pm 18)h^{-1}{\rm Mpc}$, which gives a total binding mass of 1.65 times the sum of the virial masses. Donegan (2001) considered three pairs, A2061/A2067, A999/A1016, and A399/A401 and found that minimum $M_G/(M_1 + M_2)$ is from 0.8 to 83 using the data available at the time. Wegner & Kang (2010) obtained more extensive data for A999/A1016 and A1228/A1267 and found ratios of 6 and 2.5, respectively (Fig. 18.7). Generally, the minimum of this ratio is several times larger than unity.

More elaborate and realistic analyses of merging galaxy clusters are being developed to be used in future surveys. Dawson (2013) produced a Bayesian Monte Carlo method, which has been used to study several clusters. Massey, Kitching, & Nagai (2011) utilize N-body hydrodynamical simulations and 'cluster bulleticity', which they define as the offset between ordinary matter and dark matter in a cluster.

In many cases, the galaxy clusters have complex sub-structures (e.g., Cohen *et al.*, 2014, 2015; Merten *et al.*, 2011; Medezinski *et al.*, 2016) and this complicates the analysis of their collisions and mergers. Nevertheless, there is a growing number of relatively simple double clusters and these are to be sorted into relaxed clusters with circular symmetry of all their dark matter, galaxy, and gas components, pre-mergers showing some distortion, and post-mergers with disturbed and separated components indicative of core-crossing events.

18.2.1. *Gas 'sloshing'*

Also interesting is the case of 'sloshing' where the clusters have narrowly missed each other, but distortions of the gas are set up (e.g., Markevitch & Vilklinin, 2007; Johnson *et al.*, 2010; ZuHone *et al.*, 2013) and produce cold fronts. The plasma is magnetized and adds information on the structure of the process. ZuHone, Zavala, & Vogelsberger (2019) have considered models, which include SIDM. They find that for $\sigma_{\rm DM} \gtrsim 1{\rm cm}^2/{\rm g}$, separation of peaks in the X-ray emission and the thermal SZ-effect might provide a constraint on the DM self-interaction.

18.3. Star Formation in Mergers

Colliding double galaxy clusters can provide a laboratory for investigating galaxy evolution and the interactions between the DM, and different baryonic components in the form of gas and stars under varying conditions of density. For galaxies in general, it is now well established that star formation since $z \approx 1.0$, or ~8 Gyr ago, is declining according to approximately $\Sigma{\rm SFR} \propto (1+z)^{3.9}$ (e.g., Le Floc'h *et al.*, 2005). This effect is known as 'downsizing'. Galaxy clusters are known to grow by merging and galaxies in their outer limits tend to be less relaxed and more gas rich, but those in the central cluster regions are more gas depleted, having undergone more collisions. Thus, a range of dynamical and density conditions are observable in galaxy clusters and the collision of two clusters will add additional conditions.

Compared to the general decrease of the star formation rate, galaxy clusters show a larger decline rate with $\Sigma{\rm SFR}/M_{\rm Cl} \propto (1+z)^6$, but colliding clusters can show enhanced star formation at lower redshifts (Finn *et al.*, 2005; Koyama *et al.*, 2010; Chung *et al.*, 2011; Popesso *et al.*, 2012; Webb *et al.*, 2013).

Massive galaxy clusters tend to show less star formation than lower mass clusters with lower temperatures (Laguna *et al.*, 2008). Some colliding clusters exhibit higher or lower degrees of enhanced star formation (e.g.,

Finn *et al.*, 2005; Hwang & Lee, 2009; Dawson *et al.*, 2012; Wegner, Chu, & Hwang, 2015) and it is interesting to look for relations between enhanced star formation and the distributions of the DM and the baryonic matter distributions.

Figure 18.8 presents the double galaxy cluster Abell 2465. The X-ray emission from the hot baryonic gas, the galaxies, and the dark matter contours are shown. This pair appears to be a pre-collision example that shows enhanced star formation (Wegner *et al.*, 2017).

Star formation rates in galaxies can be estimated a number of ways, but the two that are widely used employ measured optical emission line strengths and NIR data. The first method utilizes optical measurements

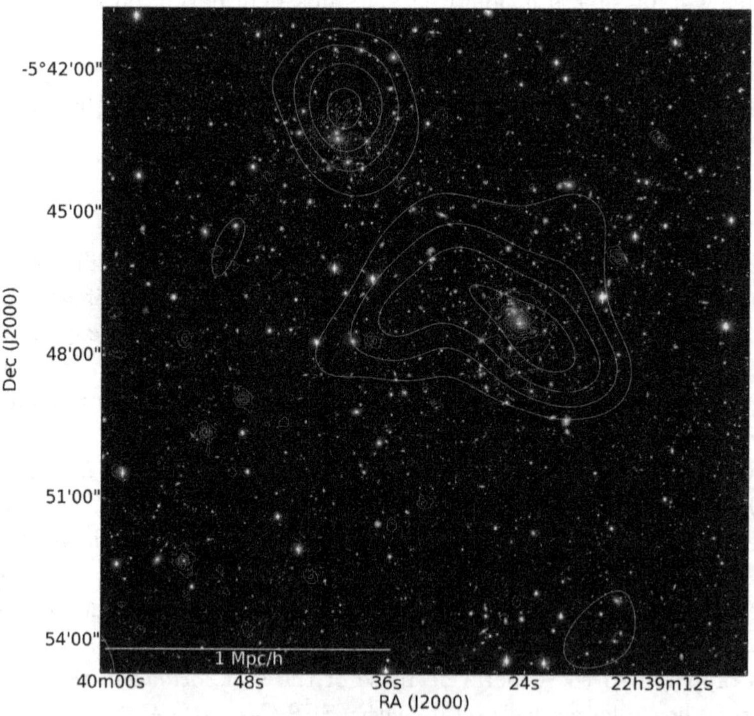

Fig. 18.8. The optical appearance of the pre-collision double galaxy cluster Abell 2465 from deep Subaru images. Superimposed are the X-ray distribution from Chandra (red contours) and the weak lensing mass contours (light curves) (*Source*: Wegner G. A., Umetsu K., Molnar S. M., Nonino M., Medezinski E., Andrade-Santos F., Bogdan A., Lovisari L., Forman W. R., Jones C. (2017), *ApJ*, 844, 67, "The double galaxy cluster A2465. III. X-ray and weak-lensing observations," Fig. 7, ©AAS Reproduced with permission.)

of the Hα emission. This requires a reddening correction, which can be determined, if spectra are available, from the ratio $L_{H\alpha}/L_{H\beta}$. Kennicut's relation for the star formation rate (Kennicut, 1998) is often used, where $L_{H\alpha}$ is in cgs units,

$$\mathrm{SFR} = L_{H\alpha}/1.26 \times 10^{41}\ M_{\odot}\mathrm{yr}^{-1}. \tag{18.18}$$

The second method is to determine the SFR from infrared dust emission. For example, this can be satellite data, such as from the WISE satellite (Wright *et al.*, 2010; Jarrett *et al.*, 2011; Chung *et al.*, 2011; Stern *et al.*, 2012; Hwang *et al.*, 2012a), but other satellite data have been used. The total infrared dust luminosity, L_{IR}, can be estimated from templates provided by Chary & Elbaz (2001) and Kennicut (1998):

$$\mathrm{SFR} = 1.72 \times 10^{-10} L_{IR}/L_{\odot}\ M_{\odot}\mathrm{yr}^{-1}. \tag{18.19}$$

Hwang *et al.* (2012b) find that the IR data from WISE and the Hα agree well.

18.3.1. *Limits on self-interacting dark matter*

The separation of the baryonic and dark matter in the collisions of two galaxy clusters has been utilized by a number of investigators. (e.g., Spergel & Steinhardt, 2000; Massey *et al.*, 2011; Harvey *et al.*, 2013, 2014; Rocha *et al.*, 2013; Peter *et al.*, 2013; Kahlhoefer *et al.*, 2014; Harvey, 2015) to study the non-gravitational interactions of the dark matter in these collisions and to provide limits on the collisional cross-sections of the dark matter if it consists of particles. If a dark matter halo passes through another body of DM particles, and if the dark matter is self-interacting (SIDM), this can produce a deceleration and evaporation of the halo. The deceleration will cause a displacement between the centres defined by the galaxies and that of the DM.

Figure 18.9 is a schematic of the collision of two galaxy clusters. The separations of the three components can be measured. The dark matter, D_a and D_b is found from gravitational lensing, the baryonic gas, G is found from X-ray images, and the centres of the galaxy distributions, S_a and S_b, are obtained from optical imaging and redshifts. They will all be affected by the gravitation, but if the DM is not perfectly collisionless, the gas will drag the DM and slow its motion. The drag has to be modelled and the precise physics can only be estimated. Another problem for this measurement is to

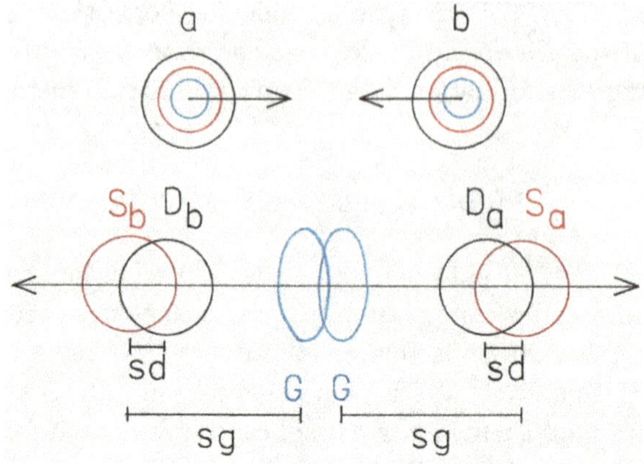

Fig. 18.9. Schematic of the collision of two equal mass galaxy clusters, a and b, along a linear trajectory seen perpendicular to the line of sight. G, S, and D represent gas, stellar, and dark matter components, respectively. The top row represents the arrangement of gas in blue, dark matter in black and galaxies in stars in red. The lower represents the displacement of the components after the collision if the DM is not perfectly collisionless.

determine the projection angle in three dimensions and whether the clusters have collided off axis.

Several investigators have found collisional cross-sections using the separation of the components for individual pairs, including Randall *et al.* (2008) who get $\sigma_{\rm DM}/m < 1.25\,{\rm cm}^2/{\rm g}$ for the 'Bullet cluster' and Merten *et al.* (2011) find $\sigma_{\rm DM} < 3 \pm 1\,{\rm cm}^2/{\rm g}$ in Abell 2744, which are typical measurements. In a large compilation, Harvey *et al.* (2015) have used 30 cluster pairs for which X-ray and optical data were available. The set a limit for the cross-section of the DM as $\sigma_{\rm DM}/m < 0.47{\rm cm}^2/{\rm g}$. Wittman, Golovich, & Dawson (2018) have reanalyzed these results using better data and correcting for large errors they found in the offset measurements and conclude that $\sigma_{\rm DM}/m \lesssim 2{\rm cm}^2/{\rm g}$ is the best upper limit.

Similar limits can also be set using the profiles of dwarf galaxies and galaxy clusters (e.g., Rocha *et al.*, 2013; Peter *et al.*, 2013; Ebert *et al.*, 2015). Kaplinghat, Tulin, & Yu (2014) have compared the astrophysical limits for SIDM with laboratory experiments in Fig. 18.10.[2] Tulin & Yu (2018) have discussed further cross-section measurements.

[2]A useful conversion between astrophysical and laboratory measurements is 1 barn/GeV = $0.6\,{\rm cm}^2/{\rm g}$.

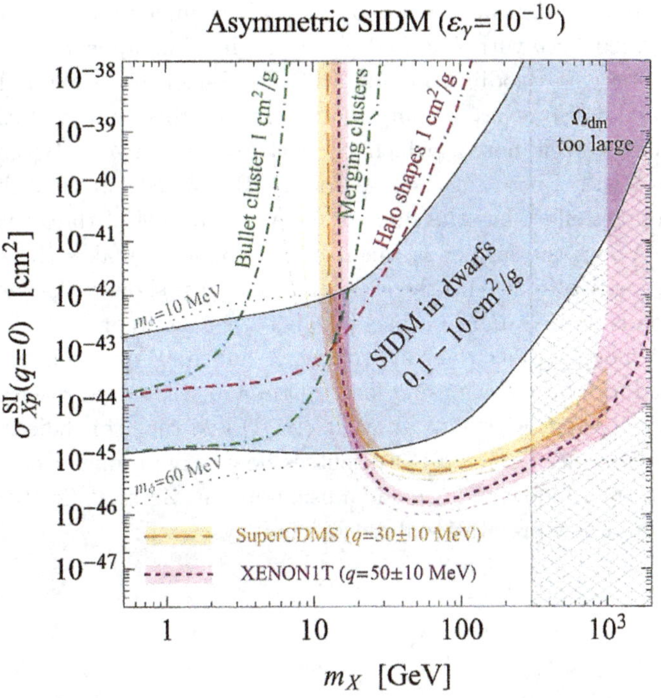

Fig. 18.10. Estimated detection limits for assymetric SIDM. Astrophysical limits for the Bullet Cluster, galaxy cluster halo shapes, and the shaded area corresponds to the lowering of dwarf galaxy central densities for $\sigma/m = 0.1 - 1.0\,\mathrm{cm}^2/\mathrm{g}$. The sensitivity for direct detection by the XENON1T experiment is the dashed red curve. (*Source*: Kaplinghat M., Tulin S., Yu H.-B. (2014), *Phys Rev.* D, 89, 5009, Fig. 3 (right half), Reprinted by permission. ©2014 by the American Physical Society.)

18.4. Infall Peculiar Velocities

Peculiar velocities produced by the velocity fields of galaxy groups and clusters can be studied. This can be compared with the concordance ΛCDM predictions and models (e.g., Ceccarelli *et al.*, 2005; Privato *et al.*, 2006; Albæk *et al.*, 2018).

18.5. Gravitational Redshifts of Galaxy Clusters

Although not directly related to double galaxy clusters, Wotjak, Hansen, & Hjorth (2011) combined velocity data from the Sloan Digital Sky Survey for 7800 galaxy clusters to measure the gravitational redshift of galaxy clusters. Although they found large errors, they compared the deviation of the measured redshift on scales 0–6 Mpc and concluded that their measurements agreed best with GR and $f(R)$ gravity, but appeared to exclude TeVeS theory, a modification of MOND (Bekenstein, 2004) described in Chapter 6. However, a some refinements to their interpretation have subsequently been noted, which do not exclude TeVeS (Bekenstein & Sanders, 2012). Zhao, Peacock, & Li (2013) pointed out that the transverse Doppler effect has the same order of magnitude of the gravitational redshift. When the former is taken into account, it makes the measurement insensitive to gravity theories. Kaiser (2013) showed that additional correction is needed due to a bias produced because more galaxies will be observed moving away than approaching. Sadeh, Feng, & Lahav (2015) conducted a new analysis using later SDSS and BAO data releases. Their result is consistent with Wotjak *et al.* (2011) and they concluded that GR gives the best fit. Croft (2013) discusses how gravitational redshifts have opened a new window and can add a useful probe. Zu *et al.* (2019) consider the design of new gravitational redshift surveys.

Chapter 19

Gravitational Radiation

One of the greatest achievements of the 21st century has been the long awaited direct measurement of gravitational radiation. This is opening a remarkable new window to the universe. Although Einstein predicted gravitational waves in 1915, several investigators had proposed their existence by the analogy between the inverse square laws of the electric and gravitational forces. Details of their reality remained obscure and controversial into the 1950s. The analogy with electromagnetic waves is that as moving electrons will radiate electromagnetic waves, so will a moving mass generate gravitational waves. However, unlike electromagnetic radiation from a dipole, which produces plane waves, gravitational radiation is quadrupole radiation. Compared to electromagnetic radiation, the gravitational radiation is very weak and moving massive systems are necessary to produce detectable amounts of it.

Non-spherical motions of matter will produce gravitational waves. Examples of possible astronomical sources include orbiting objects such as planets and satellites orbiting stars, binary stars, especially short period systems of compact objects such as white dwarfs, neutron stars, and black holes, non-spherical rotating objects, asymmetric supernova events and stellar pulsations. Inflation models of the early universe with asymmetric expansion could produce gravitational waves. As discussed in Chapter 15, efforts are being conducted to detect the circular polarization or the B-mode in the CMB from the inflation, but despite earlier discovery claims it has not been detected.

Unlike electromagnetic radiation (radio, visible, X-ray, etc.) that are affected by absorption and scattering processes, the universe is expected to be completely transparent for gravity waves. So far gravitational radiation has been detected from compact binaries as described below, but not from cosmological sources.

19.1. Detection of Gravitational Waves

Following, e.g., Hartle (2003), Adler, Bazin & Schiffer (1965), Misner, Thorne, & Wheeler (1970), Landau & Lifshitz (1975) and many other authors, gravitational waves can be demonstrated by linearizing Einstein's gravitational field equations

$$R_{\alpha\beta} - \frac{1}{2}g_{\alpha\beta}R = 8\pi T_{\alpha\beta}. \tag{19.1}$$

Assuming that $T_{\alpha\beta}$ is small, the metric is written as a small perturbation on the flat case

$$g_{\alpha\beta} = \eta_{\alpha\beta} + h\alpha\beta(x). \tag{19.2}$$

Setting

$$\bar{h}_{\alpha\beta} = h_{\alpha\beta} - \frac{1}{2}\eta_{\alpha\beta}h, \tag{19.3}$$

where $h = h^{\gamma}_{\gamma}$. From this, the linearized field equation is

$$\Box\bar{h}_{\alpha\beta} = -16\pi T_{\alpha\beta}, \tag{19.4}$$

which shows that each component follows a separate flat-space wave equation with a source $j(x)$ of the form

$$-\frac{\partial^2 f(x)}{\partial t^2} + \nabla^2 f(x) = j(x), \tag{19.5}$$

for which the general solution is

$$\bar{h}_{\alpha\beta}(t,x) = 4\int d^3x' \frac{[T^{\alpha\beta}(t'x')]_{\mathrm{ret}}}{|x-x'|}, \tag{19.6}$$

where $T_{\alpha\beta}$ is evaluated at the retarded time $t_{\mathrm{ret}} = t - |x - x'|$.

19.2. Gravitational Waves Detected in Binary and Mergers

The first evidence for gravitational radiation was from the orbital decay of binary pulsars. For weak gravity the orbits of stars can be explained by the post-Newtonian approximation to the General Relativistic equations of motion (Will 1993, 2014). In a binary system of two masses, m_1 and m_2, separated by r, energy is radiated by gravitational radiation, which produces the orbit to decay, and the rate of decrease of the orbital period can be shown (e.g., Landau & Lifshitz, 1975) to be

$$\frac{dP}{dt} = -\frac{64}{5}\frac{G^3}{c^5}\frac{(m_1 m_2)(m_1 + m_2)}{r^3}, \tag{19.7}$$

which leads to a merger time

$$t = \frac{5}{256}\frac{c^5}{G^3}\frac{r^4}{(m_1 m_2)(m_1 + m_2)}. \tag{19.8}$$

For most stellar binaries, these times exceed the age of the universe and a combination of very large masses and small distances is required to produce observable effects.

Observations of binary pulsars using radio telescopes provide indirect confirmation of gravitational radiation. The decay of the orbital period of the binary pulsar PSR B1913+16 (Hulse & Taylor, 1975; Weisberg & Taylor, 2002). This system has total mass, $2.83M_\odot$ and each has $1.4M_\odot$ and orbital period $P = 7.75$ hours. It follows equation (19.7) to better than $\sim$0.2%. Similar results have now been found for at least nine binary pulsars (Weisberg & Huang, 2016). Brown *et al.* (2017) have also observed these effects in WD systems. After restarting observations April 1, 2019, LIGO and VIRGO have reported observing mergers involving additional neutron stars and black holes.

Direct detection of gravitational radiation requires strong gravity processes involving massive relativistic systems. Even so, the expected movement produced by a passing gravity wave is $\sim 10^{-18}$ m, smaller than the diameter of a typical atom (Thorne, 1994). Consequently, producing instruments sufficiently sensitive to detect these small displacements required the development of new technology and types of detectors requiring over fifty years of technical development involving large teams of colleagues. The earliest attempts to detect gravitational radiation were pioneered by Weber (1967) employing 'bar detectors.' These apparatuses typically consisted of 2 m × 1 m diameter aluminium cylinders with piezoelectric sensors. Subsequently, these were improved using cryogenic systems with other types of electromechanical transducers. None of the devices of this type have proven sensitive enough (Trimble, 2017).

19.3. Numerical Models

The theoretical calculation of the mergers of binary black holes and neutron stars needs to be studied using *numerical relativity* (Baumgarte &

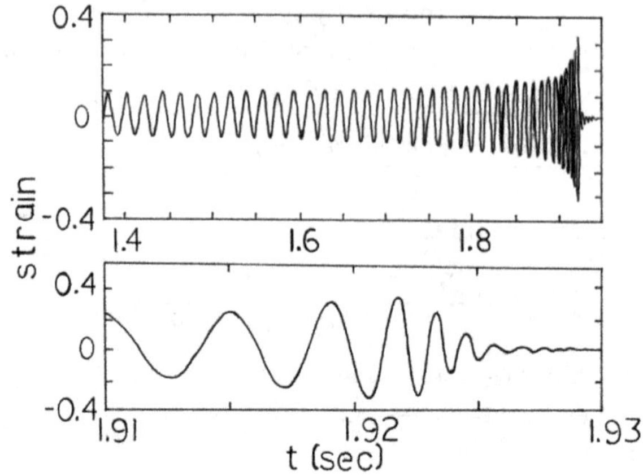

Fig. 19.1. Schematic of a hypothetical black hole merger signal observed by a gravity detector. The top panel illustrates the inspiral part of the merger characterized by the decreasing period and the increase in signal up to the time of the merger near 1.9 seconds. The lower panel is an enlargement of the signal covering the last 0.02 seconds of the merger showing the coalescence segment followed by the ringdown after about 1.925 seconds. (*Based on*: Wardell, Barry, Hinder, Ian, & Bentivegna, Eloisa. (2016). Simulation of GW150914 binary black hole merger using the Einstein Toolkit [Data set]. Zenodo. doi:10.5281/zenodo.155394.)

Shapiro, 2011) and leads to the events illustrated in Figs. 19.1 and 19.2. The approximation used is the ADM formulation (Arnowitt, Deser, & Misner, 1962), employing a 3+1 decomposition of Einstein's equations. The complications of gravity stem from it being nonlinear and that gravitational radiation scales as the second time derivative of the quadrupole moment. These equations are solved numerically by difference equation techniques, but relaxing spherical symmetry, needed for binary neutron star and black hole inspiral and merger events, led to numerical instabilities requiring a new reformulation. These techniques include the Baumgarte–Shapiro–Shibata–Nakamura (BSSN) formulation. The analysis methods continue to be developed. McWilliams (2019) has devised analytical solutions for a large range of parameters for mergers.

The signal expected from these mergers is sketched in Fig. 19.1. The inspiral stage as the two masses approach is characterized by an increasing orbital circular frequency Ω (e.g., Tiwari *et al.*, 2016; Holz, Hughes, & Schutz, 2018),

$$\dot{\Omega} = \frac{96}{5}\left(\frac{GM_c}{c^3}\right)\Omega^{11/3}, \tag{19.9}$$

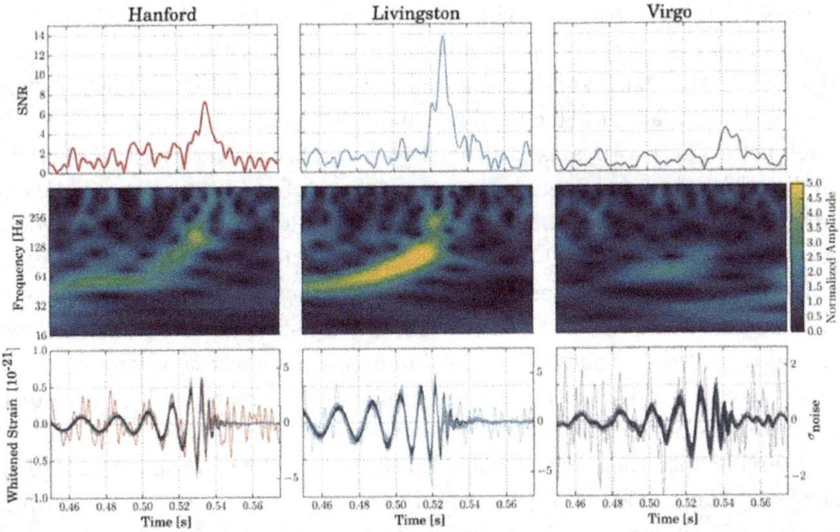

Fig. 19.2. Details of the black hole merger GW170814 from three independent gravity wave observatories. The two LIGO devices are located at Hanford, WA and Livingston, LA and Virgo is at Santo Stefano a Macerata, Italy. Top panels have the observed raw signals. Middle panels show the frequency changes with time. The central panels show the change in frequency or chirp. The bottom panels are fits to the observed merger signals, indicating merger masses of approximately 30 and 25 $M_\odot$. These are also typical of signals from merging black holes. (*Source*: Abbott B. P. *et al.* (2017b), *PhRvL*, 119, 141101, "GW170814: A three-detector observation of gravitational waves from a binary black hole coalescence," Fig. 1.)

where M_c is the 'chirp mass',

$$M_c = \frac{(m_1 m_2)^{3/5}}{(m_1 + m_2)^{1/5}}, \tag{19.10}$$

m_1 and m_2 are the individual masses, G is the gravitational constant, and c is the speed of light. The inspiral section reaches a maximum amplitude at the time of the merger with the coalescence followed by the ringdown.

19.4. Some Observed Properties of Mergers

The development of considerably more sensitive interferometric detectors has produced a class of instruments capable of detecting the mergers of black hole and neutron star binaries. As a result, the number of known sources is rapidly growing. These make use of the Michelson interferometer principle. The LIGO (Laser Interferometer Gravitational-wave

Observatory) detectors in Livingston, LA and Hanford, WA, each contain two 4 km tunnels arranged perpendicularly in an 'L' with a laser beam, split at their centre. Each sub-beam is reflected by suspended mirrors at their ends and follow about 280 trips down and back each leg after which they are combined. The two beams are tuned to destructively interfere. If there is no disturbance to the detector the beams cancel, but a passing gravity wave creates a resonance signal. LIGO has been followed by similar interferometers, Virgo and GEO600 in Europe, and additional devices, e.g., LIGO India and KAGRA in Japan are under construction. Future experiments in space such as LISA (Laser Interferometric Space Antenna) planned to be launched by the European Space Administration, some time around 2034 would have 2.5×10^6 km arms and promises even greater ability to detect more sources peaking at lower frequencies.

Starting with the detection of GW150914 by LIGO (Abbot *et al.*, 2016) involving black holes of approximately 36 $M_\odot$ and 31 $M_\odot$, an increasing number of mergers have been found. These are primarily mergers of two black holes. Figure 19.2 shows the observed gravitational wave signal from the GW170814 black hole merger event from three independent observatories illustrating the three phases of the event. Gravity wave signals have also been detected from neutron star mergers, notably from GW170817 (e.g., Abbott *et al.*, 2017 a–c), which was from about 1.5 and 1.3 $\mathrm{M}_\odot$ objects and had an optically observed counterpart that has interesting implications for nucleosynthesis of heavy elements. Not all neutron star mergers have been observed in the optical.

19.5. Standard Sirens

In addition to the masses, measurements of the gravity waves during the inspiral and merger can be utilized to determine the luminosity distance, D_L, of the sources termed 'standard sirens' (Schutz, 1986; Holz & Hughes, 2005; Holz *et al.*, 2018). This promises to be an important cosmological distance indicator analogous to the standard candle method. The observed amplitude of the waveform depends on the chirp mass and distance and for the two polarizations is given by

$$h_+ = \mathcal{A}(1 + \cos^2 i) \cos 2\Phi(t), \tag{19.11}$$

$$h_\times = \mathcal{A} \cos i \sin 2\Phi(t), \tag{19.12}$$

where $\mathcal{A} = \frac{2c}{D}\left(\frac{GM_c}{c^3}\right)^{5/3} \Omega^{2/3}$, i is the inclination, Ω is the circular frequency, and $\Phi(t)$ is the orbital phase accumulated with time. Measurements in both polarizations can then yield the inclination and the parameters of the whole event can be determined in principle. This cannot be used to directly determine H_0 from gravity waves alone. A redshift measurement is required, which necessitates an optical identification. The measured redshift requires corrections due to the internal dynamics of its host galaxy and its peculiar motion.

The binary neutron star merger GW170817 was the first event to be identified and have its redshift measured yielding $H_0 = 70^{+12}_{-8}$ km s^{-1}Mpc^{-1} (Abbott *et al.*, 2017d), while Guidorzi *et al.* (2017) indicate that D_L is highly correlated with the inclination of the system's orbit and derive $H_0 = 74.0^{+11.5}_{-7.5}$ km s^{-1}Mpc^{-1}. The situation should improve when more sources are discovered. Chen, Fishbach, & Holz (2018) predict that H_0 should be determined from standard sirens with a precision of $\sim$2% within five years.

19.6. Heavy Element Production in BH and Neutron Star Mergers

The discovery that the binary neutron star merger, GW170817, that could be observed spectroscopically for element abundances (Abbott *et al.*, 2017b,c) has provided some answers to puzzles about the formation of elements heavier than Fe in the universe. LIGO and VIRGO have reported detecting additional mergers involving neutron stars. Only the lightest elements and their isotopes (H^1, H^2, He^3, He^4, Li^6, L^7, etc.) could be formed by nucleosynthesis in the early universe (Chapter 15). Heavier atoms reaching iron are cooked up in the interiors of stars. The classical formulation of the different nuclear processes producing elements was given by Burbidge *et al.* (1957) in the famous B^{2}FH study. Elements heavier than iron can be produced by bombarding iron nuclei with neutrons in the slow (s) and rapid (r) processes. The accumulation of slow electrons (s-process) produces Sr, Ba, etc., which are observed in barium and related stars. Heavier nuclei, e.g., rare earths, Au, Pt, Th, U, and even transuranic nuclei are built up by the r-process by reactions in the formation of neutron stars. In the core collapse Type II supernovae, the inverse β-process ($pe^- \rightarrow n\nu_e$) produces neutrons using electrons and protons from disintegrating Fe nuclei. Nuclei heavier than Fe can be produced by bombarding a 'seed' nucleus, e.g., Fe, with neutrons. In a plot of the numbers of neutrons (n) and protons (p) in the atomic nuclei table, nuclei fall near a stability line of specific numbers of neutrons and protons, e.g., the stable ^{56}Fe nucleus has 26 p and 30 n, whilst ^{69}Fe with 26 p and 43 n is unstable rapidly decays to ^{68}Co. If the density of neutrons is sufficiently high, however, more neutrons can be added to an unstable nucleus before it decays and through a chain of decays and adding neutrons, increasingly heavy nuclei can be built as long as the high density and temperature conditions for the r-process persist. Seeger, Fowler, & Clayton (1965), Clayton (1968), Wallerstein *et al.* (1997) and many others review the observational evidence for nucleosynthesis producing many of the elements.

The high density and temperature conditions required for the r-process can be realized in SNe II or core collapse supernovae with masses, $M > 8$–$20 M_\odot$. Although this has been generally accepted as a plausible origin for the heavy elements, it does not explain the r-process elements observed in some of the oldest metal poor stars in nearby dwarf galaxies. From what is known of the evolution of these low mass galaxies from semi-analytic models incorporating a standard initial mass function, like the Salpeter mass function, $N(M) \sim M^{-3.5}$, could not produce sufficient numbers of supernovae from the rare massive stars to enrich the interstellar environment with the r-process elements observed in these systems (Ji *et al.*, 2016). Of order 10^3 supernovae would be needed. A merger of two neutron stars, producing even higher energies, might do it but the amounts of production are still unclear (e.g., Li & Paczynski, 1998; Qian, 2000; Agast, Samland, & Thielemann, 2004; Matteucci *et al.*, 2014; van de Voort *et al.*, 2015; Ishimaru, Wanjo, & Prantzos, 2015; Kobayshi, Karakas, & Lugaro, 2020). This process is known as a 'kilonova'.

The merging of two neutron stars or neutron stars merging with black holes have been proposed as producers of r-process elements (Lattimer & Schramm, 1974; Rosswog, Freburghaus, & Thielemann, 2001; Metzger *et al.*, 2010), but this has not been explored as intensively until recently. The coalescing double neutron star event, GW170817 produced heavy elements including lanthanides along with Au and Pt (Drout *et al.*, 2017; Shappee *et al.*, 2017). Doubtless, further details will emerge as more of these objects are discovered.

How close binaries of neutron stars and black holes form is an interesting problem. A related question is whether the hypothetical Thorne–Żytkow stars comprised of a neutron star in an extended envelope exist. A possible candidate is HV2112 in the SMC (Levesque *et al.*, 2014).

Stars with enhancement of elements heavier than iron, are known, but disentangling the situation is confusing as diffusion processes can act instead of nuclear reactions. The Ap and Am stars show enrichments of lanthanides and heavier elements, but are thought to have had these heavier elements pushed upwards from deeper layers towards the star's surface by their radiation fields (Adelman, 1973; Michaud, 1970; LeBlanc *et al.*, 2009). Some of these stars have strong magnetic fields and star-spots of metal enrichment. Nevertheless, stars formed from r-process enriched material, the remains of earlier neutron star mergers in the Milky Way, are to be expected. In energetic gravitational wave events where heavy nuclei are produced, some peculiar stars might be the place to search.

A possible suspect might be like the star HD 101065,[1] although a handful of somewhat similar stars at the extreme end of the cool Ap stars have been found (Ryabchikova *et al.*, 2000; Ryabchikova & Ramonovskaya, 2017), where heavy element enrichments have been interpreted to be explained by the diffusion process. These unusual stars have very complicated spectra and atmospheric analyses show large enhancements of heavier elements compared to the sun. Figure 19.3 presents different abundance analyses for this object and shows the progress using new spectroscopy, improved atomic data such as oscillator strengths and opacities, and stratified model atmospheres, and confirms the enhancement of the heaviest elements, including Th and U. Cowley *et al.* (2004) have indicated evidence for the presence of radioactive Pm and Tc in this star and HD 965. Gopka *et al.* (2008) have claimed the presence of weak lines of actinides Np, Pu, Am, Cm, Bk, Cf, and Es (atomic numbers 89, 91, 93–99). However, this seems controversial due to the complex spectrum of HD 101065, which likely has many unclassified lines of lighter elements, notably different ionized lanthanides. These actinides are complex to interpret with some large isotopic shifts.

Some of the transuranic elements have short half-lives (e.g., for ^{252}Es it is 472 days) and were they present in a star, it would require an ongoing energetic process to generate them or the decay of even heavier nuclei. Some isotopes have longer half-lives, and might conceivably be detected if they were produced in a relatively recent merger event such as, e.g., Np^{93}, Pu^{94}, and Cm^{96} that have half-lives of 2×10^6, 81×10^6, and 156×10^6 years, respectively. Interesting speculations, which cannot be detailed here, relate to the well-known 'island of stability' for long-lived heavy nuclei, which has yet to be found. Nuclear isomers in metastable states might be produced with greatly lengthened half-lives. The nuclear isomer ^{180m}Ta with HL $= 45 \times 10^{15}$ years is known to exist and the hyperdeformed doubly magic nucleus ^{271m}Hs^{108} has been proposed to possibly have a long half-life. Thornton & Burdette (2019) have reviewed the difficulties of finding such elements. There have been many searches for signs of superheavy elements using a variety of techniques, e.g., Ter-Akopian & Dmitriev (2015)

[1]Also known as Przybylski's star.

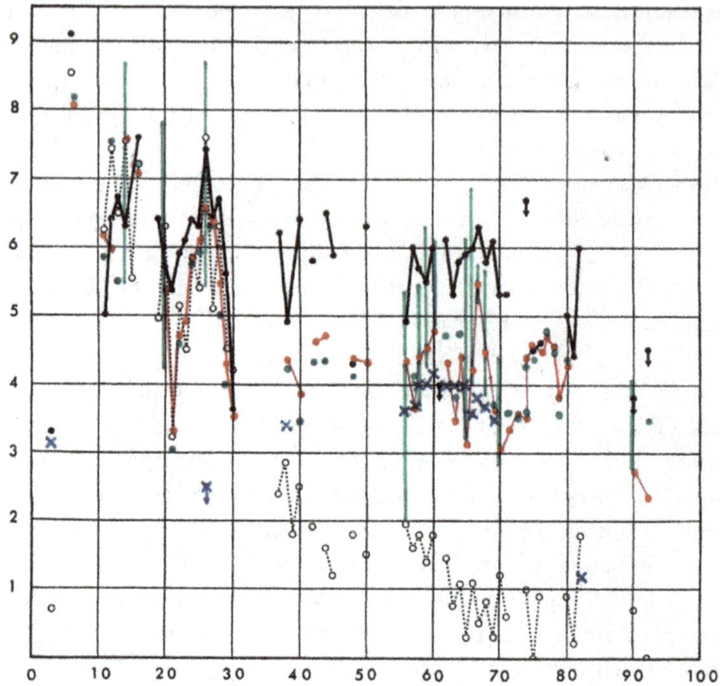

Fig. 19.3. Results of some abundance analyses of elements in HD101065. These have been normalized to $\log N = 12.0$ for hydrogen. The abscissa is the atomic number. Blue crosses: Przybylski (1966); black: curve of growth analysis (Wegner & Petford, 1974); red: standard model atmosphere (Cowley *et al.*, 2000); green: stratified model atmosphere; vertical bars indicate ranges of stratified elements (Shuylak *et al.*, 2010). Open dotted circles are solar abundances from Allen (1973). (*Source*: Wegner G., Petford A. D. (1974), *MNRAS*, 168, 557, "Abundance analysis of Przybylski's star (HD 101065)," Fig. 4 (referenced abundance analyses overlain), Reproduced by permission of Oxford University Press on behalf of the Royal Astronomical Society.)

and Ludwig *et al.* (2012) who used a mass spectrometer to intensely scan samples of Pt and Os for superheavy elements, but none were detected and they set an upper limit of $\sim 10^{-15}$ for the abundance of Hs relative to chemically similar Os.

In the future large spectroscopic surveys like the RAVE project may find more examples of peculiar stars with enhanced heavy elements.

Chapter 20

21st Century Peculiar Motions Studies

Efforts to measure peculiar motions continue in the 21st century utilizing new generations of instruments and analyses. The structures and natures of regions like the Great Attractor and the derived cosmological parameters have been brought into sharper focus. Comparing and combining peculiar motion data with the CMB, BAO, and weak-lensing results provide new opportunities. The primary goals for peculiar motion investigations remain, although with the advances in cosmology, the emphasis has shifted: (1) Mapping of mass structures and flow patterns; as coverage improves and extends to greater distances, the motions are studied to see if they reveal new large attractors, such as the Great Attractor and Perseus–Pisces regions, and understand their properties including dark matter; (2) Extracting cosmological data; variables $\sigma_8, \beta, \Omega, \gamma$, etc. are determined and the bulk flow amplitudes, apex directions, and convergence consistency with the CMB dipole and ΛCDM or other models are verified; (3) Using peculiar velocities to constrain theories of gravity can be explored. In addition, peculiar motions provide an important correction to redshifts in Hubble constant measurements.

In general, as the motions of the matter are driven by gravity, the simplification of linear theory can be applied. Peculiar motions have the advantage of being independent of other methods. Their analysis does not require assumptions about the physics of the early high-z universe and the validity of ΛCDM. Peculiar motions do not currently extend to large redshifts and are relatively insensitive to the cosmological model. The properties of peculiar motions of nearby galaxies can be compared with other methods at higher redshifts and provide a consistency check or combined with them.

Advances in peculiar velocity measurements have increased the numbers and quality of observational data. This permits development of new analysis algorithms. The accuracy of distance indicators have benefited from improvements in the astronomical distance scale, e.g., by Gaia and other experiments that are used to determine zero points of the relations. Compared to redshift, surveys determining accurate distances and peculiar velocities are more labourious. Both high quality spectroscopic and photometric measurements are required for each source, and the number of measurements does not increase with the speed of lower resolution redshift surveys where millions of objects have been observed. Galaxies with peculiar velocities still number in the order of tens of thousands. In the following, the latest work on peculiar motions is sketched.

20.1. Larger V_{pec} Databases

The available data for studying peculiar motions have been increased by larger observational programmes and compiling the available measurements. Although all distance determination methods outlined earlier are utilized, new ones are promising. The TF and FP methods still dominate, particularly for larger distances and work to understand their nature and calibration is active. The sparser SNeIa distances are more accurate and growing, so in the future this promises to be the preferred distance indicator. Also, new statistical methods for analysis have been proposed by several investigators. Awaiting the completion of larger peculiar motion surveys, recent databases have been applied to re-reducing older data with new analyses.

20.2. Recent Observational Surveys

20.2.1. *6dF galaxy survey*

The 6dF galaxy survey (6dFGS) employed the 1.2-m UK Schmidt telescope in Australia with a 150 multi-fibre spectrograph (Jones *et al.*, 2004, 2009). Campbell (2009) and Campbell *et al.* (2014) used the 6dFGS to construct distances and $v_{\rm pec}$ for 8,885 early-type galaxies with redshifts $z < 0.055$ with distance errors of $\sim \pm 25\%$ for each measured distance. This work has generated several subsequent analyses of the data or in combination with other surveys, e.g., Springob *et al.* (2014) found strong flow towards the Shapley–Vela clouds, but a more negative flow from the Pisces–Cetus clusters.

20.2.2. *2MASS TF survey*

The 2MASS TF (2MTF) survey is described by Huchra *et al.* (2012). Observations of 44,599 near infrared magnitudes ($K_S \leq 11.75$) from the 2MASS all-sky survey (Skrutskie *et al.*, 2006) were combined with spectroscopic observations of about 11,000 galaxies from several observatories and published sources. The resulting dataset is the Two Micron All-Sky Survey or 2MRS (Erdoğdu *et al.*, 2006a) containing 23,200 galaxies. The 2M++ catalogue is an improved dataset from the 2MRS assembled by Lavaux & Hudson (2011).

20.2.3. *SFI++ survey*

Additional enlarged databases employing the TF relation include a galaxy cluster survey (Dale *et al.*, 1999a–c) and the SFI++ (Springob *et al.*, 2005, 2007, 2014). This has enlarged and improved the SFI project described in Chapter 14 and produced new good quality data, which have expanded the available data for more objects.

20.2.4. *Supernova peculiar motions*

The number of available supernova SNe Ia distances continues to grow from several groups. Turnbull *et al.* (2012) utilized 245 supernovae in the 'First Amendment' compilation. Boruah, Hudson, & Lavaux (2020) derived and discussed additional objects from a 'Second Amendment' list of 1,164 supernovae from Pan-STARRS and 174 from the Foundation Supernova Survey (Jones *et al.*, 2019). This now surpasses the 400 $D_n - \sigma$ distances of the early 7S survey (Lynden-Bell *et al.*, 1988) and since these provide higher accuracy distances ($\sim\pm5\%$) compared to ($\sim\pm20\%$) for scaling relations, this is becoming one of the methods of choice.

20.3. Compilations of V_{pec} Data

As in the case of the Mark III catalogue, which was used earlier for several peculiar motion studies, Tully and collaborators assembled databases, 'Cosmicflows — 1, 2, 3'. (Tully *et al.*, 2008, 2013, 2016), which have been used by several investigations based on peculiar motions. Tully *et al.* (2013) compile distances and v_{pec} for 8,315 galaxies using TF spirals and the FP for E/S0 galaxies. They combine the literature and add new infrared observations from the Spitzer satellite. Using the Cepheid P–L and TRGB relations, they conclude $H_0 = 74.4 \pm 3.0\,\text{km s}^{-1}\,\text{Mpc}^{-1}$. Tully *et al.* (2016) added 2,257 new TF measures using 3.6μ Spitzer data and the 8,885 6dF objects to get 17,669 objects with v_{pec} data. They obtain a new calibration of the Cepheid P–L relation based on the top of the red giant branch (TRGB) and 2MASS K-band data, and construct a new 3.6μ ITF relation.

20.4. Mapping and Structures

The mapping of the mass and flow distributions has continued and utilized larger databases. Tully *et al.* (2014) confirm the Local Group's peculiar velocity is decomposed into three major components: (1) the expansion of the Local Void, (2) attraction of the Virgo cluster, and (3) beyond $V \approx 3{,}000\,\text{km s}^{-1}$, the flow is towards Centaurus. They conclude the best value for the Hubble constant is $H_0 = 75 \pm 2\,\text{km s}^{-1}\,\text{Mpc}^{-1}$, which is consistent with $H_0 = 73.52 \pm 1.62\,\text{km s}^{-1}\,\text{Mpc}^{-1}$ from the Cepheid P–L relation (Riess *et al.*, 2018), but not the CMB values $H_0 = 67.4 \pm 0.5$ and $69.32 \pm 0.80\,\text{km s}^{-1}\,\text{Mpc}^{-1}$ for Planck (Planck Collaboration, 2018) and WMAP (Bennett *et al.*, 2013), respectively.

Although estimates of H_0 have been determined by several groups using TF and FP scaling relations, as already mentioned, one should note that these are not wholly independent. They are based on secondary distance indicators, calibrated with primary distance indicators, including the Cepheids.

Finer details of the flow patterns have been revealed, although the two major regions known from early studies (Chapters 13 and 14) affecting the nearby flow, Hydra–Centaurus Norma towards the GA and Perseus–Pisces remain much the same. Tully *et al.* (2014) have used the Wiener filter with the least action method to determine flow trajectories for ~8000 galaxies, confirming the dominant regions. This is shown in Fig. 20.1, giving the major mass concentrations. The structures of the main density regions and the peculiar motions reveal more details and extend to greater distances compared to the earlier results shown in Chapters 13 and 14. The local supercluster, Virgo, and Perseus–Pisces regions appear. The earlier Great Attractor and Norma region reach to greater distances and are denoted by Tully *et al.* as the Laniaka supercluster. Other groups have constructed similar flow pattern maps.

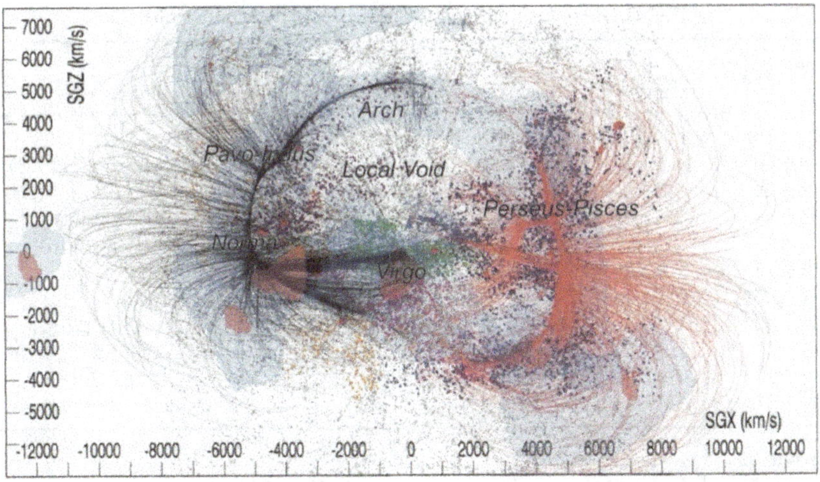

Fig. 20.1. Details of the flows and structures to $z = 6000\,\text{km s}^{-1}$ (about 80 Mpc from the sun) according to Tulley *et al.* (2014) in their Laniaka supercluster. This is a map of structure and flows due to mass within $cz < 6000\,\text{km s}^{-1}$ ($\sim$80 Mpc). Shaded patches are clusters (red) and filaments (blue). Coloured threads represent flow paths, black indicates flows terminating near the Norma Cluster and red shows flows directed at the Perseus–Pisces region. The Arch and Antilla Wall are shown in black. (*Source*: Tully R. B., Cortois H., Hoffman Y., Pomarède (2014), *Nature*, 513, 71, "The Laniakea supercluster of galaxies," Extended data Fig. 2, ©2014 Reprinted by permission from Nature.)

20.5. Comparison with the CMB and Cosmological Models

Several peculiar motion investigations have been published based on the newest databases. As in the earlier Chapters 13 and 14, efforts have focused on quantities that can be compared with the CMB, BAO, RSD, etc. and combined with them. The parameters that are derived can in turn test ΛCDM and other cosmologies. One expects the answers from the newer results to converge to the same answers as larger datasets grow, but a number of outstanding questions remain. The three general areas where peculiar motions are employed and compared with cosmology are: (1) bulk flow dipole, (2) determination of $\beta = f\Omega_m^\gamma$, yielding the matter density, and (3) tests of gravity.

20.5.1. *Bulk flow measurements*

An important goal for peculiar velocity measurements, as already discussed, is to test whether the convergence of the bulk flow amplitude and the dipole

Table 20.1. Comparison of some dipole determinations.

V (km s^{-1})	l (degrees)	b (degrees)	Sample	Reference
369.0 ± 0.9	263.99 ± 0.14	48.26 ± 0.03	CMB	Hinshaw *et al.* (2009)
570 ± 60	307 ± 11	9 ± 8	D_n–σ	Lynden-Bell *et al.* (1988)
612 ± 22	268 ± 3	27 ± 3	IRAS PSCz	Rowan-Robinson *et al.* (2000)
407 ± 81	287 ± 9	8 ± 6	Pec vels	Watkins *et al.* (2009)
627 ± 22	276 ± 3	30 ± 3	2MRS	Lavaux *et al.* (2010)
249 ± 76	319 ± 18	7 ± 14	SNe	Turnbull *et al.* (2012)
281 ± 25[a]	296 ± 16	19 ± 6	2MTF	Hong *et al.* (2014)
540 ± 40	268 ± 4	38 ± 6	2M++	Carrick *et al.* (2015)
248 ± 58[b]	318 ± 20	40 ± 13	6dFGSv	Scrimgeour *et al.* (2016)
288 ± 24	296 ± 6	21 ± 25	2MTF, 6dFGS	Qin *et al.* (2018)
246 ± 11	292 ± 11	14 ± 5	SNe, SFI++	Boruah *et al.* (2020)

[a]Inside $30h^{-1}$Mpc shell.
[b]Inside $50h^{-1}$Mpc shell.

apex to the CMB values are explained by the known mass structures including the GA and Pisces–Perseus regions. The inability to achieve this could indicate the existence of more distant large attractors or that some basic assumptions are invalid. Several investigators have pursued this aspect.

Table 20.1 lists some of the dipole determinations from peculiar velocities and compares them to the CMB and other methods. Earlier measurements were discussed in Chapter 14. In the distance range covered, the dipole apex shown in Fig. 20.2 and the velocity changes slightly with distance as in Fig. 20.3, (Qin *et al.*, 2018, 2019a) which are given in the original publications.

Watkins *et al.* (2009) investigated how local mass concentrations affect the measured dipole direction. They combined peculiar motion data from nine catalogues available at that time and concluded that the observed dipole velocity **B** is high inside a $50h^{-1}$ Mpc window compared to the WMAP5 normalization for the power spectrum. Thus, some disagreement exists between the observations and predictions from ΛCDM, which requires knowledge of the mass distributions of more distant and less studied superclusters.

Lavaux *et al.* (2010) derived motions from the 2MRS catalogue and considered dipole convergence. Although they recovered agreement with the WMAP5 to about the 1σ level, they estimated that less than half of the velocity is generated by the Hydra–Centaurus–Norma supercluster and that inclusion of additional mass regions beyond would be needed.

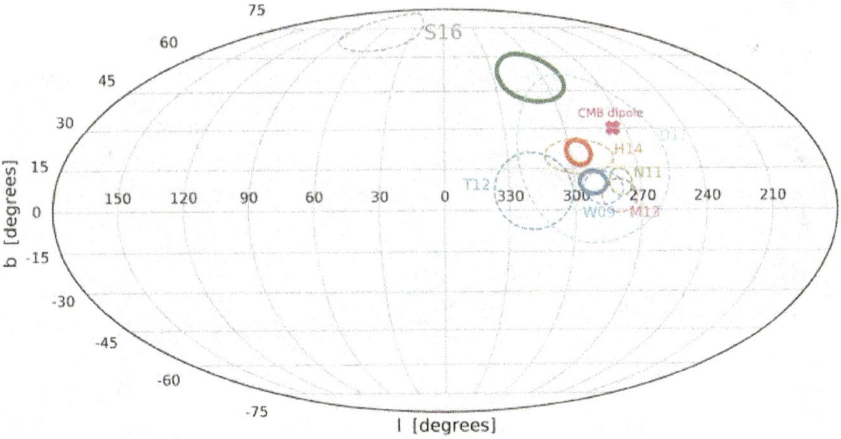

Fig. 20.2. Bulk flow directions in galactic coordinates found from several recent investigations compared with 2MTF + CF3 data plotted on the celestial sphere in galactic coordinates. The abbreviations denote: H14: Hong *et al.* (2014), M13: Ma & Scott (2013), S16: Scrimgeour *et al.* (2016), T12: Turnbull *et al.* (2012), N11: Nusser & Davis (2011), W09: Watkins *et al.* (2009), D11: Dai *et al.* (2011), and C11: Colin *et al.* (2011). The solid blue, green, and red circles denote data from 2μ TF, 6dF, and combination of the two. (*Source*: Qin F., Hoelett C., Staveley-Smith L., Hong T. (2019), *MNRAS*, 482, "Bulk flow and shear in the local Universe: 2MTF and COSMICFLOWS-3," Fig. 8, Reproduced by permission of Oxford University Press on behalf of the Royal Astronomical Society.)

Turnbull *et al.* (2012) used supernovae from the 'First Amendment' compilation and found the dipole is consistent with ΛCDM. However, compared with the IRAS PSCz, the density does not provide about $150 \pm 43\,\text{km s}^{-1}$of the observed bulk motion. Feindt *et al.* (2013) also utilized the 112 Union2 SNe and studied the dipole and flow with distance produced in the direction of the Shapley supercluster. They also found that at large distances this does not produce the observed motions. Carrick *et al.* (2015) reach somewhat similar results. They employed galaxies in the 2M++ catalogue and from SNe distances and compared with results from the SFI++ catalogue. The 2M++ catalogue is an improved dataset from the 2MRS assembled by Lavaux & Hudson (2011) and near infrared TF distances were constructed. They concluded as Turnbull *et al.* (2012) that there is an additional external bulk flow of $159 \pm 23\,\text{km s}^{-1}$in the direction $(l, b) = (304 \pm 11,\ 6 \pm 13)$ degrees, which is still unexplained. Although this could possibly be from distant structures such as the Vela supercluster, it might be an effect of the data. Boruah *et al.* (2020) also find that about this same

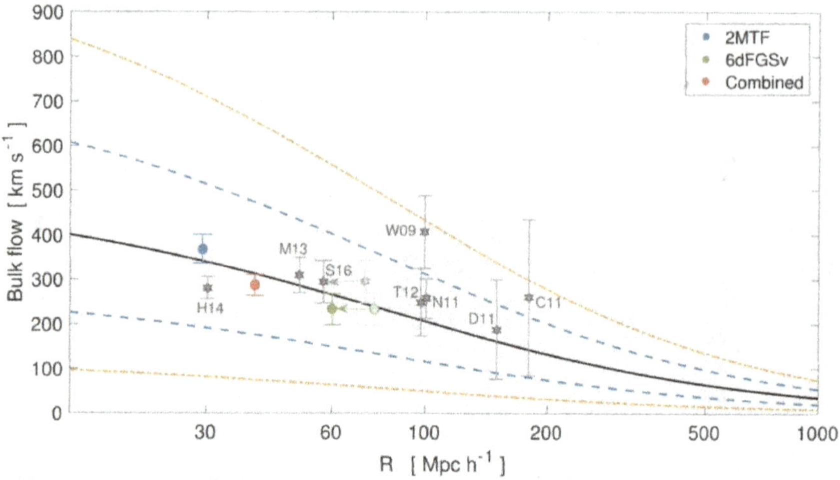

Fig. 20.3. Comparison of bulk flow amplitudes vs. depth from recent peculiar motion studies including 2MTF + CF3 data. The black curve is the prediction of ΛCDM. The blue and yellow curves are one and two standard deviations from it, respectively. The labelled stars denote the results reported by several investigations; the abbreviations are the same as in Fig. 20.2. The dashed contours and dash-dotted contours denote one and two σ contours. (*Source*: Qin F., Hoelett C., Staveley-Smith L., Hong T. (2019), *MNRAS*, 482, "Bulk flow and shear in the local Universe: 2MTF and COSMICFLOWS-3," Fig. 7, Reproduced by permission of Oxford University Press on behalf of the Royal Astronomical Society.)

amount of velocity is still unexplained. Rameez *et al.* (2018) examined the dipole anisotropy using AllWISE galaxies and conclude that a large source at $z > 0.03$ remain unexplained.

Most recent studies conclude that there is agreement or at least consistency with ΛCDM theory, although this is not to the precision of the CMB measurements. Hong *et al.* (2014) used 2018 galaxies selected from the 2MASS and 2MTF surveys. They used J, H, and K photometry and TF template relations from Masters *et al.* (2008)[1] and the maximum likelihood and minimum variance methods to derive the dipole parameters. They concluded that the properties of the dipole agree with ΛCDM to about the 1 σ level. Scrimgeour *et al.* (2016) applied the MV method to the 6dFGSv data consisting of 8885 galaxies to determine the bulk flow inside 50 and $70h^{-1}$ Mpc radius windows. This result was compared to the MLE method and Malmquist and selection bias were studied using

[1]Masters *et al.* (2014, AJ, 147, 124) have later revised these relations.

mock catalogues. The amplitude and directions of the bulk flows are found consistent with ΛCDM. Their paper contains a useful compilation of bulk flows.

Qin *et al.* (2018, 2019a) also find results consistent with ΛCDM theory. They used 10194 measurements of galaxies from combining the 2MTF and 6dFGSv surveys and developed a Maximum Likelihood estimator. Figure 20.2 plots some of the recent dipole apex directions on the celestial sphere in galactic coordinates compared to their result. Figure 20.3 compares the corresponding observed bulk flow velocities or amplitudes with depth for ΛCDM according to Qin *et al.* (2018, 2019a) for several investigations. Inspection of earlier results in Figure 13.2 indicates the directions of the derived dipoles are contained in a smaller region of the sky, while the bulk amplitudes give a tighter appearance than in Figure 14.7. The scatter is mostly reduced by the elimination of some earlier apparently spurious determinations and the agreement is good to $\sim 1\sigma$. Although the most outlying determinations of the dipole, such as the amplitude of the dipole shown in Fig. 14.7, seem to have gone away, the newer determinations still give somewhat different results.

20.5.2. *Determinations of β and Ω_m*

These parameters have been estimated from peculiar velocities related to several of the dipole solutions already discussed, but they can also be estimated by other methods besides using peculiar velocities. For example, Bond *et al.* (2005) discussed obtaining σ_8 and β from CMB and ΛCDM models. Juszkiewicz *et al.* (2010) compare determining σ_8 from several estimators: cosmic flows, Lyα forest, the cluster number density, weak lensing, and the galaxy clustering power spectrum. From this they conclude that for flat ΛCDM, the most likely value is $0.8 < \sigma_8 < 0.95$. The value of $\beta = \sigma_8 \Omega_m^\gamma$ as explained earlier can be measured directly from the peculiar velocities. If σ_8 is known, Ω_m can be found. For the CMB, using $\Omega_m = 0.30$ as an example, one expects β to be 0.39 if $\gamma = 0.6$ or $\beta = 0.42$ if $\gamma = 0.55$ using σ_8 from Table 20.2. Table 20.2 compares some of the available parameters (most authors do not give all of the three parameters) mostly from peculiar velocity data.

In addition to analyzing the flows as in the case of the dipoles, several investigators have utilized the mean pairwise velocity pairs, v_{12}, in the BBGKY hierarchy (Chapter 13). Wang *et al.* (2018) used the 2-point peculiar velocity correlation function. Dupuy *et al.* (2019) estimated the local

Table 20.2. Some representative β, σ_8, and Ω_m determinations.

β	σ_8	Ω_m	Sample	Reference
	0.811 ± 0.006	0.315 ± 0.007	CMB	Planck Collaboration (2018)
0.5 ± 0.1			pec vel	Nusser *et al.* (2001)
0.42 ± 0.04			pec vel	Branchini *et al.* (2001)
	$1.13^{+0.22}_{-0.23}$	$0.30^{+0.17}_{-0.07}$	pec vel	Feldman *et al.* (2003)
	$0.83^{+0.03}_{-0.03}$	$0.31^{+0.05}_{-0.04}$		Reichardt (2008)
0.785 ± 0.043	0.771 ± 0.029	0.248 ± 0.029	wk lens	Fu *et al.* (2008)
	$1.02^{+0.16}_{-0.18}$			Juszkiewicz *et al.* (2010)
0.40 ± 0.07			SNe	Turnbull *et al.* (2012)
0.401 ± 0.024				Carrick *et al.* (2015)
	$1.01^{+1.07}_{-0.58}$		pec vel	Scrimgeour *et al.* (2016)
$0.51^{+0.09}_{-0.08}$			pec vel	Howlett *et al.* (2017)
	$0.92^{+0.440}_{-0.295}$	$0.315^{+0.205}_{-0.135}$	vel cor	Wang *et al.* (2018)
0.43 ± 0.11			pec vel	Dupuy *et al.* (2019)
0.341 ± 0.024			pec vel	Said *et al.* (2020)

growth rate, $f\sigma_8$, from the Cosmicflows-3 data and compared several published determinations of $f\sigma_8$. Nusser (2017) also used the Cosmicflows-3 data and studied velocity–density correlations. He concluded the $0.32 < \Omega^{0.55}\sigma_8 < 0.48$ and $0.34 < \Omega^{0.55} < 0.52$, which is in the range of other determinations.

One complication for these methods is cosmic variance (Chapter 9). These methods are applied by constructing mock catalogues, which needs additional assumptions. Although these investigators conclude that these results support ΛCDM cosmology, the data still cannot accurately constrain the parameters to high precision.

20.5.3. *Constraints on theories of gravity and cosmologies*

Peculiar velocities provide a test of gravity and cosmologies through the growth rate of the perturbations

$$f(\Omega) = \frac{d \ln D}{d \ln S}, \tag{20.1}$$

relative to the scale factor, S (Huterer *et al.*, 2015). This appears in the peculiar velocities via the parameter β,

$$\beta = \sigma_8 \Omega_m^{\gamma}, \tag{20.2}$$

where γ, is the gravitational growth index. Often, the approximation $\gamma \approx 0.6$ described in Chapter 13 has been used. Wang & Steinhardt (1998) reference earlier literature and discussed using galaxy cluster abundances and large-scale structures to derive specific growth relations for ΛCDM and different models. These produce differing values for γ. For ΛCDM, $\gamma = 0.55$ (Linder, 2005) and modified gravity theories give different values. Linder & Cahn (2007) developed a Minimal Modified Gravity (MMG) theory, which resembles the method of Parameterized Post Newtonian (PPN) theory (Will 1993, 2014) to derive γ for some modified gravity theories, e.g., Braneworld (e.g., Dvali *et al.*, 2000; Brax & van de Bruck, 2003) that predicts $\gamma = 0.6875$

Several studies have estimated γ from β derived using peculiar velocities. Table 20.3 gives some representative values of γ. This requires separate information on Ω_m and σ_8. Some examples are: Abate & Lahav (2008) used SNeIa peculiar motions and assumed $\Omega_m = 0.25$ and $w = -1$ to find $\gamma = 0.72 \pm 0.21$. Nusser & Davis (2011) used TF distances to derive peculiar motions and obtained $\gamma = 0.495 \pm 0.096$. Beutler *et al.* (2012) found $\gamma = 0.547 \pm 0.088$. Hudson & Turnbull (2012) compiled literature measurements and give a weighted mean of $\gamma = 0.619 \pm 0.054$.

Further estimates of γ continue to appear. Huterer *et al.* (2017) analyzed the SNe and 6dFGS. Qin, Howlett, & Staveley-Smith obtained

Table 20.3. Some representative determinations of the gravitational growth index γ.

γ	Sample	Reference
0.72 ± 0.21	SNeIa V_{pec}	Abate & Lahav (2008)
0.495 ± 0.096	TF V_{pec}	Nusser & Davis (2011)
0.619 ± 0.054	V_{pec} compilation	Hudson & Turnbull (2012)
0.547 ± 0.088	RSD	Beutler *et al.* (2012)
$0.55^{+0.10}_{-0.09}$[a]	SNeIa + 6dFGS	Huterer *et al.* (2017)
0.566 ± 0.058	BAO	Mueller *et al.* (2018)
0.54 ± 0.19	BAO	Gil-Marín *et al.* (2018)
$0.62^{+0.20}_{-0.15}$[a]	BAO	Hou *et al.* (2018)
$0.56^{+0.17}_{-0.14}$[a]	BAO	Zarrouk *et al.* (2018)
0.60 ± 0.03	V_{pec}	Qin *et al.* (2019b)
$0.76^{+0.07}_{-0.07}$[a]	V_{pec} 6dFGS + SDSS	Said *et al.* (2020)

[a]Converted from $f(z)\sigma_8$ using Planck 2018 $\Omega_m = 0.315$ and $\sigma_8 = 0.811$.

$\gamma = 0.60 \pm 0.03$ from 2MTF + 6dF data, which they consider fully consistent with ΛCDM and GR. Said *et al.* (2020) used 6dFGS and SDSS data and employ the parameter $S_8 = \sigma_8(\Omega_m/0.3)^{0.5}$. If one adopts some of the PLANCK 2018 CMB parameters, e.g., $\Omega_m = 0.315 \pm 0.007$ and $\sigma_8 = 0.811 \pm 0.006$, this yields values of γ that favor GR. Mueller *et al.* (2018) have used acoustic oscillation and employed redshift data from the Baryon Oscillation Spectroscopic Survey (BOSS). They obtain $\gamma = 0.566 \pm 0.058$ and also set limits on $f(R)$ gravity. When observations can further reduce the errors on γ, this promises to provide an interesting test of modified gravity.

20.5.4. *Comparison of measurements*

If one assumes the standard ΛCDM values for the parameters $\Omega_m = 0.30$ and $\gamma = 0.55$, β can be re-written as

$$S_8 = \beta/(0.3)^{0.55} = \sigma_8(\Omega_m/0.3^{0.55}) \tag{20.3}$$

derived from different types of cosmological investigations and compared to simulations. Boruah *et al.* (2020) have compiled determinations from the literature from different methods including CMB, cluster abundances, weak lensing, and RSD and renormalized them in the form of S_8. These are compared to results from peculiar velocities (including their own from SNe Ia) in Fig. 20.4. This summary shows that here is reasonable agreement to about ±10%, although for the error weighted mean this is 4%. This corresponds to a depth of $\sim 40h^{-1}$Mpc, and indicates that to within this level peculiar velocities are consistent with other methods and ΛCDM.

20.5.5. *Kinetic SZ effect peculiar velocities*

Utilizing peculiar motions at large distances becomes increasingly difficult and new distance methods need to be considered. Kashlinsky *et al.* (2008, 2009) pioneered using the kinetic SZ effect. They determined peculiar velocities of galaxy clusters at distances beyond those achieved using conventional distance indicators and claimed evidence for a large motion called 'dark flow'. Analysis of 3-year WMAP data for nearly 700 galaxy clusters to redshift $z \approx 0.1$ yielded a dipole direction of $(l = 283, b = 11)$ and an unusually high velocity $B = 600$–$1000\,\text{km s}^{-1}$. As seen from Fig. 20.3, this is inconsistent with ΛCDM models. This flow would suggest a primordial CMB dipole of non-kinematical origin, presumably from a large attractor

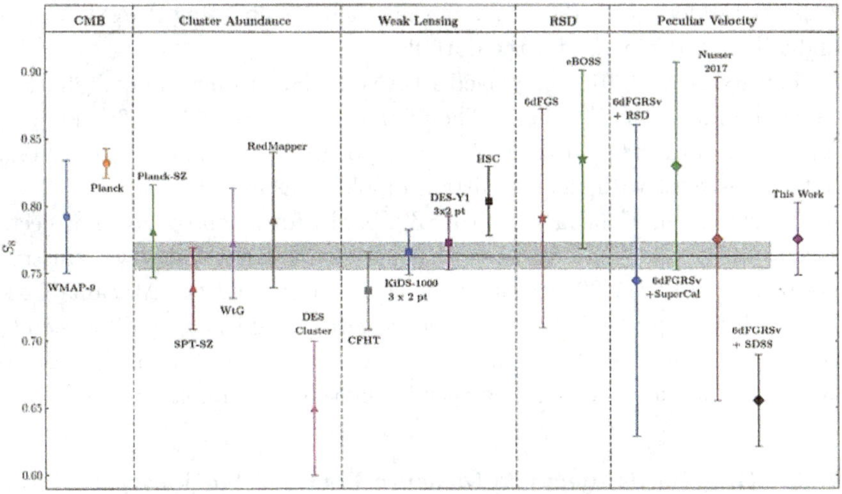

Fig. 20.4. Results of several methods for determining the $S_8 = \sigma_8(\Omega_m/0.3)^{0.55}$ parameter according to Boruah *et al.* (2020) where the data are discussed. References to the original sources: WMAP-9 (Hinshaw *et al.*, 2013), Planck (Planck Collaboration, 2019), Planck-SZ (Bocquet *et al.*, 2019), SPT-SZ (Mantz *et al.*, 2015), WtG (Constanzi *et al.*, 2018), RedMapper (Rykoff *et al.*, 2016), CFHT (Heymans *et al.*, 2013), KiDs-450 (Hildebrandt *et al.*, 2017), DES-Y1 (Abbott *et al.*, 2018), HSC (Hamana *et al.*, 2019), 6dFGS (Beutler *et al.*, 2012), BOSS (Alam *et al.*, 2017), 6dFGRSv (Adams & Blake, 2017), 6dFGRSv+SuperCal (Huterer *et al.*, 2017), CosmicFlows (Dupuy *et al.*, 2019), Work at end (Boruah *et al.*, 2020). The peculiar velocity determinations in the right panel are compared to cosmological constraints from the CMB and cluster abundances, several weak lensing results, and redshift space distortions (RSD). The horizontal line near 0.78 is the uncertainty weighted mean and the horizontal shaded bar indicates the weighted uncertainty. (*Source*: Boruah S. S., Hudson M. J., Lavaux G. (2020), *MNRAS*, 498, 2703, "Cosmic flows in the nearby Universe: new peculiar velocities from SNe and cosmological constraints," Fig. 9, Reproduced by permission of Oxford University Press on behalf of the Royal Astronomical Society.)

beyond the horizon of the visible universe, proof for the existence of a multiverse.

These results have proven controversial. Keisler (2009) analyzed 7-year WMAP data and 736 ROSAT clusters, which place limits ~3× more sensitive than Kashlinsky *et al.* and found no kinematical SZ signal for the dark flow. The data are consistent with ΛCDM, and the earlier results are instrumental. Osborne *et al.* (2011) obtained almost an order of magnitude better signal-to-noise and got no dark flow signal. Likewise, the Planck Collaboration (2014) found no significant detection of a kinematical SZ effect.

Atrio-Barandela *et al.* (2015) reanalyzed their results using PLANCK data and still claim evidence for the dark flow.

Lavaux *et al.* (2013) employed PLANCK data to determine bulk flow using the kinematic SZ effect. They found a velocity of 533±263 km s^{-1} directed toward $(l, b) = (324\pm27, -7\pm17)$ degrees, which has higher errors, but is consistent with the less distant dipole measures.

Measurement of the kinematical SZ effect by future and ongoing projects such as the South Pole Telescope (SPT), a 10-m diameter microwave instrument (Chang *et al.*, 2009; Carlstrom *et al.*, 2011) and the Atacama Cosmology Telescope (ACT Pol), which is getting underway (Hilten *et al.*, 2018; Miyatake *et al.*, 2019), in conjunction with optical data, promise to eventually provide data for new peculiar motion determinations.

20.6. Ongoing Projects to Measure Peculiar Motions

Several large surveys underway or planned will increase the numbers, accuracy, and depths of peculiar velocities. Some preliminary results of these have been discussed above.

Wallaby: This is a Tully–Fisher survey of ∼40,000 galaxies (Johnston *et al.*, 2008; Howlett *et al.*, 2017; Koribalski *et al.*, 2020). This project is a portion of the Australian Square Kilometer Array (AuSKA) Pathfinder H I survey in the southern hemisphere and can be combined with the Westerbork Northern Sky H I survey (WNSHS) in the north.

Taipan: Employing the Fundamental Plane, measurements of distances and peculiar motions of ∼45,000 galaxies to $z \sim 0.1$ are planned. This survey effort follows on the 6dF survey by utilizing the 1.2 m UK Schmidt telescope, but with an improved MOS spectrograph and the StarBugs positioner. Taipan will cover declination $\delta \lesssim 10°$ in the southern hemisphere and will secure redshifts for about 2×10^6 galaxies (da Cunha *et al.*, 2017).

Supernovae: In the future, supernovae may provide the best peculiar velocity estimates because of the higher accuracy of their distances, but the time and location of an outburst is unpredictable and they cannot be rushed. The number of Sne Ia distances will continue to build up and several projects are designed to find them. The Full Foundation Supernovae Sample is constructed from the Panoramic Survey Telescope and Rapid Response System (Pan-STARRS) in Hawaii that currently employs two

1.8-m Ritchey–Chrétien telescopes with large CCD arrays giving a wide 3 deg^2 field. Currently, ~800 SN Ia with $z \lesssim 0.1$ are available (Foley, *et al.*, 2018; Jones *et al.*, 2019). The Sloan digital sky supernova survey (SDSS-II) located 4607 confirmed supernovae and suspects (Sako *et al.*, 2018). The Zwicky Transient Facility (ZTF) using the 48-inch (1.2 m) Schmidt telescope on Mount Palomar (Bellm *et al.*, 2019; Yao *et al.*, 2019)) covers 47 deg^2 and has added 336 supernovae as part of its goals. The Large Synoptic Survey Telescope (LSST) and related surveys will also add to the supernova database (Garcia *et al.*, 2020), while they are intended to provide new results in many areas and reach higher redshifts.

20.7. Summary and Further Questions

Peculiar velocities will continue to be important among the developing new techniques to study the universe. Understanding matter in motion has advanced rapidly in just the past few decades through remarkable discoveries. These range from the fundamental forces and particles through solar system motions, stellar dynamics in galaxies, and galaxy clusters, to flows in the large-scale structure. Key to this is the interplay of gravitation and dark matter on increasing size scales from subatomic particles to cosmological dimensions and from weak gravity to collapsed systems with strong gravity. The peculiar motions have been described and include the Local Group, the local supercluster, and the more remote galaxies and structures. They can be combined with redshift surveys and CMB data to obtain cosmological information. At the present time, one can draw four general conclusions from peculiar motions.

Firstly, the properties of the flows and how they are generated are now clearer. The nature of the Great Attractor (GA) is seen to be centred toward the Hydra–Cen–Norma supercluster. Other mass concentrations also contribute to the peculiar motions. These include the Perseus–Pisces region and the more distant Shapley cloud. They have complicated structures and are outlined by galaxy clusters. Whether or not the motions in these regions contribute to the Hubble constant problem is still unsettled.

Secondly, large unseen mass concentrations that require new forms of matter and produce significant deviations from the CMB dipole direction, such as dark flow, appear ruled out. Seeing if the gravitational attraction of the observed mass regions fully explains the flows revealed by the peculiar motions needs further verification. New investigations with larger and higher quality data reach better agreement, but several investigators find small residuals of the order of 150 km s^{-1} that might be caused by more distant attractors.

Thirdly, the bulk flow convergence of the dipole direction and velocity amplitude with redshift is related to this. In addition parameters, e.g. β, γ, Ω_m, and σ_8 that are determined probe cosmology and ΛCDM theory. This is a test of the power spectrum and these can be compared with the predictions from theory. Whether the bulk flow disagrees with standard ΛCDM cosmology depends on how it is calculated. Given the different peculiar velocity catalogs, the required corrections to the data, and analysis schemes, it is not surprising that some disagreement lingers at present. Currently, most investigations find that the measurements of these parameters are consistent with ΛCDM, but the agreement is to about one σ or $\pm 68\%$.

Fourthly, peculiar velocities show promise for obtaining and testing some of the modified gravity theories. Parameter estimates such as $\sigma_8(\Omega_m/0.3)^\gamma$, derived from independent methods, again, agree within about one σ and are consistent with ΛCDM. The expected values of $\gamma \approx 0.55$ are also consistent within the errors and can exclude some theories of modified gravity. For future determinations, higher accuracies are to be desired. A straight mean of many estimates gives $\gamma \approx 0.60 \pm 0.02$.

Peculiar velocities can be compared with other cosmological models and observations. Investigative methods based on independent physical assumptions are important. Different surveys and samples of galaxies give varying results and further work is needed. In astrophysics and cosmology, there are very few 'clean' measurements free of a large number of contributing factors that require correction.

Recalling the complexity of the measurements of distances needed for deriving peculiar motions, there is the complicated chain of measurements and corrections. These also need to be applied to other cosmological measurements so it is likely that some results will change and, eventually, all of these can be brought into better agreement. However, if discrepancies continue, they could yield clues leading to new interpretations of the data. Peculiar motions are limited to relatively nearby distances, but some nurse hope that peculiar motions at great distances employing new methods, e.g., by the kinetic SZ effect, standard sirens, and BAOs, etc., could provide new cosmological information.

Chapter 21

Concluding Remarks

The developments in the measurement of peculiar motions have been described. Knowledge of the motions of matter have been expanded by several new discoveries. How do these connect with contemporary astrophysics and cosmology? The picture that one sees, rather than being a sharp focus is a broad patchwork of phenomena. Whilst new questions have arisen in the 21st century, others are over a century old. Some even originated in antiquity. The accumulation of empirical data has accelerated due to the wide availability of modern instruments. This is complimented by the application of detailed computer modelling and data analyses. Investigations have grown from one or a few individuals drinking coffee all night, peering into eyepieces at objects near the limits of human vision, to projects employing 300–400 or more colleagues.

Major funding is required for projects based on astronomical satellites and large earth-based observatories and their homes. This stems from the complexity of the instrumentation and the data reduction programs available that nurse delicate signals from immense amounts of complicated data. These efforts and their interpretation have spawned new fields of study that barely existed at the end of the last century.

Many theoretical endeavours strive to develop new physical theories, involving, e.g., quantum gravity, string theory, and higher dimensional and non-isotropic cosmological models. Any of these might give the impetus for a jump in fundamental physics. At present, however, this seems to have reached a plateau accompanied by the rapid growth in observational data. The older problems of the intertwined natures of dark matter, dark energy, and gravitation seem unsolved. There appears to be no observation that provides an overwhelming clue for any of these yet. One might conclude that astronomical sciences are still in a data collection mode.

The nature of gravity lies at the centre of understanding the universe. General Relativity is a remarkably accurate theory of gravity in all tests so far devised. In astrophysics, two observational regimes for studying motions, will produce progress. The first is in motions under the control of strong gravity, of neutron stars and black holes revealed by gravitational radiation and other non-classical effects. This is a relatively new subject. Gravitational waves, and the structure of black holes as observed with the Event Horizon Telescope and elsewhere appear to confirm GR predictions. The second involves size scales related to the large-scale structure revealed by weak lensing, galactic dynamics and galaxy clusters. This is where peculiar motions fit in.

After over two centuries of measuring the Newtonian gravitational constant, G, it is still the most poorly determined fundamental constant. This illustrates the difficulties with dealing with gravity on earth. Although the literature contains disprovable theories of gravity, it is impossible to conclude whether any simple theories other than GR can account better for known physical phenomena.

Combining GR with the notions of cold dark matter and dark energy explains many extragalactic and cosmological data using ΛCDM simulations. The Planck Collaboration VI (2018) concludes that with six parameters, Planck data are well fit. This includes the large-scale structure, galaxy formation and evolution, the BAOs, and other fundamental observations, a remarkable achievement. Nevertheless, despite intensive effort at great expense, science is still powerless to explain dark matter and dark energy and to connect quantum to large-scales. So one is left with a great edifice of accumulating astronomical data, and interpretations that are mostly focused on ΛCDM cosmology.

ΛCDM appears to be an excellent theory for many applications, but one is obliged to asks how good it is. Problems bubble below the surface. Some investigators consider that they will go away with larger and better N-body simulations and the accumulation of greater quantities of more detailed and accurate observational data.

Although the ΛCDM model accounts impressively for many observed phenomena, in some areas there is disagreement with N-body cosmological simulations. Many of these are considered small-scale challenges that may not be serious and the resolution could require a better formulation of the baryonic physics, or the complex physics of DM might be involved (Weinberg *et al.*, 2015). Some of the outstanding discrepancies include:

(1) *The cuspy halo problem*: In many instances, the simulated halos reach higher central densities than observed.
(2) *The missing satellite problem*: Simulations predict more small mass satellite halos surrounding larger objects than is observed around the Milky Way and M31.
(3) *Satellite planes*: Satellites of M31 and the Milky Way appear to lie in planes, while CDM predicts random distributions for satellites.
(4) *Morphology problem*: If massive galaxies grow from mergers, the predicted number of bulgless disk galaxies is too large (Kormendy *et al.*, 2010).
(5) *The too big to fail problem*: The simulations predict subhalos that are more massive than observed.

Searches for new MW satellites and satellite galaxies around MW galactic analogs (Geha *et al.*, 2017; Nadler *et al.*, 2020; Mao *et al.*, 2021) indicate that issues (2) and (3) may not be serious and new results are broadly consistent with the ΛCDM model. The CMB anomalies discussed in Chapter 15 also might hint at new phenomena in the early universe, but these are still controversial and could change with new data.

Everyone anticipates new and revolutionary discoveries in physics and astronomy that would enhance or even change our world view. Gradual changes from the accumulation of new facts more likely will answer these questions rather than one large quantum jump. Along with the derivation of gravitational models, a tremendous energy has been devoted to determining H_0.

Setting aside looming questions, already mentioned, the different Hubble constants derived from the CMB and nearby optical data (Freedman, 2017; Riess, 2019) might be the sought clue. Some investigators believe there are real inconsistencies in H_0, a fundamental basis for understanding the universe. Any universe theory must account for it. The tendency is nearby H_0 tends to lie higher than that found from the CMB value. From Cepheid distances Riess *et al.* (2019) obtain $H_0 = 74.03 \pm 1.42\,\mathrm{km\ s^{-1}\,Mpc^{-1}}$ for the local universe, while the CMB, at an earlier era gives $H_0 = 67.4 \pm 0.5\mathrm{km\ s^{-1}\,Mpc^{-1}}$ (Planck Collaboration, 2018). Figure 21.1 is a compilation by Riess *et al.* (2019) that compares H_0 measured in the nearby universe to the earlier era CMB values. Verde, Treu, & Riess (2019) have summarized some of the situation for the nearby universe and CMB H_0 values. It is important to note that these are derived from different methods requiring corrections including peculiar velocities,

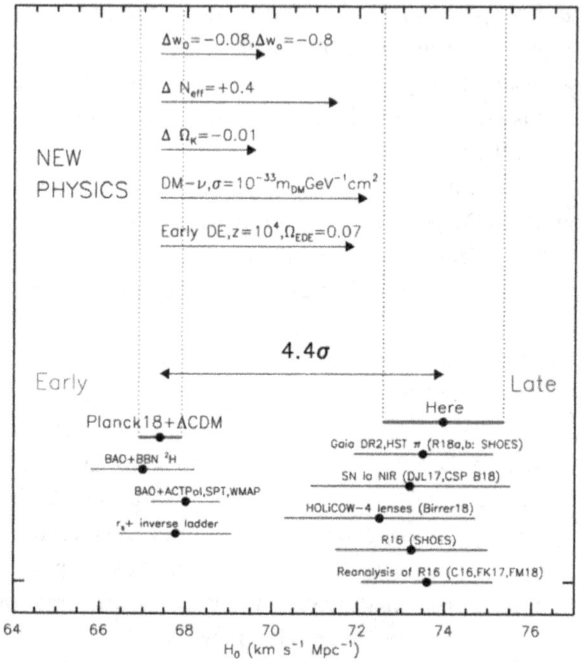

Fig. 21.1. Compilation of determinations of the Hubble constant based on determinations from the CMB for a redshift in a relatively young universe age $\approx 0.2\,\mathrm{Gyr}$ (blue) and revised optical data nearer the present age $\approx 10\,\mathrm{Gyr}$ (red). This shows a 4.4 standard deviation in the two determinations. The effects of possible new physics in bringing the values of H_0 together are indicated in the upper portion of the figure. (*Source*: Riess A. G., Casertano S., Yuan W., Macri L. M., Scolnic D. (2019), *ApJ*, 876, 85.)

an important correction to the observed redshifts of each galaxy. If this range in the Hubble constant is confirmed by future high precision determinations, modifications to the ΛCDM model would be indicated.

The determination of the Hubble constant has long remained a classical problem. Remembering the large ranges with small errors of the past century, with H_0 ranging from $100 \pm 10\,\mathrm{km\ s^{-1}\,Mpc^{-1}}$ (de Vaucoulers *et al.*, 1981) to $50 \pm 7\,\mathrm{km\ s^{-1}\,Mpc^{-1}}$ (e.g., Sandage & Tammannn, 1982; Rowan-Robinson, 1985; Tammann, Sandage, & Reindl, 2008). Considerable progress has been achieved in the measurement of H_0 in the past decades, but one question persists: is this current difference still due to observational errors or hints at new physics? A considerable number of estimates of the Hubble constant continue to be made using different datasets and approaches.

As already described, there are new methods for determining distances. The TGB method yields an intermediate $H_0 = 69.8 \pm 1.9\,\mathrm{km\ s^{-1}\,Mpc^{-1}}$ (Freedman *et al.*, 2019). Other new techniques, independent of the same distance indicators and assumptions, include the standard siren method, which gives $H_0 = 70^{+12}_{-8}\,\mathrm{km\ s^{-1}\,Mpc^{-1}}$ and the γ-ray attenuation method, also independent of distance indicators (Domínguez *et al.*, 2019) that finds $H_0 = 67.4^{+6}_{-6.2}\,\mathrm{km\ s^{-1}\,Mpc^{-1}}$ but yields $\Omega_M = 0.14^{+0.06}_{-0.07}$. These fall between the earlier sets of measurements but also point to changes in other parameters. Lensing time delays give $82 \pm 8\,\mathrm{km\ s^{-1}\,Mpc^{-1}}$ (Jee *et al.*, 2019).

Explanations of the different measures of H_0 range from theoretical to observational. Theoretically this would entail modification to DM, the cosmology changing the scale factor, gravity on large scales, etc. Poulin *et al.* (2019) have proposed that adding a second form of dark energy as a scalar field that decays could modify the early universe acceleration and explain this discrepancy. Many other theories are coming forth. However this reminds one that 'a good theoretician can always come up with at least two or three explanations for most observations'.[1]

Some investigators consider the differences in H_0 measured nearby and far as effects of systematic errors in the measurement methods. This has been checked without finding any errors. The role of peculiar motions in corrections to the redshift measurements used in determining the Hubble constant is an important consideration because this is the right order of magnitude. Boruah, Hudson, & Lavaux (2021) considered how different velocity and density field models change the derived H_0 and do not find large differences. They find that using megamaser distances, the 2M++ velocity field gives the best result of $H_0 = 69^{+2.9}_{-2.8}\mathrm{km\ s^{-1}\,Mpc^{-1}}$. Rameez & Sarkar (2019) indicate that systematic differences in the publicly available supernova databases and the importance of the V_{pec} corrections to their redshifts are sufficient to explain the H_0 'tension'.

The location of the Milky Way in a large local void or under-density extending to ~150 Mpc has been revived. This would produce an enhanced expansion rate relative to the mean of the universe of about the right order of magnitude (Whitbourn & Shanks, 2016 using 6dF, 2MASS, and SDSS redshift data; Hoscheit & Barger, 2018 using the linear kinetic SZ-effect; Lombriser, 2020 from SNe Ia and Cepheids; Böringer, Chon, & Collins,

[1] I do not know the origin of this quote, but it was first told to me by Donald Lynden-Bell.

2020 from X-ray galaxy clusters), although some earlier work (Giovanelli *et al.*, 1998b, 1999 using peculiar motions; Wu & Huterer, 2017 using SNe Ia distances and ΛCDM modelling) question whether a void is able to solve the H_0 discrepancy.

There are additional challenges for the standard ΛCDM model. Although these results may not stand up to further scrutiny, they represent the kinds of tests that should continue. Recent investigations of the lensing of the CMB and the luminosity dependence of SNe Ia with redshift lead to claims of no or reduced Λ and non-zero curvature. Betoule *et al.* (2014) analyzed 740 SNe Ia collected by the SDSS-II and SNLS collaborations and confirmed earlier cosmological parameters. Several investigators have examined whether the properties of the SNe Ia remain constant with redshift. Kang *et al.* (2020 and references therein) as well as Tutusaus *et al.* (2019) studied these objects in galaxies covering a range of population-age, and metallicity using Lick indices in their spectra. Significant differences were found in the SNe Ia luminosities. Sources in low mass younger galaxies are fainter than those in high mass older galaxies by ~0.14mag./4 Gyr. Correcting for this effect, Kang *et al.* (2020) find that in the redshift-luminosity Hubble diagram this eliminates the need for Λ and get a best fit with $\Omega_M = 0.27$ and $\Omega_\Lambda = 0$. Rose *et al.* (2020) question the robustness of this result after rejecting 12% of the sample.

Colin *et al.* (2019) used supernovae to study the bulk flow. They found a significant dipole component of the deceleration parameter, q_0. This could be considered evidence that the cosmic acceleration interpreted as dark energy could be an artefact.

Di Valentino, Melchiorri, & Silk (2020a,b) interpret enhanced lensing amplitude in the CMB cosmic shear power spectrum from the PLANCK-2018 data compared with ΛCDM predictions. Their best correlation provides a closed universe, rather than a flat one, with curvature $\Omega_k = -0.04$ corresponding to a radius $R_c \sim 10\,\text{Gpc}$. However, Efstathiou & Gratton (2020) raise questions about this analysis and support a flat universe. Whether results like these will be confirmed remains to be tested, but this shows that conclusions about the cosmological model for the universe are still in play.

Extrapolating back to $t = 0$ in the Big Bang in the nearby universe, at present, $t \approx 14 \times 10^9$ years, despite discrepancies such as those mentioned, the ΛCDM Big Bang model seems highly successful; the current large-scale structure along with the peculiar motions, discussed in the previous Chapter 20, can be explained fairly well.

More serious is that looking back beyond the CMB time $t \sim 3 \times 10^5$ years and the nucleosynthesis time $t \sim 1$ sec, direct evidence supportive of this extrapolation becomes increasingly vague. Thus in trying to peer even earlier, the possibility for new physics has an opening. It is well known that to get past the classical problems of the Big Bang, e.g., the horizon and flatness problems, some form of inflation, extremely rapid expansion by ~100 orders of magnitude at about 10^{-32} sec is required. Many versions of inflation theory have been proposed. Inflation predicts the initial scale invariant power spectrum of the fluctuations with $n \sim 1$ from which the large-scale structure evolves, but other than this, there is no direct evidence for inflation. While one test is that it should produce circular polarization in the CMB, this has not been observed.

From a philosophical standpoint, Zinkernagel (2002, 2014) has argued mirroring the concerns of a number of other investigators, that although it is a highly popular and plausible notion, which many accept, there is really no direct evidence that particle physics shapes the large-scale structure of the universe and that the alleged connection between particle physics and early cosmology is weaker than popularly claimed. This likewise calls into question that the structure of the universe is ultimately governed by quantum mechanics, such as in the Wheeler–DeWitt equation (DeWitt, 1967; Rovelli, 2010; Shestakova, 2018).

Burbidge (2008) and collaborators long defended non-Big-Bang cosmologies and the need for new physics. Whilst these ideas are currently unpopular, Burbidge pointed out that, in the general case of data fitting if there are enough free parameters, almost any model can be accommodated. He maintained that this is the case for ΛCDM models,

Peebles (2015 and references therein) has discussed how well the ΛCDM model agrees with observations including all observational constraints. If one is willing to accept that ~5–10% is a good fit and brush away the number of problems that have been mentioned, the ΛCDM model can be limited to six empirically derived parameters: ρ_b, ρ_{DM}, H_0, tilt, n_s, and Λ, baryonic matter density, dark matter density have been checked independently using different methods, but they have not been predicted from first principles. This seems to be consistent with current data, but probably needs to be modified if higher precision is to be obtained.

Brandenberger (2018) and Brandenburger & Peter (2017) consider that the ΛCDM paradigm is not the only one that is consistent with available observations. Although the ΛCDM Big Bang is highly successful, there are also early universe models that do not require inflation and the initial singularity to explain the observed large-scale structure. These are based on superstring theory and thus require more complicated and unexplored physics that are not presently understood. These include bouncing and string-gas cosmologies, pre-big-bang and ekpyrotic scenarios. Some investigators propose that other cosmologies can eliminate the need for dark energy or other of the six empirical parameters, e.g., inhomogeneous cosmologies such as the Lemaître–Tolman–Bondi (LTB) or Stephani cosmologies (Mustapha & Hellaby, 2001; Enqvist, 2008) and provide the differences in H_0. However such models have a number of difficulties explaining all of the observational constraints (Balcerzak *et al.*, 2015; Ong *et al.*, 2018).

Of course, there are also other questions about the universe: what is it and why is it here? These will not be approached here and answers to them are left to the experts.

Many authors have discussed the Scientific Method; almost innumerable volumes have been written about it. I do not attempt to elaborate on this, but it is important to keep in mind that the Scientific Method lies at the core of scientific research and always incorporates comparing observations, which rely on reproducible facts, and theoretical explanations. Progress is made by obtaining better data and further improving or rejecting theories, but in the 21st century some extensive changes have transformed astrophysics and cosmology.

The developments of space borne instruments, large ground-based observatories, supercomputers, and particle accelerators have become costly and complex so that one nation, or even a consortium of them, can only construct a few. The quantity of data is mind boggling and hundreds or even thousands of devoted colleagues are sometimes needed to complete a project and boil this down to a few fundamental facts. While it is envisioned that this approach will answer several pressing current questions, one analyzes the measured massive amounts of data and fits many parameters simultaneously to models that have been already been prescribed, e.g., the ΛCMB models are examples. Although many arguments support this approach, in some circles, it seems to be felt that the agreement between observation and theory through numerical simulations has become so good that the border between observed and computer generated data is becoming blurred and observations are unnecessary, seemingly a risky modification to the Scientific Method when reaching the fringes of knowledge.

Coming to the end of this book, after all of the facts and details that were examined, we cannot give a fully satisfactory answer to the question asked in the first chapter: "Where are we and where are we heading?" There has been progress. It seems possible to state that while peculiar motions are consistent with ΛCDM, they do not prove it; this needs to be considered along with all the other types of methods and their approximations. We appear to be at the edge of the local void region near the Virgo galaxy cluster and between the larger Perseus–Pisces and Norma–Centaurus regions. Our galaxy heads in the original GA direction driven by the gravitational attraction of masses at about $400\,\mathrm{km\ s^{-1}}$. It remains uncertain whether visible masses or more distant unseen ones account for all of the motion. It is harder to say what lies farther. One can expect significant improvements and widening of the vast panorama coming in the future.

One needs to continue to rely on observational data and consequently, the well-known conclusion to Hubble's (1936) book, *The Realm of the Nebulae*: "The search will continue. Not until the empirical resources are exhausted, need we pass on to the dreamy realms of speculation" remains equally compelling today as nearly a century past.

Bibliography

Aaronson M., Huchra J., Mould J. (1979), *ApJ*, 229, 1.

Aaronson M., Dawe J. A., Dickens R. J., Mould J. R., Murray J. B. (1981), *MNRAS*, 195, 1p.

Aaronson M., Huchra J., Mould J., Scechter P. L., Tully R. B. (1982), *ApJ*, 258, 64.

Aaronson M. *et al.* (1982b), *ApJS*, 50, 241.

Aaronson M., Bothun G., Mould J. R., Huchra J., Schommer R. A. (1986), *ApJ*, 302, 536.

Aaronson M. *et al.* (1989), *ApJ*, 338, 654.

Aarseth S. J. (2003), *Gravitational N-body Simulations: Tools and Algorithms.* Cambridge: Cambridge University Press.

Abate A., Lahav O. (2008), *MNRAS*, 389, 47.

Abe K. *et al.* (2018), *PhLB*, 787, 153.

Abbott B. P. *et al.* (2016), *PhRvL*, 116, 061102.

Abbott B. P. *et al.* (2017a), *PhRvL*, 119, 141101.

Abbott B. P. *et al.* (2017b), *PhRvL*, 119, 161101.

Abbott B. P. *et al.* (2017c), *ApJ*, 848, L13.

Abbott B. P. *et al.* (2017d), *Nature*, 551, 85.

Abbott T. M. C. *et al.* (2018), *PhRvD*, 98, 043526.

Abell G. (1958), *ApJS*, 3, 211.

Abraham R., van den Bergh S. (2001), *Sci*, 293, 1273.

Adams C., Blake C. (2017), *MNRAS*, 471, 839.

Ade P. A. R. *et al.* (2014), *PhRvL*, 112, 241101.

Ade P. A. R. *et al.* (2016), *A&A*, 594, A21.

Ade P. A. R. *et al.* (2021), *PhRvL*, 127, 151301.

Adelberger E. G., Heckel B. R., Nelson A. E. (2003), *ARNPS*, 53, 77.

Adelman S. J. (1973), *ApJ*, 183, 95.

Adler R., Bazin M., Schiffer M. (1965), *Introduction to General Relativity.* New York: McGraw-Hill Book Company.

Aerts C., Christian-Dalsgaard J., Kurtz D. (2010), *Astroseismology.* New York: Springer.

Agast D., Samland M., Thielemann F. K., Qian Y.-Z. (2004), *A&A*, 416, 997.
Aghamousa A. *et al.* (2016a), arXiv:1611.00036.
Aghamousa A. *et al.* (2016b), arXiv:1611.00037.
Aghanim N. *et al.* (2018), arXiv:1807.06209.
Aihara H. *et al.* (2018), *PASJ*, 70, 8.
Akiyama K. *et al.* (2019), *ApJ*, 875, 1.
Akrami Y. *et al.* (2019), arXiv:1906.02552.
Alam S. *et al.* (2017), *MNRAS*, 470, 2617.
Alam S. *et al.* (2021), *PhRvD*, 103, 083533.
Albæk L., Hansen S. H., Martizzi D., Moore B., Teyssier R. (2017), *MNRAS*, 472, 3486.
Alcock C. *et al.* (2000), *ApJ*, 542, 281.
Allen C. W. (1973), *Astrophysical Quantities*, 3rd ed., London: Athalone Press.
Allen P. D. *et al.* (2005), *MNRAS*, 360, 1244.
Alonso D. *et al.* (2015), 449, 670.
Alonso M. V., Bernardi M., da Costa L. N., Wegner G., Willmer C. N. A., Pellegrini P. S., Maia M. A. (2003), *AJ*, 125, 23017.
Alpher R. A., Herman R. C. (1948), *Nature*, 162, 774.
Amanullah R. *et al.* (2010), *ApJ*, 716, 712.
Amaré J. *et al.* (2019), *PhRvL*, 123, 031301.
Amon A. *et al.* (2021), arXiv:2105.13543.
Antoniazzi A., Califano F., Fanelli D., Ruffo S. (2007), *PhRvL*, 98, 0602.
Apaslan M. *et al.* (2014), *MNRAS*, 440, L106.
Appelby S., Shafielo A. (2014), *JCAP*, 10, 070.
Aprile E. *et al.* (2017a), *PhRvL*, 119, 181301.
Aprile E. *et al.* (2017b), *EPJC*, 77, 881.
Aragon-Calvo M. A., Szalay A. S. (2013), *MNRAS*, 428, 3409.
Argelander F. W. A. (1852–62), *Bonner Durchmusterung.*
Arnaud M. (2009), *A&A*, 500, 103.
Arnold C., Matteo L., Li B. (2019), *NatAs*, 3, 945.
Atrio-Barandela F., Kashlinsky A., Ebeling H., Fixsen D. J., Kocevski D. (2015), *ApJ*, 810, 143.
Baade W. (1953), *Trans IAU*, 8, 398.
Babichev E., Deffayet C. (2013), *CQGra*, 30, 400.
Balcerzak A., Dabrowski M. P.,Denkiewicz T., Polarski D., Puy D. (2015), *PhRvD*, 91, 083506.
Balona L. A., Martin W. L. (1978), *MNRAS*, 184, 1.
Banse K., Grosbøl P., Baade D. (1992), *ASPC*, 25, 120.
Barbon R., Ciatti F., Rosino L. (1973), *A&A*, 25, 241.
Bardeen J. M., Bond J. R., Kaiser N., Szalay A. S. (1986) *ApJ*, 304, 15.
Bardeen J. M., Bond J. R., Efstathiou G. (1987), *ApJ*, 321, 28.
Bartelmann M., Schneider P. (2001), *PhRvL*, 340, 291.
Bassett G., Hlozek R. (2010), in *Dark Energy: Observational and Theoretical Approaches*, ed. P. Ruiz-Lapuente. Cambridge: Cambridge Press University, p. 246.
Baumgarte T. W., Shapiro S. L. (2011), *PhT*, 64, 32.

Beers T. C., Geller M. J., Huchra J. P. (1982), *ApJ*, 257, 23.
Bekenstein J. D. (2004), *PhRvD*, 90, 3509.
Bekenstein J. D., Sanders R. H. (2012), *MNRAS*, 421, 59.
Bellm E. C. *et al.* (2019), *PASP*, 131, 018002.
Belokurov V. (2013), *NewAR*, 57, 100.
Bender R. (1990), *A&A*, 229, 441.
Bender R., Faber S. M., Burstein D. (1992), *ApJ*, 399, 462.
Bender, R., Burstein, D., Faber, S. M. (1997), in *Galaxy Scaling Relations: Origins, Evolution and Applications", Proceedings from the ESO Workshop*, November, 1996, eds. Luiz da Costa and Alvio Renzini, Springer-Verlag, p. 95.
Bengaly Jr. C. A. P., Novaes C. P., Xavier H. S., Bilicki M., Bernui A., Akaniz J. S. (2017), *MNRAS*, 464, 768.
Bennett C. L. *et al.* (2013), *ApJS*, 208, 20.
Bernardi M., Alonso M. V., da Costa L. N., Wilmer C. N. A., Wegner G., Pellegrini P. S., Rite C., Maia M. A. G. (2002a), *AJ*, 123, 2990.
Bernardi M. *et al.* (2003), 125, 1866.
Bertin G., Stiavelli M. (1984), *A&A*, 137, 26.
Bertotti B., Iess L., Tortora P. (2003), *Nature*, 425, 374.
Bergé Métris G., Pernot-Borràs M., Touboul P., Uzan J.-P. (2018), *PhRvL*, 120, 141101.
Bernardi M., Alonso M. V., da Costa L. N., Wilmer C. N. A., Wegner G., Pellegrini P. S., Rité C., Maia M. A. G. (2002a), *AJ*, 123, 2159.
Bernardi M., Alonso M. V., da Costa L. N., Wilmer C. N. A., Wegner G., Pellegrini P. S., Rité C., Maia M. A. G. (2002b), *AJ*, 125, 2307.
Bertschinger E., Dekel A. (1989), *ApJ*, 336, L5.
Bertschinger E. (1998), AR*A&A*, 36, 599.
Bertschinger E., Gelb J. M. (1991), *ComPh* 5, 164.
Besla G. *et al.* (2012), *MNRAS*, 421, 2109.
Bessell F. W. (1838), *AN*, 16, 65.
Bessell M. S. (2005), *ARA&A*, 43, 293.
Betoule M. *et al.* (2014), *A&A*, 568, A22.
Beutler F. *et al.* (2011), *MNRAS*, 416, 3017.
Beutler F. *et al.* (2012), *MNRAS*, 423, 3430.
BICEP2/Keck Collaboration (2015), *PhRvL*, 114, 101301.
Binney J., Merrifield M. (1998), *Galactic Astronomy*, Princeton: Princeton University Press.
Binney J., Tremaine S. (2008), *Galactic Dynamics*, 2nd ed., Princeton: Princeton University Press.
Birkinshaw M. (1999), *PhR*, 310, 97.
Birkinshaw M., Gull S. F. (1983), *Nature*, 302, 315.
Birkinshaw M., Hughes J. P., Arnaud K. A. (1991), *ApJ*, 379, 466.
Blaauw A., Gum C. S., Pawsey J. L., Westerhout G. (1960), *MNRAS*, 121, 123.
Blake C. *et al.* (2011), *MNRAS*, 418, 1707.
Blakeslee J. P., Davis M., Tonry J. L., Dressler A., Ajhar E. A. (1999), *ApJ*, 52, 527, L73.

Blakeslee J. P., Lucey J. R., Tonry J. L., Hudson M. J., Naraynanan V. K., Barris B. J. (2002), *MNRAS*, 330, 443.
Blakeslee J. (2012), *Ap&SS*, 341, 179.
Blandford R. D., Narayan R. (1992), *ARA&A*, 30, 311.
Blanton M. R., Roweis S. (2007), 133, 734.
Boehle A. *et al.* (2016), *ApJ*, 830, 17.
Böringer H., Chon G., Collins C. A. (2020), *A&A*, 633, A19.
Bolejko K., Hellaby C. (2008), *GReGr*, 40, 1771.
Bolokurov V. *et al.* (2017), *MNRAS*, 466, 4711.
Bolzonella M., Miralles J.-M., Pelló (2000), *A&A*, 363, 476.
Bonamente M., Joy M. K., LaRoque S., Carlstrom J. E., Reese E. D., Dawson K. S. (2006), *ApJ*, 645, 27.
Bond J. R., Efstathiou G. (1984), *ApJ*, 285, L45.
Bond J. R. *et al.* (2005), *ApJ*, 626, 12.
Borgani S., *et al.* (1997), *ApJ*, 482, L121.
Borgani S. (2000), *ApJ*, 537, 1.
Boruah S. S., Hudson M. J., Lavaux G. (2020), *MNRAS*, 498, 2703.
Boruah S. S., Hudson M. J., Lavaux G. (2021), arXiv:2010.01119.
Bottinelli L., Gouguenheim L., Paturel G., Teerikorpi P. (1988), *ApJ*, 328, 4.
Bothun G. D., Schommer R. A., Williams T. B., Mould J. R., Huchra J. P. (1992), *ApJ*, 388, 253.
Boutan C. *et al.* (2018), *PhRvL*, 121, 261302.
Boylan-Kolchin M., Springel V., White S. D. M., Jenkins A., Lemson G. (2009), *MNRAS*, 398, 1150.
Boyle W., Smith G. E. (1970), *Bell Sys Tech J*, 49, 587.
Brammer G. B., van Dokkum P. G., Coppi P. (2008), *ApJ*, 686, 1503.
Branch D. (1998), *ARA&A*, 36, 17.
Branchini E. *et al.* (2001), *MNRAS*, 326, 1191.
Brault J. M., White O. R. (1971), *A&A*, 13, 169.
Briggs F. H. (1990), *ApJ*, 352, 15.
Broadhurst T. J., Ellis R. S., Koo D. C., Szalay A. S. (1990), *Nature*, 343, 726.
Bromley B. C. (1994), The large-scale structure of the universe: wavelet analysis of the distribution of galaxies. PhD thesis, Dartmouth College.
Brown W. R., Kilic M., Kosakowski A., Gianninas A. (2017), *ApJ*, 847, 10.
Bruzual G., Charlot S. (2003), *MNRAS*, 344, 1000.
Buchdal H. A. (1970), *MNRAS*, 150, 1.
Buchler J. R. (2010), *AIP Conf Proc* 1170, 51.
Burbidge G. (2008), arXiv:0811.2402.
Burbidge E. M., Burbidge G. R., Fowler W. A., Hoyle F. (1957), *RvMP*, 29, 547.
Burke B. E. (1957), *AJ*, 62, 90.
Burstein D., Heiles C. (1978), *ApJ*, 225, 40.
Burstein D., Heiles C. (1982), *AJ*, 87, 1165.
Burstein D., Heiles C. (1984), *ApJS*, 54, 33.
Burstein D., Rubin V. C. (1985), *ApJ*, 297, 423.
Burstein D., Davies R. L., Dressler A., Faber S. M., Stone R. P. S., Lynden-Bell D., Terlevich R. J., Wegner G. (1987), *ApJS*, 64, 601.

Burstein D., Bender R., Faber S. M., Nolthenius R. (1997), *ApJ*, 114, 1365.
Burstein D. (1990), *RPPh*, 53, 421.
Campbell L. (2009), PhD thesis, Australian National University.
Campbell L. A. (2014), *MNRAS*, 443, 1231.
Cannon A. J., Pickering E. C. (1918–1924), *AnnHar*, 91–99, 1.
Capitaine N., Klioner S., McCarthy D. (2012), *IAUJD*, 7, 40.
Cappellari M. *et al.* (2006), *MNRAS*, 366, 1126.
Cappellari M. *et al.* (2013a), *MNRAS*, 432, 1709.
Cappellari M. *et al.* (2013b), *MNRAS*, 432, 1862.
Carlberg R. G. (2019), *ApJ*, 885, 17.
Carlberg R. G., Grillman C. J. (2013), *ApJ*, 775, 90.
Carlstrom J. J. *et al.* (2011), *PASP*, 123, 568.
Carrick J., Turnbull S. J., Lavaux G., Hudson M. J. (2015), *MNRAS*, 450, 317.
Cautun M. *et al.* (2014), *MNRAS*, 441, 2923.
Cavendish H. (1798), *Phil Trans R Soc B*, 88, 469.
Cawson M. G. M. (1987), GRASP Documentation. Tucson: University of Arizona.
Ceccarelli C., Valotto C., Lambas D. G., Padilla N., Giovanelli R., Haynes M. P. (2005), *ApJ*, 622, 853.
Chang C. L. *et al.* (2009), *AIP Conf Proc*, 1185, 475.
Charlier C. V. L. (1922a), *MeLuF*, 98, 1.
Charlier C. V. L. (1922b), *Arkiv Mat Astron Phys*, 16, 1.
Chen H.-Y., Fishbach M., Holt D. E. (2018), *Nature*, 562, 545.
Chen X., Wang S., Deng L., de Grijs R., Liu C., Tian H. (2019), *NatAs*, 202C.
Christian-Dalsgaard J. *et al.* (1996), *Science*, 272, 1286.
Christianson J. R. (2003), On Tycho's Island. Cambridge: Cambridge University Press.
Clarke A. M. (1908), *A Popular History of Astronomy during the Nineteenth Century*, 4th Edition. London: Adam & Charles Black.
Clampitt J., Cai Y.-C., Li B. (2013), *MNRAS*, 431, 749.
Clayton D. D. (1968), *Principles of Stellar Evolution and Nuclear Synthesis.* New York: McGraw-Hill.
Clifton T., Ferreira P. G., Padilla A. (2012), *PHR*, 513, 1.
Clowe D. *et al.* (2006), *ApJL*, 648, L109.
Clutton-Brock M., Peebles P. J. E. (1981), *AJ*, 86, 1115.
Cohen S. A., Hickox R. C., Wegner G. A., Einasto M., Vennik J. (2014), *ApJ*, 783, 186.
Coil A. L. *et al.* (2011), *ApJ*, 767, 741, 8.
Cole S. *et al.* (2005), *MNRAS*, 362, 2005.
Cole S., Lacey C. G., Baugh C. M., Frenk C. S. (2000), *MNRAS*, 319, 168.
Colin J., Mohayaee R., Rameez M., Sarkar S. (2019), *A&A*, 631, L13.
Colless M., Burstein D., Wegner G., Saglia R. P., McMahan R., Davies R. L., Bertschinger E., Baggley G. (1993), *MNRAS*, 262, 475.
Colless M. (1995), *AJ*, 109, 1937.
Colless M., Burstein D., Davies R. L., McMahan Jr. R. K., Saglia R. P., Wegner G. (1999), *MNRAS*, 303, 813.

Colless M., Saglia R. P., Burstein D., Davies R. L., McMahan Jr. R. K., Wegner G. (2001a), *MNRAS*, 321, 277.
Colless M. M. *et al.* (2001b), *MNRAS*, 328, 1039.
Collister A. A., Lahav O. (2004), *PASP*, 116, 345.
Collister A. *et al.* (2007), *MNRAS*, 375, 68.
Contreras C. *et al.* (2013), *MNRAS*, 430, 924.
Cooke R. J., Pettini M., Steidel C. C. (2018), *ApJ*, 855, 102.
Cooke R. J., Fumagalli M. (2018), *NatAs*, 2, 957.
Cool R. J. *et al.* (2013), *ApJ*, 767, 118.
Corollo C. M., de Zeeuw P. T., van der Marel R. P. (1995), *MNRAS*, 276, 1131.
Corsini E. M., Wegner G., Saglia R. P., Thomas J., Bender R., Thomas D. (2008), *ApJS*, 175, 462.
Corsini E. M., Wegner G. A., Thomas J., Saglia R. P., Bender R., Pu S. B. (2013), *IAUS*, 295, 225.
Corsini E. M., Wegner G. A., Thomas J., Saglia R. P., Bender R. (2017), *MNRAS*, 466, 974.
Courteau S., Faber S, M., Dressler A., Willick J. A. (1993), *ApJ*, 412, L51.
Courtois H. M., Hoffman Y., Tully R. B., Gottlöber S. (2012), *ApJ*, 744, 43.
Courtois H., Pomarède D., Tully R, B., Hoffman Y., Courtois D. (2013), *AJ*, 146, 69.
Cousins A. W. J. (1973), *MmRAS*, 77, 223.
Cousins A. W. J. (1976), *MmRAS*, 81, 25.
Covington M. D. *et al.* (2010), *ApJ*, 710, 279.
Covington M. D., Primak J. R., Porter L. A., Croton D. J., Somerville R. S., Dekel A. (2011), *MNRAS*, 415, 3135.
Cowley C. R., Ryabchikova T., Kupka F., Bord D. J., Mathys G., Bidelman W. P. (2000), *MNRAS*, 317, 299.
Cowley C. R., Bidelman W. P., Hubrig S., Mathys G., Bord D. J. (2004), *A&A*, 419, 1087.
Cox J. P. (1980), *Theory of Stellar Pulsation.* Princeton: Princeton University Press.
Croft R. (2013), *MNRAS*, 434, 3008.
Croft R. A. C., Efstathiou G. (1994), *MNRAS*, 268, L23.
Cyburt R. H. *et al.* (2016), *Rev Mod Phys*, 88, 015004.
da Costa L. N., Freudling W., Wegner G., Giovanelli R., Haynes M. P., Salzer J. J. (1996), *ApJ*, 468, L5.
da Costa L. N., Nusser A., Freudling W., Giovanelli R., Haynes M. P., Salzer J. J., Wegner G. (1998), *MNRAS*, 299, 425.
da Costa L. N., Bernardi M., Alonso M. V., Wegner G., Willmer C. N. A., Pellegrini P. S., Rité C., Maia M. A. G. (2000a), *AJ*, 120, 95.
da Costa L. N., Bernardi M., Alonso M. V., Wegner G., Willmer C. N. A., Pellegrini P. S., Rité C., Maia M. A. G., Zaroubi S. (2000b), *ApJ*, 537, 81.
da Cunha E. *et al.* (2017), *PASA*, 34, 47.
Dahlen T. *et al.* (2010), *ApJ*, 754, 425.
Dale D. A., Giovanelli R., Haynes M. P., Campusano L. E., Hardy E., Borgani S. (1999a), *ApJ*, 510, L11.

Dale D. A., Giovanelli R., Haynes M. P., Hardy E., Campusano L. E. (1999b), *AJ*, 118, 1468.
Dale D. A., Giovanelli R., Haynes M. P., Hardy E., Campusano L. E. (1999c), *AJ*, 118, 1489.
Dale D. A., Giovanelli R. (2000), in *Towards an Understanding of Large-Scale Structure*, ASP Conference Series, Vol. 201, 25, eds. S. Courteau, M. A. Strauss, J. A. Willick.
Dalton G. B., Efstathiou G., Maddox S. J., Sutherland W. J. (1992), *ApJ*, 390, L1.
Davies M., Peebles P. J. E. (1983), *ARA&A*, 21, 109.
Davies R. L., Burstein D., Dressler A., Faber S. M., Lynden-Bell D., Terlevich R. J., Wegner G. (1987), *ApJS*, 64, 581.
Davis M., Nusser A., Willick J. A. (1996), *ApJ*, 473, 22.
Davis M., Peebles P. J. E. (1977), *ApJS*, 34, 425.
Davis T. M., Scrimgeour M. I. (2014), *MNRAS*, 442, 1117.
Da Vinci L. (1998), *The Notebooks of Leonardo Da Vinci*, ed. I. A. Richter. Oxford: Oxford's World Classics Paperback, pp. 52f, 379.
Dawson W. A. *et al.* (2012), *ApJ*, 747, L42.
Dawson W. A. (2013), *ApJ*, 772, 131.
Deason A. J., Van der Marel R. P., Guhathakurta P., Sohn S. T., Brown T. M. (2013), *ApJ*, 766, 24.
de Blok W. J. G. (2010), *AdAst*, 2010, 789293.
de Jong R. S., Simard L., Davies R. L., Saglia R. P., Burstein D., Colless M., McMahan R., Wegner G. (2004), *MNRAS*, 355, 1155.
Dekel A., Bertschinger E., Faber S. M. (1990), *ApJ*, 364, 349.
Dekel A. (1994), *ARA&A*, 32, 371.
Dekel A. (2000), *ASPC*, 201, 420.
Dekel A., Eldar A., Kolatt T., Yahil A., Willick J. A., Faber S. M., Courteau S., Burstein D. (1999), *ApJ*, 522, 1.
Delabrouille J. *et al.* (2013), *A&A*, 553, A96.
De Lucia G. *et al.* (2006), *MNRAS*, 366, 499.
Delva P. *et al.* (2018) *PhRvL*, 121, 231101.
Desjacques V., Jeong D., Schmidt F. (2018), *PhR*, 733, 1.
de Lapparent V., Geller M. J., Huchra J. P. (1986), *ApJ*, 302, 1.
de Vaucouleurs G. (1948), *AnAp*, 11, 247.
de Vaucouleurs G. (1958), *AJ*, 63, 253.
de Vaucouleurs G., Malik G. M. (1969), *MNRAS*, 142, 387.
de Vaucouleurs G., Capaccioli M. (1979), *ApJS*, 40, 699.
DeWitt B. S. (1967), *PhRv*, 160, 1113.
Diaz J. D., Bekki K. (2012), *ApJ*, 750, 36.
Dieter N. H. (1965), *AJ*, 70, 552.
Dirac P. A. M. (1937), *Nature*, 139, 323.
Disney M., Davies J., Philipps S. (1989), *MNRAS*, 239, 939.
Di Valentino E., Melchiorri A., Silk J. (2020a), *NatAs*, 4, 196.
Di Valentino E., Melchiorri A., Silk J. (2020b), *JCAP*, 01, 013D.
Domínguez A. *et al.* (2019), *ApJ*, 885, 137.

Donegan M. M. (2001), Constraints on the total binding masses of three double clusters of galaxies. PhD thesis, Dartmouth College.
Dong F., Yu y., Zhang J., Yang X., Zhang P. (2020), *MNRAS*, 500, 3838.
Dooley G. A., Peter A. H. G., Yang T., Willman B., Griffen B. F., Frebel A. (2017), *MNRAS*, 471, 4894.
Doran M. (2005), *JCAP*, 10, 001.
Dossett J., Hu B., Parkinson D. (2014), *JCAP*, 3, 46.
Douglas K. A. *et al.* (2018), *ApJ*, 864, 144.
Dressler A., Lynden-Bell D., Burstein D., Davies R. L., Faber S. M., Terlevich R. J., Wegner G. (1987a), *ApJ*, 313, 42.
Dressler A., Faber S. M., Burstein D., Davies R. L., Lynden-Bell D., Terlevich R. J., Wegner G. (1987b), *ApJ*, 313, L37.
Dressler A., Faber S. M. (1990a), *ApJ*, 354, 13.
Dressler A., Faber S. M. (1990b), *ApJ*, 354, 45.
Dressler A., Faber S. M., Burstein D. (1991), *ApJ*, 368, 54.
Dreyer J. L. E. (1888), *MmRAS*, 49, 1.
Dreyer J. L. E. (1895), *MmRAS*, 51, 185.
Dreyer J. L. E. (1910), *MmRAS*, 59, 105.
Drinkwater M. J. *et al.* (2010), *MNRAS*, 401, 1429.
Driver S. P., Robotham A. S. G. (2010), *MNRAS*, 407, 2131.
Drout M. R. *et al.* (2017), *Sci*, 358, 1570.
Duffey A. R. *et al.* (2012), *MNRAS*, 426, 3385.
Dupuy A., Cortois H. M., Kubik B. (2019), *MNRAS*, 486, 440.
Dvali G., Gabadadze G., Porrati M. (2000), *PhLB*, 485, 208.
Dyson F. W., Eddington A. S., Davidson C. (1920), *RSPTA*, 220, 291.
Ebert O. D., Bullock J. S., Garrison-Kimmels S., Rocha M., Oñorbe J., Peter A. H. G. (2015), *MNRAS*, 453, 29.
Eddington A. S. (1917), *Obs*, 40, 290.
Eddington A. S. (1930), *The Internal Constitution of the Stars*, Cambridge: Cambridge Univ. Press; 2nd edition reprinted 1959, New York: Dover Publications.
Eddington A. S. (1932), *MNRAS*, 92, 471.
Edge D. O., Shakeshaft J. R., McAdam W. B., Baldwin J. E., Archer S. (1959), *MmRAS*, 68, 37.
Efstathiou G. (1990), peu conf, 361.
Efstathiou G. (1993), *PNAS*, 90, 4859.
Efstathiou G., Sutherland W. J., Maddox S. J. (1990), *Nature*, 348, 705.
Efstathiou G., Bond J. R., White S. D. M. (1992), *MNRAS*, 258, 1P.
Efstathiou G., Gratton S. (2020), arXiv:2002.06892.
Einasto J. *et al.* (1997a), *Nature* 385, 139.
Einasto J. *et al.* (1997b), *MNRAS*, 289, 801.
Einasto J. *et al.* (2011), *A&A*, 534, A128.
Einasto J. *et al.* (2018), *ARA&A*, 56, 1.
Einstein A. (1916), *Ann. Phys.*, 49, 769.
Einstein A., (1936), *Sci*, 84, 506.
Eisenstein D. J., Hu W., Silk J., Szalay A. S. (1998a), *ApJ*, 494, L1.

Eisenstein D. J., Hu W., Tegmark M. (1998b), *ApJ*, 504, L57.
Eisenstein D. J., Seo H.-J., White M. (2007), *ApJ*, 664, 660.
Ekholm T., Teerikorpi P. (1994), *A&A*, 284, 369.
Enqvist K. (2008), *GReGR*, 40, 451.
Erdoğdu P. *et al.* (2006a), *MNRAS*, 368, 1515.
Erdoğdu P. *et al.* (2006b), *MNRAS*, 373, 45.
Faber S. M., Burstein D. (1988), in *Large Scale Motions in the Universe*, eds. V. C. Rubin and G. V. Coyne. Princeton: Princeton University Press, p. 115.
Faber S. M., Jackson R. E. (1976), *ApJ*, 204, 668.
Faber S. M., Wegner G., Burstein D., Davies R. L., Dressler A., Lynden-Bell D., Terlevich R. J. (1989), *ApJS*, 69, 763.
Fabricant D., Herz E. (1990), *SPIE*, 1235, 747.
Fairall A. P. (1985), *MNSSA*, 44, 114.
Fairall A. P., Vettolani G., Chincarini G. (1989), *A&AS*, 78, 269.
Fairall A. P., Maurellis A., Jones A., Kaufmann G., Matavers B., Ellis G. F. R. (1991), *ASPC*, 15, 149.
Fairall A. P. *et al.* (1992), *AJ*, 103, 11.
Falcón-Borroso J. *et al.* (2011), *A&A*, 532, 95.
Fan X., Carilli C. L., Keating B. (2006), *ARA&A*, 44, 415.
Fang Y. *et al.* (2019), *MNRAS*, 490, 3573.
Feldman H. A., Watkins R. (1994), *ApJ*, 430, L17.
Feldman H. A. *et al.* (2003), *ApJ*, 596, L131.
Feldman H. A., Watkins R., Hudson, M. J. (2010), *MNRAS*, 407, 2038.
Fernández L. M., Cepa J., Bongiovanni A., Pérez García A. M., Lara-López M. A., Pović M., Sánchez-Portal M. (2010), *A&A*, 521, 27.
Fernández L. M. *et al.* (2011), *A&A*, 526, 72.
Finn R. A. *et al.* (2005), *ApJ*, 630, 206.
Fisher K. B., Lahav O., Hoffman Y., Lynden-Bell D., Zaroubi S. (1995), *MNRAS*, 272, 885.
Fixen D. J., Cheng E. S., Wilkinson D. T. (1983), *PhRvL*, 50, 620.
Fomalont E., Kopeiken S., Lanyi G., Benson J. (2009), *ApJ*, 699, 1395.
Foley R. J. *et al.* (2018), *MNRAS*, 475, 193.
Foqué P., Solanes J. M., Sanchis T., Balkowski C. (2001), *A&A*, 375, 770.
Ford W. K. Jr. (1968), *ARA&A*, 6, 1.
Forman W., Jones C. (1982), *ARA&A*, 20, 547.
Freedman W. L. *et al.* (2001), *ApJ*, 553, 47.
Freedman W. L. (2017), *NatAs*, 1, 0121.
Freedman W. L., Madore B. F. (2010), *ARA&A*, 48, 673.
Freedman W. L. *et al.* (2019), *ApJ*, 882, 34.
Freeman K. C. (1970), *ApJ*, 160, 811.
Frenk C. S. (1991), *PhST*, 36, 70.
Freudling W., da Costa L. N., Wegner G., Giovanelli R., Haynes M. P., Salzer J. J. (1995), *AJ*, 110, 920.
Freudling W. *et al.* (1999), *ApJ*, 523, 1.
Fried D. L. (1966), *JOSA*, 56, 1372.

Fu L. *et al.* (2008), *A&A*, 479, 9.
Gaitskell R. J. (2004), *ARNPS*, 54, 315.
Garcia K., Quartin M., Siffert B. B. (2020), *PDU*, 29, 100519.
Garcia-Berro E. *et al.* (2006), *IJMPD*, 15, 1163.
Gaztañaga E., Baugh C. M. (1998), *MNRAS*, 294, 229.
Gaztañaga E., Norberg P., Baugh C. M., Croton D. J. (2005), *MNRAS*, 364, 620.
Ge J., Mao S., Lu Y., Cappelari M., Yan R. (2019), *MNRAS*, 485, 1675.
Geha M. *et al.* (2017), *ApJ*, 847, 4.
Geller M. J., Huchra J. P. (1989), *Sci*, 246, 897.
Geller M. J., Dell'Antonio I. P., Kurtz M. J., Ramella M., Fabricant, D. G., Caldwell N., Tyson J. A., Wittman D. (2005), *ApJ*, 635, L125.
Geller M. J., Kurtz M. J., Dell'Antonio I. P., Ramella M., Fabricant D. G. (2010), *ApJ*, 709, 832.
Geller M. J., Hwang H. S., Diaferio A., Kurtz M. J., Coe D., Rines K. J. (2014), *ApJ*, 783, 52.
Gerdes D. M., Sypniewski A. J., McKay T. A., Hao J., Weis M. R., Wechsler R. H., Busha M. T. (2010), *ApJ*, 715, 823.
Ghez A. M. (2004), *GCNew*, 17, 11.
Giannantonio T. *et al.* (2008), *PhRvD*, 77, 123520.
Gibson B. K. *et al.* (1999), *ApJ*, 512, 48.
Gillies G. T. (1987), *Metro* 24, Supplement.
Gillies G. T. (1997), *RPPH*, 60, 151.
Gil-Marín S. *et al.* (2018), *MNRAS*, 477, 1604.
Gingerich O. (1999), *PASP*, 111, 254.
Giovanelli R., Haynes M. P. (1985), *AJ*, 90, 2445.
Giovanelli R., Haynes M. P., Salzer J. J., Wegner G., da Costa L. N., Freudling W. (1994), *AJ*, 107, 2036.
Giovanelli R., Haynes M. P., Salzer J. J., Wegner G., da Costa L. N., Freudling W. (1995), *AJ*, 110, 1059.
Giovanelli R., Haynes M. P., Wegner G., da Costa L. N., Freudling W., Salzer J. J. (1996a), *ApJ*, 464, 99.
Giovanelli R., Haynes M. P., Chamaraux P., da Costa L. N., Freudling W., Salzer J. J., Wegner G. (1996b), *IAUS*, 168, 183.
Giovanelli R., Haynes M. P., da Costa L. N., Freudling W., Salzer J. J., Wegner G. (1997a), *ApJL*, 477, L1.
Giovanelli T., Haynes M. P., Herter T., Vogt N. P., da Costa L.N., Freudling W., Salzer J. J., Wegner G. (1997b), *AJ*, 113, 53.
Giovanelli R., Haynes M. P., Salzer J. J., Wegner G., Da Costa L. N., Freudling W. (1998b), *AJ*, 116, 2632.
Giovanelli R., Dale D. A., Haynes M. P., Hardy E., Campusano L. E. (1999), *ApJ*, 525, 25.
Giovanelli R. (2013), *Proc. IAU Symposium No. 289, Advancing the Physics of Cosmic Distances*, ed. R. de Grijs, p. 296.
Glass I. S., Evans T. L. (1981), *Nature*, 291, 303.
Gnedin O. Y., Kravtsov A. V., Klypin A. A., Daisuke N. (2004), *ApJ*, 616, 16.
Gonzales M. E. *et al.* (2011), *ApJ*, 743, 102.

Gould A. (1993), *ApJ*, 412, L55.
Gopka V. F., Yuschenko A. V., Yuschenko V. A., Panov I. V., Kim C. H. (2008), *KPCB*, 24, 89.
Gordon K. D., Clayton G. C., Misselt K. A., Landolt A. U., Wolf M. J. (2003), *ApJ*, 594, 279.
Gott J. R., III *et al.* (2005), *ApJ*, 624, 463.
Gough D. O. *et al.* (1996), *Sci*, 272, 1296.
Granett B. R., Neyrinck M. C., Szapudi I (2008), *ApJ*, 683, L99.
Greene J. E. *et al.* (2015), *ApJ*, 807, 11.
Grogin N. A., Geller M. J. (2000), *AJ*, 119, 32.
Groth E. J., Peebles P. J. E. (1975), *ApJ*, 196, 647.
Groth E. J., Peebles P. J. E. (1977), *ApJ*, 217, 385.
Guedes J., Callegari S., Madau P., Mayer L. (2011), *ApJ*, 742, 76.
Guidorzi C. *et al.* (2017), *ApJ*, 851, L36.
Guinan E., Wasatonic R., Calderwood T., Carona D. (2020), *ATel*, No. 13512.
Guzzo L. *et al.* (2008), *Nature*, 451, 541.
Hamana T. *et al.* (2019), arXiv:1906.06041.
Hamilton A. J. S. (1998), *ASSL*, 231, 185.
Hammer F. *et al.* (2013), *MNRAS*, 431, 3543.
Han M., Mould J. (1990), *ApJ*, 360, 448.
Han M., Mould J. (1992), *ApJ*, 396, 453.
Hanimeli E., Tutusaus I., Lamine B., Blanchard A. (2020), *MNRAS*, 497, 4407.
Hartle J. B. (2003), *Gravity*. San Francisco: Addison-Wesley.
Harvey D. *et al.* (2014), *MNRAS*, 441, 404.
Harvey D., Massey R., Kitching T., Taylor A., Tittley E. (2015), *Sci*, 347, 1462.
Hawkins E. *et al.* (2003), *MNRAS*, 346, 78.
Haynes M. P., Giovanelli R. (1984), *AJ*, 89, 758.
Haynes M. P., Giovanelli R., Salzer J. J., Wegner G., Freudling W., da Costa L. N., Herter T., Vogt N. P. (1999a), *AJ*, 117, 1668.
Haynes M. P., Giovanelli R., Chamaraux P., da Costa L. N., Freudling W., Salzer J. J., Wegner G. (1999b), *AJ*, 117, 2039.
Hearnshaw J. B. (1996), *The Measurement of Starlight: Two Centuries of Astronomical Photometry*. Cambridge: Cambridge University Press.
Heiles C., 1975, *A&AS*, 20, 37.
Heisler J., Tremaine S., Bahcall J. N. (1985), *ApJ*, 298, 8.
Hernandez X., Jiménez A., Allen C. (2014), *ASSP*, 38, 43.
Herschel W. (1783), *Phil Trans*, lxxiii, 247.
Heymans C. *et al.* (2013), *MNRAS*, 432, 2433.
Hildebrandt H. *et al.* (2017), *MNRAS*, 465, 1454.
Hillebrandt W., Kromer M., Röpke F. K., Ruiter A. J. (2013), *FrPhy*, 8, 116.
Hilten M. *et al.* (2018), *ApJ*, 235, 20.
Hinshaw G. *et al.* (2009), *ApJS*, 180, 225.
Hoell J., Priester W. (1991), *A&A*, 251, L23.
Hoessel J. G. (1980), *ApJ*, 241, 493.
Hoffmeister C. (1962), *ZA*, 55, 46.
Hogg D. W. (2000), arXiv:9905116.

Holmberg E. (1940), *ApJ*, 92, 200.
Holmberg E. (1941), *ApJ*, 94, 385.
Holmberg E. (1958), *MeLuS*, 136, 1.
Holz D. E., Hughes S. A. (2005), *ApJ*, 629, 15.
Holz D., Hughes S., Schutz B. (2018), *PhT*, 71, 12, 34.
Holzapfel W. L., Ade P. A. R., Church S. E., Mauskopf P. D., Raphaeli Y., Willbanks T. M., Lange A. E. (1997), *ApJ*, 481, 35.
Hong T. *et al.* (2014), *MNRAS*, 445, 402.
Hopkins P. F., Cox T. J., Robertson B. E., Hernquist L. (2008), *MNRAS*, 391, 481.
Hoscheit B. L., Barger A. J. (2018), *ApJ.*, 854, 46.
Hoskin M. (1980), *JHA*, 11, 153.
Hou J. *et al.* (2018), *MNRAS*, 480, 2521.
Howlett C. *et al.* (2017), *MNRAS*, 464, 2517.
Howlett C., Staveley-Smith L., Blake C. (2017), *MNRAS*, 464, 2517.
Hoyle F., Vogeley M. S., Pan D. (2012), *MNRAS*, 426, 3041.
Hsien B. C., Yee H. K. C., Lin H., Gladders M. D. (2005), *ApJS*, 158, 161.
Hu W., Dodelson S. (2002), *ARA&A*, 40, 171.
Hu W., Sawicki I. (2007), *PhRvD*, 76, 064004.
Hubble E. P. (1925), *PA*, 33, 252.
Hubble E. P. (1929), *PNAS*, 15, 168.
Hubble E. P. (1930), *ApJ*, 71, 231.
Hubble E. P. (1934), *ApJ*, 79, 8.
Hubble E. P. (1936), *Realm of the Nebulae.* New Haven: Yale University Press.
Huchra J., Geller M., Clemens C., Tokarz S., Michel A. (1992), *BICDS*, 41, 31.
Huchra J. P. *et al.* (2012), *ApJS*, 199, 26.
Hudson M. J. (1994a), *MNRAS*, 266, 468.
Hudson M. J. (1994b), *MNRAS*, 266, 475.
Hudson M. J., Lucey J. R., Smith R. J., Steel J. (1997), *MNRAS*, 291, 488.
Hudson M. J., Smith R. J., Lucey J. R., Schlegel D. J., Davies R. L. (1999), *ApJ*, 512, 79L.
Hudson M. J., Lucey J. R., Smith R. J., Schlegel D. J., Davies R. L. (2001), *MNRAS*, 327, 265.
Hudson M. J., Smith R. J., Lucey J. R., Branchini E. (2004), *MNRAS*, 352, 61.
Hudson M. J., Stevenson J. B., Smith R. J., Wegner G. A., Lucey J. R., Simard L. (2010), *MNRAS*, 409, 405.
Huggins W. (1868), *RSPT*, 158, 529.
Huggins W., Miller W. A. (1864), *RSPT*, 154, 437.
Hulse R. A., Taylor J. H. (1975), *ApJL*, 195, L51.
Humason M. L., Mayall N. U., Sandage A. R. (1956), *AJ*, 61, 97.
Humphreys R. M., Davidson K. (1979), *ApJ.*, 232, 409.
Huterer D., Shafer D. L., Scolnic D. M., Schmidt F. (2017), *JCAP*, 05, 015.
Hwang H. S., Lee M. G. (2009), *MNRAS*, 397, 2111.
Hwang H. S., Geller M. J., Diaferio A., Rines K. J. (2012a), *ApJ*, 752, 64.
Hwang H. S., Geller, M. J., Kurtz, M M. J., Dell'Antonio I. P., Fabricant D. G. (2012b), *ApJ*, 758, 25.

Ibata R. A., Gilmore G., Irwin M. J. (1994), *Nature*, 370, 1941.
Ibata R. A. *et al.* (2013), *Nature*, 493, 62.
Ishimaru Y., Wanajo S., Prantzos N. (2015), *ApJ*, 804, L35.
Ivanov V. D. *et al.* (2019), *A&A*, 629, 100.
Jacoby G. H. *et al.* (1992), *PASP*, 104, 599.
Jaeger S. (2008), *ASPC*, 394, 623.
Jasche J., Wandelt B. D. (2013), *ApJ*, 779, 15.
Jeans J. H. (1915), *MNRAS*, 76, 70.
Jeans J. H. (1919), *Problems of Cosmogony and Stellar Dynamics*. Cambridge: Cambridge University Press.
Jee I., Suyu S. H., Komatsu E., Fassnacht C. D., Hilbert S., Koopmaris L. V. E. (2019), *Sci*, 365, 1134.
Jensen J. B. *et al.* (2003), *ApJ*, 583, 712.
Ji A. P. *et al.* (2016), *Nature*, 531, 610.
Jones D. H. *et al.* (2004), *MNRAS*, 355, 747.
Jones D. H. *et al.* (2009), *MNRAS*, 399, 683.
Jones D. O. *et al.* (2019), *ApJ*, 881, 19.
Johnson H. L., Morgan W. W. (1953), *ApJ*, 117, 313.
Johnson R. E., Markevitch M., Wegner G. A., Jones C., Forman W. R. (2010), *ApJ*, 710, 1776.
Johnston S. *et al.* (2008), *Ex As*, 22, 151.
Jorden P. R., Jorden D., Jerram P. A. (2014), *Proc SPIE*, 9154.
Jørgensen I., Franx M., Kjaergaard P. (1995), *MNRAS*, 276, 1341.
Jørgensen I., Franx M., Kjaergaard P. (1996), *MNRAS*, 280, 167.
Juszkiewicz R., Ferreira P. G., Feldman H. A., Jaffe A. H., Davis M. (2000), *Sci*, 287, 109.
Juszkiewicz R., Feldman H. A., Fry J. N., Jaffe A. H. (2010), *JCAP*, 2, 021.
Kahlhoefer F., Schmidt-Hoberg K., Frandsen M. T., Sarkar S. (2014) *MNRAS*, 437, 2865.
Kahn F. D., Woltjer L. (1959), *ApJ*, 130, 705.
Kaiser N. (1988), *MNRAS*, 231, 149.
Kaiser N. (2013), *MNRAS*, 435, 1278.
Kaiser N., Squires G. (1994), *ApJ*, 404, 441.
Kaiser N., Stebbins A. (1991), *ASPC*, 15, 111.
Kandrup H. E. (1998), *ApJ*, 500, 120.
Kant Imm. (1755), *Algemeine Naturgeschichte und Theorie des Himmels*. Königsberg und Leipzig: Johann Friedrich Petersen.
Kang Y., Lee Y.-W., Chung C., Ree C. H. (2020), *ApJ*, 889, 8.
Kaplinghat M., Tulin S., Yu H.-B. (2014), *PhRvD*, 89, 5009.
Kashlinsky A., Atrio-Barandela F., Kocevski D., Ebeling H. (2008), *ApJ*, 686, L49.
Kashlinsky A., Atrio-Barandela F., Kocevski D., Ebeling H. (2008), *ApJ*, 691, 1479.
Kassin S. A. *et al.* (2007), *ApJL*, 660, L35.
Keeler J. E. (1890), *PASP*, 2, 284.
Keisler R. (2009), *ApJ*, 707, L42.

Kellermann K. I. (1996), *PASP*, 108, 729.
Kerr F. J., Henning P. A. (1987), *ApJ*, 320, L99.
Kinney A. L., Calzetti D., Bohlin R. C., McQuade K., Storchi-Bergmann T., Schmitt H. R. (1996), *ApJ*, 467, 38.
Kirchoff G. (1861), *Untersuchungen über das Sonnenspectrum und die Spectren der chemischen Elemente*. Berlin: Königl. Akad. der Wissenschaften zu Berlin.
Kirschner R. P., Kwan J. (1974), *ApJ*, 193, 27.
Kirshner R. P. *et al.* (1981), *ApJ*, 248, L57.
Kirshner R. P. *et al.* (1983), *IAU Symposium 104, Early Evolution of the Universe and its Present Structure*, p. 197.
Kirshner R. P. *et al.* (1987), *ApJ*, 314, 493.
Klypin A. A., Trujillo-Gomez S., Primack J. (2011), *ApJ*, 740, 102.
Klypin A., Yepes G., Gottlober S., Prada F. (2016), *MNRAS*, 457, 4340.
Kobayashi C., Karakas A. I., Lugaro M. (2020), *ApJ*, 900, 179.
Kocevski D. D., Ebeling H. (2006), *ApJ*, 645, 1043.
Koda J. *et al.* (2014), *MNRAS*, 445, 4267.
Koen C., Lombard F. (2009), *MNRAS*, 395, 1657.
Kogut A. *et al.* (1993), *ApJ*, 419, 1.
Kolatt T., Dekel A. (1997), *ApJ*, 479, 592.
Kolb E. W., Turner M. S. (1990), *The Early Universe*. Redwood City: Addison-Weseley.
Koribalski B. S. (2020), *MNRAS*, submitted.
Kormendy J., Drory N., Bender R., Cornell M. E. (2010), *ApJ*, 723, 54.
Kovac J. M., Leitch E. M., Pryke C., Carlstrom J. E., Halverston N. W., Holzapfel W. L. (2002), *Nature*, 420, 772.
Kraan-Korteweg R. C., Woudt P. A., Fairall A. P., Balkowski C., Henning P. A. (1996), *Nature*, 379, 519.
Kraan-Korteweg R. C., Stavely-Smith L., Donley J., Koribalski B., Henning P. A. (2005), *IAUS*, 216, 203.
Kraan-Korteweg R. C. *et al.* (2017), *MNRAS*, 466, 29.
Kraus J. D. (1977), VA, 20, 445.
Kreckel K. *et al.* (2011), *AJ*, 141, 204.
Kreckel K. *et al.* (2012), *AJ*, 144, 16.
Kreckel K. *et al.* (2015), *ApJ*, 798, L15.
Kroupa P. (2001), *MNRAS*, 322, 231.
Kroupa P. (2012), *PASA*, 29, 395.
Kunder A. *et al.* (2013), *AJ*, 146, 119.
Kunder A. *et al.* (2017), *AJ*, 153, 75.
Kurtz D. W. (1978), *Inf Bull Var Stars*, No. 1436.
Kurtz D., Wegner G. (1979), *ApJ*, 232, 510.
Kurtz M. J., Geller M. J., Utsumi Y., Miyazki S., Dell'Antonio I. P., Fabricant D. G. (2012), *ApJ*, 750, 168.
Kurucz R. L. (1992), *IAUS*, 149, 225.
La Barbera F., Carvalho R. R., de La Rosa I. G., Lopes P. A. A. (2010), *MNRAS*, 408, 1335.

Lacey C., Cole S. (1993), *MNRAS*, 262, 627.

Lacey C. *et al.* (2001), *ASPC*, 222, 273.

Lagos C. del P., Tobar R. J., Robotham A. S. G., Obreschkow G., Mitchell P. D., Power C., Elahi P. J. (2018), *MNRAS*, 481, 3573.

Lahav O., Santiago B. X., Webster A. M., Strauss M. A., Davis M., Dressler A., Huchra J. P. (2000), *MNRAS*, 312, 166.

Lallemand A., Duchesne M., Walker M. F. (1960), *PASP*, 72, 268.

Lam T. Y. *et al.* (2015), *MNRAS*, 428, 3409.

Lambert J. H. (1761), *Cosmologische Briefe.* Augsburg: Eberhard Klettes

Landau L. D., Lifshitz E. M. (1975), *The Classical Theory of Fields*, 4th ed., Oxford: Pergamon Press.

Landolt A. U. (1983), *AJ*, 88, 439.

Landy S. D., Szalay A. (1992), *ApJ*, 391, 494.

Lattimer J. M., Schramm D. N. (1974), *ApJ*, 192, L145.

Lauer T. R., Postman M. (1994), *ApJ*, 425, 418.

Lavaux G., Tully R. B., Mohayaee R., Colombi S. (2010), *ApJ*, 709, 483.

Lavaux G., Hudson M. J. (2011), *MNRAS*, 416, 2840.

Lavaux G., Ashfordi N., Hudson M. J. (2013), *MNRAS*, 430, 1617.

Lawrence A. *et al.* (1999), *MNRAS*, 308, 897.

Leavitt H. S., Pickering E. C. (1912), *HarCi*, 173, 1.

Leavitt H. S., Pickering E. C. (1917), *AnHar*, 71, 47.

Lee C.-H., Riffeser A., Seitz S., Bender R., Koppenhoefer J. (2015), *ApJ*, 806, 161.

Le Floc'h E. *et al.* (2005), *ApJ*, 632, 169.

Lelli F., McGaugh S. S., Schombert J. M. (2016), *ApJL*, 816, L14.

Lelli F., McGaugh S. S., Schombert J. M., Desmond H., Katz H. (2019), *MNRAS*, 484, 3267.

Lemaître G. (1927), *AASB*, 47, 49.

Lesser M. (2015), *PASP*, 127, 1097.

Levesque E. M., Massey P., Żytkow A. N., Morrell N. (2014), *MNRAS*, 443, L94.

Li, B., Zhao G.-B., Koyama K. (2012), *MNRAS*, 421, 3481.

Li Q. *et al.* (2018), *Nature*, 560, 582.

Li L.-X., Paczyński B. (1998), *ApJ*, 507, L59.

Liebscher D.-E., Priester W., Hole J. (1992), *A&A*, 261, 377.

Liebeskind N. I. *et al.* (2018), *MNRAS*, 473, 1195.

Limber D. N. (1954), *ApJ*, 119, 655.

Lin D. N. C., Lynden-Bell D. (1977), *MNRAS*, 181, 59.

Lin D. N. C. (1996), in *Gravitational Dynamics*, eds. O. Lahav, E. Terlevich, R. J. Terlevich. Cambridge: University Cambridge Press, p. 15.

Linder E. V. (1997), *First Principles of Cosmology.* Harlow: Addison-Wesley.

Linder E. V. (2005), *PhRvD*, 72, 043529.

Linder E. V., Cahn R. N. (2007), *APh*, 28, 481.

Lockyer J. N. (1890), *The Meteoritic Hypothesis.* London: MacMillan & Co.

Łokas E. L., Mamon G. A. (2001), *MNRAS*, 321, 155.

Lombriser L., Koyama K., Zhao L.-B., Li B. (2012), *PhRvD*, 85, 124054.

Lombriser L. (2020), *PhLB*, 803, 135303.

Longair M. S. (1993), QJRAS, 34, 157.
Longair M. S. (2011), *High Energy Astrophysics*, 3d ed., Cambridge: Cambridge University Press, p. 257.
Lorimer D. R. (2008), *LRR*, 11, 8.
Lotz J. M. *et al.* (2017), *ApJ*, 837, 97.
Lu J. *et al.* (2019), *AnnPhys*, 531, 1900013.
Lubin P. M., Villela T., Epstein G. L., Smoot G. F. (1985), *ApJ*, 298, L1.
Lucey J. R., Guzmán R., Carter D., Terlevich R. J. (1991), *MNRAS*, 253, 584.
Ludwig P. *et al.* (2012), *PhRvC*, 85, 024315.
Lundmark K. (1925), *MNRAS*, 85, 125.
Lynam P. D., Collins C. A., James P. A., Böhringer H., Neumann D. M. (2000), *ASPC*, 201, 146.
Lynden-Bell D. (1967), *MNRAS*, 136, 101.
Lynden-Bell D. (1975), *VA*, 19, 299.
Lynden-Bell D. (1976), *MNRAS*, 174, 695.
Lynden-Bell D. (1986), *QJRAS*, 27, 319.
Lynden-Bell D. (2004), *ASPC*, 327, 243.
Lynden-Bell D., Faber S. M., Burstein D., Davies R. L., Dressler A., Terlevich R. J., Wegner G. (1988), *ApJ*, 326, 19.
Lynden-Bell D., Lynden-Bell R. M. (1995), *MNRAS*, 275, 429.
Ma C.-P., Greene J. E., McConnell N., Janish R., Blakeslee J. P., Thomas J., Murphy J. D. (2014), *ApJ*, 795, 158.
Maartens R., Clarkson C., Chen S. (2018), *JCAP*, 01, 013.
Macaulay E., Feldman H., Ferreira P. G., Hudson M. J., Watkins R. (2011), *MNRAS*, 414, 621.
Macaulay E., Feldman H. A., Ferreira P. G., Jaffe A. H., Agarwal S., Hudson M. J. (2012), *MNRAS*, 425, 1709.
MacDonald E. C. *et al.* (2004), *MNRAS*, 352, 1255.
Madau P., Pozzetti L., Dickinson M. (1998), *ApJ*, 498, 106.
Madau P., Dickenson M. (2014), *ARA&A*, 52, 415.
Maddox S. J., Efstathiou G., Sutherland W. J., Loveday J. (1990), *MNRAS*, 242, 43P.
Magoulas C. *et al.* (2012), *MNRAS*, 427, 245.
Makarov D. *et al.* (2006), *AJ*, 132, 2729.
Malmquist K. G. (1920), *Medd. Lunds Obs Ser* II, No. 22, 1.
Mao Q. *et al.* (2017), *ApJ*, 835, 161.
Mao Y.-Y. *et al.* (2021), *ApJ*, 907, 85.
Markevitch M., Vikhlinin A. (2007), *Ph R*, 443, 1.
Maskelyne N. (1775), *Phil Trans R Soc*, 65, 500.
Massey R., Kitching T., Nagai D. (2011), *MNRAS*, 413, 1709.
Masters K. L., Springob C. M., Haynes M. P., Giovanelli R. (2006), *ApJ*, 653, 861.
Masters K. L., Springob C. M., Huchra J. P. (2008), *AJ*, 135, 1738.
Masters K. L., Springob C. M., Huchra J. P. (2018), arXiv:0711.4305.
Mathews G. J., Kusakabe M., Kajino T. (2017), *IJMPE*, 26, 1741001.
Mathewson D. S., Cleary M. N., Murray J. D. (1974), *ApJ*, 190, 291.

Mathewson D. S. (1992), *PASA*, 10, 87.
Mathewson D. S., Ford V. L. (1994), *ApJ*, 434, L39.
Mathewson D. S., Ford V. L. (1996), *IAUS*, 168, 175.
Mathewson D. S., Ford V. L., Buchhorn M. (1992a), *ApJ*, 389, L5.
Mathewson D. S., Ford V. L., Buchhorn M. (1992b), *ApJS*, 81, 413.
Matteucci F., Romano D., Areones A., Korobkino O., Rosswog S. (2014), *MNRAS*, 438, 2177.
Mattig W. (1958), *AN*, 284, 108.
Maxwell J. C. (1865), *RSPT*, 155, 459.
Maxwell J. C. (1877), *Matter and Motion.* New York: Dover Publications, Reprint.
Mazets E. P. *et al.* (1981a), *Ap&SS*, 80, 3.
Mazets E. P. *et al.* (1981b), *Ap&SS*, 80, 85.
Mazets E. P. *et al.* (1981c), *Ap&SS*, 80, 119.
McGaugh S. S. (2012), *AJ*, 143, 40.
McWilliams S. T. (2019), *PhRvL*, 122, 191102.
Medezinski E. *et al.* (2013), *ApJ*, 777, 43.
Medezinski E. *et al.* (2016), *ApJ*, 817, 24.
Meggers W. F., Corliss C. H., Scribner B. F. (1975), *Tables of Spectral-Line Intensities*, Part I, Arranged by Elements, NBS Monogr. 145.
Mehlert D., Saglia R. P., Bender R., Wegner G. (2000), *A&AS*, 141, 449.
Mehlert D., Thomas D., Saglia R. P., Bender R., Wegner G. (2003), *A&A*, 407, 423.
Mei S. *et al.* (2007), *ApJ*, 655, 144.
Merritt D. R. (1985), *AJ*, 90, 1027.
Merritt D., Graham A. W., Moore B., Diemand J., Terzić B. (2006), *ApJ*, 132, 2685.
Merten J. *et al.* (2011), *MNRAS*, 417, 333.
Messier C. (1781), *CdT*, 1784, 227.
Metzger B. D. *et al.* (2010), *MNRAS*, 406, 2650.
Michaud G. (1970), *ApJ*, 160, 641.
Milgrom M. (2015), *CaJPh*, 93, 107.
Misner C. W., Thorne K. S., Wheeler J. A. (1973), *Gravitation.* San Francisco: W. H. Freeman.
Mitchell S. A. (1932), *Eclipses of the Sun.* New York: Columbia University Press.
Miyatake H. *et al.* (2019), *ApJ*, 875, 63.
Mkrtichian D. E., Hatzes A. P., Sao H., Shobbook R. R. (2008), *A&A*, 490, 1109.
Moffat A. F. J. (1969), *A&A*, 3, 455.
Mohayee R., Mathis H., Colombi S., Silk J. (2006), *MNRAS*, 365, 939.
Mohr J. J., Wegner G. (1997), *AJ*, 114, 25.
Mohr P. J., Newall D. B., Taylor B. N. (2016), *RvMP*, 88, 035009.
Molnar S. M., Hearn S. C., Stadel J. G. (2012), *ApJ*, 748, 45.
Moore B. (1994), *Nature*, 370, 629.
Moore C. E., Minnaert M. G. J., Houtgast J. (1966), *The Solar Spectrum 2935A to 8770A*, NBS Monogr. 61.
Moore D. E. (1965), *Electron Apr* 19, 1965, p. 114.

Moorman C. M. *et al.* (2014), *MNRAS*, 344, 3559.
Moorman C. M. *et al.* (2015), *ApJ*, 810, 108.
Moorman C. M. *et al.* (2016), *ApJ*, 831, 118.
Moran, S. M., Ellis R. S., Ellis R. S., Treu T., Smith G. P., Rich R. M., Smail I. (2007), *ApJ*, 671, 1503.
Moss A., Zibin J. P., Scott D. (2011), *PhRvD*, 83, 103515.
Mossa V. *et al.* (2020), *Nature*, 587, 210.
Moster B. P., Somerville R. S., Newman J. A., Rix H.-W. (2011), *ApJ*, 731, 113.
Mould J. (2020), *FrASS*, 7, 21.
Mould J. R., Staveley-Smith L., Schommer R. A., Bothun G. D., Jall P. J. (1991), *ApJ*, 383, 467.
Mould J. R., Akeson R, L., Bothun G. D., Han M., Huchra J. P., Roth J., Schommer R. A, (1993), *ApJ*, 409, 14.
Mould J., Sakai S. (2009), *ApJ*, 694, 1331.
Mould J. *et al.* (2015), *Ap&SS*, 357, 162.
Mould J., Uddin S. A. (2014), *PASA*, 31, 15.
Müller K. R., Freudling W., Watkins R., Wegner G. (1998), *ApJ*, 507, L105.
Müller K. R., Wegner G., Raychaudury S., Freudling W. (1999), *A&AS*, 140, 327.
Mustapha N., Hellaby C. (2001), *GReGR*, 33, 455.
Mutabazi T. *et al.* (2014), *MNRAS*, 439, 3666.
Nadler E. O. *et al.* (2020), *ApJ*, 893, 4.
Narlikar J. V. (1993), *Introduction to Cosmology*, 2nd ed., Cambridge: University Cambridge Press.
Navarro J. F., Frenk C. S., White S. D. M. (1996), *ApJ*, 462, 563.
Newberg H. J., WillettB. A., Yanny B., Xu Y. (2010), *ApJ*, 711, 32.
Nelan J. E. *et al.* (2005), *ApJ*, 632, 137.
Neureiter B. *et al.* (2021), *MNRAS*, 500, 1437.
Newton I. (1687), *Philosophiae Naturalis Principia Mathematica*. London: Joseph Streeter.
Newton I. (1730), *Opticks, 4th ed.*, London: William Innys (Reprinted Dover Publications).
Neyman J., Scott E. L. (1959), *Handbuch Phys*, 53, 416.
Nicholis A., Rattazzi R., Trincherini E. (2009), 79, 064036.
Nidever D. L., Majewski S. R., Burton W. B. (2008), *ApJ*, 679, 432.
Nilson P. N. (1973), *Uppsala Obs Ann*, 6.
Nishizawa A. J. (2014), *PTEP*, 2014, 06B110.
Nusser A. (2014), *ApJ*, 795, 3.
Nusser A. (2016), *MNRAS*, 455, 178.
Nusser A. (2017), *MNRAS*, 470, 445.
Nusser A, da Costa L. N., Branchini E., Bernardi M., Alonso M. V., Wegner G., Wilmer C. N. A., Pellegrini P. S. (2001), *MNRAS*, 320, 21.
Ogle P. M., Jarrett T., Lanz L., Cluver M., Alatalo K., Appleton P. N., Mazzarella J. M. (2019), *ApJL*, 884, 11.
Oguri M. *et al.* (2018), *PASJ*, 70, 26.
Okabe N., Umetsu K. (2008), *PASJ*, 60, 345.
Ong Y. C., Hashemi S. S., An R., Wang B. (2018), *EPJC*, 78, 405.

Onofrio R, (2010), *PhRvD*, 82, 5008.
Onofrio R, (2012), *EPJC*, 72, 2006.
Onofrio R., Wegner G. A. (2014), *ApJ*, 791, 125.
Oort J. H. (1932), *BAN*, 6, 249.
Osborne S. J., Mak D. S. Y., Church E., Pierpaoli E. (2011), *ApJ*, 737, 98.
Osipkov L. P. (1979), *PAZh*, 5, 77.
Osterbrock D. E. (1989), *Astrophysics of Gaseous Nebulae and Active Galactic Nuclei.* Mill Valley, CA: University Science Books.
Ostriker J. P., Peebles P. J. E. (1973), *ApJ*, 186, 467.
Padmanabhan T. (1993), *Structure Formation in the Universe.* Cambridge: Cambridge University Press.
Pagel B. E. J. (2001), *PASP*, 113, 137.
Pahre M. A., Djorgovski S. G., de Carvalho R. R. (1998), *AJ*, 116, 1591.
Pan D. L. *et al.* (2012), *MNRAS*, 421, 926.
Papastergis E., Adams E. A. K., van der Hulst J. M. (2016), *A&A*, 593, 39.
Parkinson D. *et al.* (2012), *PhRvD*, 86, 3518.
Parrent J., Friesen B., Parthasarathy M. (2014), *Ap&SS*, 351, 1.
Partridge R. B., Wilkinson D. T. (1967), *PhRvL*, 18, 557.
Peacock J. A., Dodds S. J. (1994), *MNRAS*, 267, 1020.
Peacock J. A. *et al.* (2001), *Nat*, 410, 169.
Peebles P. J. E. (1971), *Physical Cosmology.* Princeton: Princeton University Press.
Peebles P. J. E. (1980), *The Large-Scale Structure of the Universe.* Princeton: Princeton University Press.
Peebles P. J. E. (1993), *Principles of Physical Cosmology.* Princeton University Press
Peebles P. J. E. (1995), *ApJ*, 449, 52.
Peebles P. J. E. (1996), in *Gravitational Dynamics.* eds. O. Lahav, E. Terlevich, R. J. Terlevich. Cambridge: Cambridge University Press, p. 219.
Peebles P. J. E. (2001), *ApJ*, 557, 495.
Peebles P. J. E., Hauser M. G. (1974), *ApJS*, 28, 19.
Peebles P. J. E., Nusser A. (2010), *Nat*, 465, 565.
Peebles P. J. E., Tully R. B., Shaya E. J. (2011), arXiv:1105.5596.
Peery S., Watkins R., Feldman H. A. (2018), *MNRAS*, 481, 1368.
Penzias A. A., Wilson R. W. (1965), *ApJ*, 142, 419.
Penzias A. A., Wilson R. W. (1967), *AJ*, 72, 315.
Perlmutter S. *et al.* (1999), *ApJ*, 517, 565.
Perrodin D., Sesano A. (2018), *ASSL*, 457, 95.
Persic M., Salucci P., Stel F. (1996), *MNRAS*, 281, 27.
Percival W. J. *et al.* (2007), *ApJ*, 657, 645.
Peter A. H. G., Rocha M., Bullock J. S., Kaplinghat M. (2013), *MNRAS*, 430, 105.
Peuch M. *et al.* (2008), *A&A*, 484, 173.
Phillips M. M. (1993), *ApJ*, 413, L105.
Pierce A. K., Breckenridge J. B. (1973), *The Kitt Peak Table of Photographic Solar Spectrum Wavelengths.* Tucson: Kitt Peak National Observatory, Contribution No. 559

Pierce M. J., Tully R. B. (1988), *ApJ*, 330, 579.
Pike R. W., Hudson M. J. (2005), *ApJ*, 635, 11.
Pinto M. C., Padilla N. D. Lambas D. G. (2006), *MNRAS*, 373, 1409.
Planck Collaboration (2014), *A&A*, 561, 97.
Planck Collaboration (2016), *A&A*, 594, 13.
Planck Collaboration (2018), arXiv:1807.06209.
Planck Collaboration VI (2020), *A&A*, 641, A6.
Podkletnov E., Nieminen R. (1992), *Physica C*, 203, 441.
Poggianti B. M. (1997), *A&ASS*, 122, 399.
Pogson N. R. (1856), *MNRAS*, 17, 12.
Poole G. B., Fardal M. A., Babul A., McCarthy I. G., Quinn T., Wadsley J. (2006), *MNRAS*, 373, 881.
Poole G. B., Babul A., McCarthy I. G., Sanderson A. J. R., Fardal M. A. (2008), *MNRAS*, 391, 1163.
Popescu C. C., Hopp U., Elsäuser H. (1997), *A&A*, 325, 881.
Popescu C. C., Hopp U., Rosa M. R. (1999), *A&A*, 350, 414.
Postman M. *et al.* (2012), *ApJS*, 199, 25.
Poulin V., Smith T. L., Karwal T., Kamionkowski M. (2019), *PhRvL*, 122, 221301.
Poynting J. H. (1911), in *Encyclopaedia Britannica*, Vol. XII, p. 384.
Proctor R. A. (1897), *Other Worlds than Ours*. New York: D. Appleton and Co.
Prusti T. *et al.* (2016), *A&A*, 595, 1.
Przybylski A. (1966), *Nature*, 210, 20.
Pustilnik S. A., Tepliakova A. L. (2011), *MNRAS*, 415, 1188.
Pustilnik S. A. *et al.* (2013), *MNRAS*, 432, 2224.
Qian Y.-Z. (2000), *ApJ*, 534, L67.
Qin F., Hoelett C., Staveley-Smith L., Hong T. (2018), *MNRAS*, 477, 5150.
Qin F., Hoelett C., Staveley-Smith L., Hong T. (2019a), *MNRAS*, 482, 1920.
Qin F., Hoelett C., Staveley-Smith L., Hong T. (2019b), *MNRAS*, 487, 5235.
Raghunathan S. *et al.* (2019), *PhRvL*, 123, 181301.
Ragland S. *et al.* (2006), *ApJ*, 652, 650.
Rameez M., Moyayaee R., Sarkar S., Colin J. (2018), *MNRAS*, 477, 1772.
Rameez M., Sarkar S. (2019), arXiv:1911.06456.
Randall S. W., Markevitch, M., Clowe D., Gonzales A. H., Brada M. (2008), *ApJ*, 679, 1173.
Rauch M. (1998), *ARA&A*, 36, 267.
Raychaudhury S. (1989), *Nature*, 342, 251.
Refsdahl S. (1964), *MNRAS*, 128, 307.
Reid M. J., Brunthaler A. (2004), *ApJ*, 616, 872.
Reid M. J., Menten K. M., Zheng X. W., Brunthaler A., Xu Y. (2009), *ApJ*, 705, 1548.
Renk J., Zumalacárregui M., Montanari F., Bareira A. (2017), *JCAP*, 10, 020.
Retana-Montenegro E., Van Hese E., Gentile G., Baes M., Frutos-Alfaro F. (2012), *A&A*, 540, 70.
Reynolds J. H. (1913), *MNRAS*, 74, 132.
Ricciardelli E. *et al.* (2014), *MNRAS*, 445, 4045.
Richstone D. O., Tremaine S. (1988), *ApJ*, 327, 82.

Ricker P. M., Sarazin C. L. (2001), *ApJ*, 561, 621.
Riess A. G. *et al.* (1998), *AJ*, 116, 1009.
Riess A. G., Casertano S., Yuan W., Macri L. M., Scolnic D. (2019), *ApJ*, 876, 85.
Rines K., Geller M., Diaferio A., Mohr J. J., Wegner, G. (2000), *AJ*, 120, 2338.
Ritter A. (1879), *Wiedemanns An*, 8, 172.
Rizzi L. *et al.* (2007), *ApJ*, 661, 815.
Robertson B. S. C., Feast M. W. (1981), *MNRAS*, 196, 111.
Rocha M. *et al.* (2013), *MNRAS*, 430, 81.
Röck B., Vazdekis A., Ricciardelli E., Peletier R. F., Lnapen J. H., Falcón-Borroso J. (2016), *A&A*, 589, 73.
Roettiger K., Loken C., Burns J. O. (1997), *ApJS*, 109, 307.
Rose B. M. *et al.* (2020), arXiv:2002.12382.
Rosi G., Sorrentino F., Cacciapuoti L., Prevedelli M., Tino G. M. (2014), *Nature*, 510, 518.
Ross A. J., Samushia L., Howlett H. C., Percival W. J., Burden A., Manera M. (2015), *MNRAS*, 449, 835.
Rosseland S. (1964), *The Pulsation Theory of Variable Stars*. New York: Dover Publications.
Rosswog S., Freburghaus C., Thielemann F.-K. (2001), *NuPhA*, 688, 344.
Rovelli C. (2010), *Quantum Gravity*. Cambridge: Cambridge University Press.
Rowan-Robinson M. (1985), *The Cosmological Distance Ladder*. New York: W. H. Freeman and Co.
Rowan-Robinson M., Saunders W., Lawrence A., Leech K. (1991), *MNRAS*, 253, 485.
Rowan-Robinson M. (1993), *PNAS*, 90, 4822.
Rowan-Robinson M. *et al.* (2000), *MNRAS*, 314, 375.
Rowland H. A. (1882), *Obs*, 5, 224.
Rubin V. C. (1954), *PNAS*, 40, 541.
Rubin V. C., Ford W. K. Jr., Rubin J. S. (1973), *ApJ*, 183, L111.
Rubin V. C., Ford Jr. W. K., Thonnard N., Roberts M. S., Graham J. A. (1976a), *AJ*, 81, 687.
Rubin V. C., Thonnard N., Ford Jr. W. K., Roberts M. S. (1976b), *AJ*, 81, 719.
Rubin V. C., Ford Jr. W. K., Thonnard N. (1980), *ApJ*, 238, 471.
Rubin V. C. (1983), *Sci*, 220, 1339.
Rudnicki K., Dworak T. Z., Flin P., Baranowski B., Sendrakowski A. (1973), *AcC*, 1, 7.
Ryabchikova T. A., Savanov I. S., Hatzes A. P., Weiss W. W., Handler G. (2000), *A&A*, 357, 981.
Ryabchikova T. A., Ramonovskaya A. M. (2017), *AstL*, 43, 252.
Rykoff E. S. *et al.* (2016), *ApJS*, 224, 1.
Sadeh I., Feng L. L., Lahav O. (2015), *PhRvL*, 114, 1103.
Saglia R. P., Bender R., Dressler A. (1993a), *A&A*, 279, 75.
Saglia R. P., Bertschinger E., Baggley G., Burstein D., Colless M., Davies R. L., McMahan R. K. Jr., Wegner G. (1993b), *MNRAS*, 264, 961.
Saglia R. P., Burstein D., Baggley G., Bertschinger E., Colless M. M., Davies R. L., McMahan R. K. Jr., Wegner G. (1997a), *MNRAS*, 292, 499.

Saglia R. P., Bertschinger E., Baggley G., Burstein D., Colless M. M., Davies R. L., McMahan R. K. Jr., Wegner G. (1997b), *ApJS*, 109, 79.
Saglia R. P., Colless M., Burstein D., Davies R. L., McMahan Jr. R. K., Wegner G. (2001), *MNRAS*, 324, 389.
Said K., Colless M., Magoulas C., Lucey J. R., Hudson M. J. (2020), *MNRAS*, 498, 4615.
Sakai S., Madore B. F., Freedman W. L., Lauer T. R., Ajhar E. A., Baum W. A. (1997), *ApJ*, 478, 49.
Sakai S. (1999), *IAUS*, 183, 48.
Sako M. *et al.* (2018), *PASP*, 130, 064002.
Sales L. V. *et al.* (2017), *MNRAS*, 464, 2419.
Salpeter E. (1955), *ApJ*, 121, 161.
Salzer J. J. *et al.* (2000), *AJ*, 120, 80.
Salzer J. J. *et al.* (2001), *AJ*, 121, 66.
Salzer J. J. *et al.* (2002), *AJ*, 123, 1292.
Sandage A. R. (1958), *ApJ*, 127, 513.
Sandage A. (1972), *ApJ*, 176, 21.
Sandage A., Reindl B., Tammann G. A. (2010), *ApJ*, 714, 1441.
Sandage A., Visvanathan, N. (1978), *ApJ*, 223, 707.
Sargent W. L. W., Schechter P. L., Boksenberg A., Shortridge K. (1977), *ApJ*, 212, 326.
Sarkar D., Feldman H. A., Watkins R. (2007), *MNRAS*, 375, 691.
Sawyer R. A. (1948), *Experimental Spectroscopy*. New York: Prentice-Hall, Inc.
Scaramella R., Baise-Pillastrini G., Chincarini G., Vettolani G., Zamorani G. (1989), *Nature*, 338, 562.
Scharf C., Hoffman Y., Lahav O., Lynden-Bell D. (1992), *MNRAS*, 256, 229.
Scharf C., Lahav O. (1993), *MNRAS*, 264, 439.
Schechter P. L. *et al.* (1997), *ApJ*, 475, L85.
Schechter P. L. (2005), *IAUS*, 225, 281.
Schlafly E. F., Finkbeiner D. P. (2011), *ApJ*, 737, 103.
Schlamminger S. (2014), *Nature*, 510, 478.
Schlamminger S. (2018), *Nature*, 560, 562.
Schlamminger S., Coi K.-Y., Wagner J. H., Gundlach J. H., Adelberger E. G. (2008), *PhRvL*, 100, 1101.
Schlegel D. J., Finkbeiner D. P., Davis M. (1998), *ApJ*, 500, 525.
Schmidt B. *et al.* (1994), *ApJ*, 432, 42.
Schmidt P. P. *et al.* (1994), *AJ*, 107, 1444.
Schmidt F., Lima M., Oyaizu H., Hu W. (2009), *PhRvD*, 79, 083518.
Schmidt F. (2010), *PhRvD*, 81, 103002.
Schmidt M. (1963), *ApJ*, 137, 758.
Schmittfull M., Baldauf T., Zaldarriaga M. (2017), *PhRvD*, 96, 3505.
Schmoldt I. M. *et al.* (1999), *AJ*, 118, 1146.
Schneider P., Ehlers J., Falco E. E. (1992), *Gravitational Lenses*. Berlin: Springer-Verlag.
Schumann M. (2019), *JPhG*, 46, 3003.
Scodeggio M., Giovanelli R., Haynes M. P. (1997), *AJ*, 113, 2087.

Scrimgeour M. I. *et al.* (2016), *MNRAS*, 455, 386.
Secchi A. (1866), *CRS*, 63, 621.
Seeger P. A., Fowler W. A., Clayton D. D. (1965), *ApJS*, 11, 121.
Seiler J., Parkinson D. (2016), *MNRAS*, 462, 75.
Seldner M., Siebers B., Groth E. J., Peebles P. J. E. (1977), *AJ*, 82, 249.
Seldner M., Peebles P. J. E. (1977), *ApJ*, 215, 703.
Seljak U., Zaldarriaga M. (1996), *ApJ*, 469, 437.
Selwyn E. W. H. (1950), *Photography in Astronomy*. Rochester: Eastman Kodak Co.
Sérsic J. L. (1968), *Atlas de Galaxes Australes*. Cordoba: Observatorio Astronomico.
Shakeshaft J. R., Ryle M., Baldwin J. E., Elsmore B., Thomson J. H. (1955), *MmRAS*, 67, 106.
Shane C. D., Wirtanen C. A. (1967), *Pub Lick Obs*, 22, Part 1 (SW).
Shapiro S. S., Davis J. L., Lebach D. E., Gregory J. S. (2004), *PhRvL*, 92, 121101.
Shapley H. (1914), *ApJ*, 40, 448.
Shapley H. (1918), *PASP*, 30, 42.
Shapley H. (1930), *Harvard Coll Bull*, 874, 9.
Shapley H. (1957), *The Inner Metagalaxy*. New Haven: Yale University Press.
Shappee B. J. *et al.* (2017), *Sci*, 358, 1574.
Shaya E. J., Tully R. B., Hoffman Y., Pomarède D. (2017), *ApJ*, 850, 207.
Shestakova T. P. (2018), *IJMPD*, 27, 1841004.
Shulyak D., Ryabchikova T., Kiliyarova R., Kochukhov O. (2010), *A&A*, 520, 88.
Simkin S. M. (1974), *A&A*, 31, 129.
Skrutskie M. F. *et al.* (2006), *AJ*, 131, 1163.
Smak J. (1981), *AcA*, 31, 395.
Smart W. M. (1962), *Text Book on Spherical Astronomy*, 5th ed., Cambridge: Cambridge University Press.
Smith H. A. (2003), *RR Lyrae Stars*. Cambridge: Cambridge University Press.
Smith R. J., Lucey J. R., Hudson M. J., Schlegel D. J., Davies R. L. (2000), *MNRAS*, 313, 469.
Smith R. J., Lucey J. R., Schlegel D. J., Hudson M. J., Baggley G., Davies R. L. (2001), *MNRAS*, 327, 249.
Smith R. J. (2004), *AJ*, 128, 1558.
Smith R. J. *et al.* (2004), *AJ*, 128, 1558.
Solà J. (2013), *JPhCS*, 453, 2015.
Soldner J. (1804), in *Berliner Astronomisches Jahrbuch für 1804*. Berlin: C. F. E. Späthen, p. 161.
Somerville R. S., Lee K., Ferguson H. C., Gardner J. C., Moustakas L. A., Giavalisco M. (2004), *ApJ*, 600, L171.
Somerville R. S., Hopkins P. F., Cox T. J., Robertson B. E., Hernquist L. (2008), *MNRAS*, 391, 481.
Sorce J. G. *et al.* (2013), *ApJ*, 765, 94.
Sorce J. G., Guo Q. (2016), *MNRAS*, 458, 2667.
Sotiriou T. P., Faraoni V. (2010), *RvMP*, 582, 451.
Sotiriou T. P. (2011), *JPhCS*, 283, 12034.

Spaenhauer A. M. (1978), *A&A*, 65, 313.
Spergel D. N., Steinhardt P. J. (2000), *PhRvL*, 84, 3760.
Springel V. (2005), *MNRAS*, 364, 1105.
Springel V. *et al.* (2005), *Nature* 435, 629.
Springob C. M., Haynes M. P., Giovanelli R., Kent B. R. (2005), *ApJS*, 160, 149.
Sprongob C. M., Masters K. L., Haynes M. P., Giovanelli R., Marinoni C. (2007), *ApJS*, 172, 599, erratum: Ibid. 182, 474.
Springob C. M. *et al.* (2014), *MNRAS*, 445, 2677.
Stavely-Smith L., Juraszek S., Henning P. A., Koribalski B. S., Kraan-Korteweg R. C. (2000), *ASPC*, 218, 207.
Starikova S. *et al.* (2014), *ApJ*, 786, 125.
Steigman G., (2007), *ARNPS*, 57, 463.
Stewart I. M., Blyth S.-L., Blok W. J. G. (2014), *A&A*, 567, 61.
Stiavelli M., Bertin G. (1985), *MNRAS*, 217, 735.
Strauss M. A., Cen R., Ostriker J. P., Lauer T. R., Postman M. (1995), *ApJ*, 444, 507.
Strauss M. A., Davis M. (1988), in *Large-Scale Motions in the Universe.* eds. V. Rubin & G. V. Coyne, Princeton: Princeton University Press, p. 256.
Strauss M. A., Willick J. A. (1995), *PhR*, 261, 271.
Sugiyama N. (1995), *ApJS*, 100, 281.
Sunyaev R. A., Zel'dovich Y. B. (1980), *ARAA*, 18, 537.
Tadros H. *et al.* (1999), *MNRAS*, 305, 527.
Takamoto M. *et al.* (2020), *Nat Photonics*, 14, 411.
Takizawa M. (2000), *ApJ*, 532, 183.
Tammann G. A., Sandage A., Reindl B. (2008), *A&ARv*, 15, 289.
Teegarden B. J., Cline T. L. (1980), *ApJ*, 236, L67.
Teegarden B. J., Cline T. L. (1981), *Ap&SS*, 75, 181.
Teerikorpi P. (1984), *A&A*, 141, 407.
Teerikorpi P. (1987), *A&A*, 173, 39.
Teerikorpi P. (1993), *A&A*, 280, 443.
Teerikorpi P. (2015), *A&A*, 576, A75.
Ter-Akopian G. M., Dmitriev S. N. (2015), *NuPhA*, 944, 177.
Terlevich R., Davies R. L., Faber S. M., Burstein D. (1981), *MNRAS*, 196, 381.
Terukina, A., Lombriser L., Yamamoto K., Bacon D., Koyama K., Nochol R. C. (2014), *JCAP*, 4, 13.
Thackeray A. D., Wesselink A. J. (1953), *Nature*, 171, 693.
Thomas D., Marston C., Bender R. (2003), *MNRAS*, 339, 897.
Thomas J. *et al.* (2005), *MNRAS*, 360, 1355.
Thomas J., Saglia R. P., Bender R., Thomas D., Gebhardt K., Magorrian J., Corsini E. M., Wegner G. (2007), *MNRAS*, 382, 657.
Thomas J., Saglia R. P., Bender R., Thomas D., Gebhardt K., Magorrian J., Corsini E. M., Wegner G. (2009), *ApJ*, 691, 770.
Thomas J. *et al.* (2011) *MNRAS*, 415, 545.
Thornton B. F., Burdette S. C. (2019), *NatCh*, 11, 4.
Tiley A. L. *et al.* (2016), *MNRAS*, 460, 103.
Tinsley B. M. (1972), *ApJ*, 173, L93.

Tisserand P. *et al.* (2007), *A&A*, 469, 387.
Tiwari V., Klimenko S., Necula V., Mirtselmakher G. (2016), *CQGra*, 33, 1.
Tody D. (1986), *SPIE*, 627, 733.
Tonry J., Davis M. (1979), *AJ*, 84, 1511.
Tonry J., Schneider D. P. (1988), *AJ*, 96, 807.
Tonry J. L., Davis M. (1981), *ApJ*, 246, 666.
Tonry J. L., Blakeslee J. P., Ajhar E. A., Dressler (1997), *ApJ*, 475, 399.
Tonry J. L., Blakeslee J. P., Ajhar E. A., Dressler (2000), *ApJ*, 530, 625.
Tonry J. L., Dressler A., Blakeslee J. P., Ajhar E. A., Fletcher A. B., Luppino G. A., Metzger M. R., Moore C. B. (2001), *ApJ*, 546, 681.
Tonry J. L., Onaka P. M., Burke B., Luppino G. (2005), in *Scientific Detectors for Astronomy*. eds. J. E. Beletec *et al.* pp. 53–62.
Toomre A., Toomre J. (1972), *ApJ*, 178, 623.
Touboul P. *et al.* (2017), *PhRvL*, 119, 231101.
Trager S. C., Worthey G., Faber S. M., Burstein D. (1998), *ApJS*, 116, 1.
Trenti M., Hut P. (2008), *Scolarpedie*, 3(5), 3930.
Treu T., Stiavelli M., Bertin G., Casertano S., Møller P. (2001), *MNRAS*, 326, 237.
Trimble V. (2017), *EPJH*, 42, 261.
Tripp R. (1998), *A&A*, 331, 815.
Trumpler R. J. (1930), *Lick Obs Bull*, 14, 154.
Tulin S., Yu H.-B. (2018), *PhR*, 730, 1.
Tully R. B. (2013), *Nature*, 493, 31.
Tully R. B., Fisher J. R. (1977), *A&A*, 54, 661.
Tully R. B., Fisher J. R. (1987), *Nearby Galaxies Atlas*. Cambridge: Cambridge University Press.
Tully R. B., Foque P. (1985), *ApJS*, 58, 67.
Tully R. B., Shaya E. J., Karachentsev I. D., Courtois H. M., Kocevski D. D., Rizzi L., Peel A. (2008), *ApJ*, 676, 184.
Tully R. B. *et al.* (2013), *AJ*, 146, 86.
Tully R. B., Cortois H., Hoffman Y., Pomarède D. (2014), *Nature*, 513, 71.
Tully R. B., Courtois H. M., Sorce J. G. (2016), *AJ*, 152, 50.
Turatto M. (2003), *LNP*, 598, 21.
Turnbull S. J., Hudson M. J., Feldman H. A., Hicken M., Kirshner R. P., Watkins R. (2012), *MNRAS*, 392, 743.
Tutusaus I., Lamine B., Blanchard A. (2019), *A&A*, 625, A15.
Tyson T. (2003), *PhT*, 56, 63.
Tyson J. A., Valdes F., Wenk R. A. (1990), 349, 1.
Udalski A., Szymański M., Kubiak M., Pietrzynski G., Soszyński I., Woźniak P., Żebruń K. (1999), *AcA*, 49, 201.
Udalski A. (2003), *AcA*, 53, 291.
Umetsu K. *et al.* (2012), *ApJ*, 755, 56.
Ungruhe R., Seitter W. C., Duerbeck H. W. (2003), *JAD*, 9, 1.
Unnikrishnan C. S., Gillies G. T. (2000), *PhRvD*, 61, 101101(R).
Unnikrishnan C. S., Mohapatra A. K., Gillies G. T. (2001), *PhRvD*, 63, 062002.

van Altena W. F., Lee J. T., Hoffleit E. D. (1995), *The General Catalogue of Trigonometric Parallaxes*, New Haven: Yale University Press.
van de Voort F., Quataert E., Hopkins P. F., Kere D., Faucher-Giguere C. A. (2015), *MNRAS*, 447, 140.
Van den Bergh S. (2000), *PASP*, 112, 529.
van Dokkum P. G., Stanford S. A. (2003), *ApJ*, 585, 78.
van Leeuwen F. (2007), *A&A*, 474, 653.
van Maanen A. (1916), *ApJ*, 44, 210.
van Waerbeke L. *et al.* (2013), *MNRAS*, 433, 3373.
Vargas-Magaña M. *et al.* (2018), *MNRAS*, 477, 1153.
Varela J. *et al.* (2012), *ApJ*, 744, 82.
Vainshtein A. I. (1972), *PhLB*, 39, 393.
Verde L., Treu T., Riese A. G. (2019), *NatAs*, 3, 891.
Vijayaragharan R., Ricker P. M. (2013), *MNRAS*, 435, 2713.
Vijayaragharan R., Ricker P. M. (2015), *MNRAS*, 449, 2312.
Visvanathan N. (1979), *ApJ*, 81, 1228.
van de Weygaert R., Platen E. (2011), *IJMPCS*, 41.
Vogel H. C. (1889), *AN*, 121, 241.
Vogelsberger M. *et al.* (2014), *Nature*, 509, 177.
von Eötvös, R., Pekar D., Fekte E. (1922), *AnnPhys* (*Leipzig*), 68, 11.
von Humboldt A. (1845), *Kosmos*, Bd I. Stuttgart und Tubingen: J. G. Cotta'scher Verlag.
Vorontsov-Vellyaminov B., Krasnogorskaya A. A., Arkhipova V. P. (1962–1974), *Morphological Catalogue of Galaxies (MCG)*. Moscow: Moscow State University.
Wallerstein G. *et al.* (1997), *RvMP*, 69, 995.
Wallerstein G. (2002), *PASP*, 114, 689.
Walsh D., Carswell R. F., Weymann R. J. (1979), *Nature*, 279, 381.
Wambsganss J. (1998), *Lv Rv Rel*, 1, 12.
Wang L., Steinhardt P. J. (1998), *ApJ*, 508, 483.
Wang Y., Roone C., Feldman H. A., Watkins R. (2018), *MNRAS*, 480, 5332.
Watkins R., Feldman H. A. (1995), *ApJ*, 453, L73.
Watkins R., Feldman H. A., Hudson M. J. (2009), *MNRAS*, 392, 743.
Watkins R., Feldman H. A. (2015a), *MNRAS*, 447, 132.
Watkins R., Feldman H. A. (2015b), *MNRAS*, 450, 1868.
Weber J. (1967), *PhRvL*, 18, 498.
Wegner G. (1974), *MNRAS*, 166, 271.
Wegner G., Petford A. D. (1974), *MNRAS*, 168, 557.
Wegner G. (1979), *Ap&SS*, 60, 15.
Wegner G. (1979), *AJ*, 84, 650.
Wegner G. (1989), *LNP*, 328, 401.
Wegner G., Davies R. L., Colless M. M., Burstein D., Bertschinger E., McMahan R. (1991), *ASP Conf Ser* 15, 129.
Wegner G., Haynes M. P., Giovanelli R. (1993), *AJ*, 105, 1251.
Wegner G., Colless M., Baggley G., Davies R. L., Bertschinger E., Burstein D., McMahan Jr. R. K., Saglia R. P. (1996), *ApJS*, 106, 1.

Wegner G., Colless M., Saglia R. P., McMahan R. K., Davies R. L., Burstein D., Baggley G. (1999), *MNRAS*, 305, 259.

Wegner G. *et al.* (2001), *AJ*, 122, 2893.

Wegner G. *et al.* (2003), *AJ*, 126, 2268.

Wegner G. (2007a), in *The Biographical Encyclopedia of Astronomers*, Vol. 2, eds. T. Hockey, V. Trimble, T. R. Williams. New York: Springer, p. 520.

Wegner G. (2007b), in *The Biographical Encyclopedia of Astronomers*, Vol. 2, eds. T. Hockey, V. Trimble, T. R. Williams. New York: Springer, p. 732.

Wegner G., Grogin N. A. (2008), *AJ*, 136, 1.

Wegner G. A. (2011), *MNRAS*, 413, 1333.

Wegner G. A., Corsini E. M., Thomas J., Saglia R. P., Bender R., Pu S. B. (2012), *AJ*, 144, 78.

Wegner G. A., Chu D. S., Hwang H. S. (2015), *MNRAS*, 447, 1126.

Wegner G. A., Onofrio R. (2015), *EPJC*, 75, 307.

Wegner G. A. *et al.* (2017), *ApJ*, 844, 67.

Wegner G. A., Salzer J. J., Taylor J. M., Hirschauer A. S. (2019), *ApJ*, 883, 29.

Wegner G., Reid I. N., McMahan R. K. (1989), *LNP*, 328, 378.

Wegner G., Reid I. N., McMahan R. K. (1991), *ApJ*, 376, 186.

Wegner W. (2002), *BaltA*, 11, 1.

Weinberg D. H., Bullock J. S., Governato F., de Naray R. K., Peter H. G. (2015), *PNAS*, 112, 12249.

Weinberg S. (2008), *Cosmology*. Oxford: Oxford University Press.

Weinberger R., Saurer W., Seeberger R. (1995), *A&AS*, 110, 269.

Weinberger R. *et al.* (2000), *ASPC*, 218, 27.

Weiner B. J. *et al.* (2006), *ApJ*, 653, 1027.

Weisberg J. M., Taylor J. H. (2002), *ApJ*, 576, 942.

Weisberg J. M., Huamg Y. (2016), *ApJ*, 829, 55.

Wesselink A. J. (1946), *BAN*, 10, 91.

West R. M. (1974), *ESO Bull*, 10, 25.

Westover M. (2007), Galaxy clustering in the two micron all sky redshift survey, PhD Thesis, Harvard University.

Whitbourn J. R., Shanks T. (2016), *MNRAS*, 459, 496.

Will C. M. (1993), *Theory and Experiment in Gravitational Physics*. Cambridge: Cambridge University Press.

Will C. M. (2014), *Lv Rv Rel*, 17, 4.

Williams J. C., Turyshev S. G., Boggs D. (2012), *CQGra*, 29, 184004.

Willick J. A. (1991), *PhD Thesis*, University California Berkeley.

Willick J. A. (1994), *ApJS*, 92, 1.

Willick J. A., Courteau S., Faber S. M., Burstein D., Dekel A. (1995), *ApJ*, 446, 12.

Willick J. A., Courteau S., Faber S. M., Burstein D., Dekel A., Kolatt T. (1996), *ApJ*, 457, 460.

Willick J. A., Courteau S., Faber S. M., Burstein D., Strauss M. A. (1997), *ApJS*, 109, 333.

Willick J. A., Strauss M. A. (1998), *ApJ*, 507, 64.

Wilczynska, M. R. *et al.* (2020), *SciA*, 6, 9672.

Wirtz C. (1924), *AN*, 222, 21.
Wittman D., Golovich N., Dawson W. A. (2018), *ApJ*, 869, 104.
Wolfe A. M., Gawiser E., Prochaska J. X. (2005), *ARA&A*, 43, 861.
Wollaston W. H. (1802), *RSPT*, 92, 365.
Womble D. S., Sargent W. L. W., Lyons R. S. (1996), *ASSL*, 206, 249.
Worthey, G. (1994), *ApJS*, 95, 107.
Wotjak R., Hansen S. H., Hjorth J. (2011), *Nature*, 477, 567.
Woudt P. A., Kraan-Korteweg R. C. (2001), *A&A*, 380, 441.
Woudt P. A., Kraan-Korteweg R. C., Lucey J., Fairall A. P., Moore S. A. (2007), *MNRAS*, 383, 445.
Woolf N. J. (1982), *ARA&A*, 20, 367.
Worthey G., Faber S. M., Gonzales J. J., Burstein D. (1994), *ApJS*, 94, 687.
Wright A. E. (1972), *MNRAS*, 157, 309.
Wright T. (1750), *An Original Theory or New Hypothesis of the Universe Founded Upon the Laws of Nature*. London: H. Chapelle.
Wu H.-Y., Huterer D. (2017), *MNRAS*, 471, 4946.
Yao Y. *et al.* (2019), *ApJ*, 886, 152.
Yahil A., Walker D., Rowan-Robinson M. (1986), *ApJ*, 301, 1.
Yahil A. (1988), in *Large-Scale Motions in the Universe*. eds. V. Rubin, G. V. Coyne, Princeton: Princeton University Press, p. 219.
Yasuda N., Fukugita M., Okamura S. (1997), *ApJS*, 108, 417.
Yuan W., Macri L. M., He S., Huang J. Z., Kanbur S. M., Ngeow C.-C. (2017), *AJ*, 154, 49.
Yoon M. *et al.* (2014), *MNRAS*, 445, L60.
Young C. A. (1873), *MNRAS*, 33, 248.
Zacharias N., Dorland B., Bredthauer R., Boggs K., Bredthauer G., Lesser M. (2007), *SPIE*, 669008.
Zaroubi S., Bernardi M., da Costa L. N., Hoffman Y., Alonso M. V., Wegner G., Willmer C. N. A., Pellegrini P. S. (2001), *MNRAS*, 326, 375.
Zarrouk P. *et al.* (2018), *MNRAS*, 477, 1639.
Zel'dovich Ya. B. (1970), *A&A*, 5, 84.
Zhao H., Peacock J. A., Li B. (2013), *PhRvD*, 88, 3013.
Zhu L. *et al.* (2018), *PhRvL*, 121, 11001.
Zhu W. W. *et al.* (2015), *ApJ*, 809, 41.
Zinkernagel H. (2002), *SHPMP*, 33, 493.
Zinkernagel H. (2014), *SHPMP*, 46, 1.
Zu H. *et al.* (2019), arXiv:190105616.
ZuHone J. A., Markevitch M., Ruskowski M., Lee D. (2013), *ApJ*, 762, 69.
ZuHone J., Zavala J., Vogelsberger M. (2019), arXiv:190111140.
Zwicky F. (1933), *AcHPh*, 6, 110.
Zwicky F. (1958), *Morphological Astronomy*. Berlin: Springer.
Zwicky F., Herzog E., Will P., Karpowicz M., Kowal C. (1961–1968), *Catalog of Galaxies and Clusters of Galaxies*, Vols. 1–6. Pasadena: California Institute of Technology (CGCG).
Zwicky F. (1965), in *Stellar Structure*, eds. L. H. Aller and D. B. McLaughlin. Chicago: Chicago University Press.

Index

CPSIA information can be obtained
at www.ICGtesting.com
Printed in the USA
LVHW051248031221
705026LV00001B/8

9 789811 211805